Aeolian Environments, Sediments and Landforms

Edited by

Andrew S. Goudie

School of Geography,
University of Oxford, UK

Ian Livingstone

School of Environmental Science,
University College Northampton, UK

and

Stephen Stokes

School of Geography,
University of Oxford, UK

JOHN WILEY & SONS, LTD
Chichester • New York • Weinheim • Brisbane • Singapore • Toronto

Copyright © 1999 by John Wiley & Sons Ltd,
Baffins Lane, Chichester,
West Sussex PO19 1UD, England

National   01243 779777
International (+44) 1243 779777
e-mail (for orders and customer service enquiries): cs-books@wiley.co.uk.
Visit our Home Page on http://www.wiley.co.uk
or http://www.wiley.com

Other Wiley Editorial Offices

John Wiley & Sons, Inc., 605 Third Avenue,
New York, NY 10158-0012, USA

WILEY-VCH Verlag GmbH, Pappelallee 3,
D-69469 Weinheim, Germany

Jacaranda Wiley Ltd, 33 Park Road, Milton,
Queensland 4064, Australia

John Wiley & Sons (Asia) Pte Ltd, 2 Clementi Loop #02-01,
Jin Xing Distripark, Singapore 129809

John Wiley & Sons (Canada) Ltd, 22 Worcester Road,
Rexdale, Ontario M9W 1L1, Canada

Library of Congress Cataloging-in-Publication Data

Aeolian environments, sediments, and landforms / edited by Andrew S.
Goudie, Ian Livingstone, and Stephen Stokes.
      p. cm. — (British geomorphological research group symposia series)
    Papers from the 4th International Conference on Aeolian Research
(ICAR 4), held in the School of Geography and St. Catherine's
College, University of Oxford, July 1998.
    Includes bibliographical references.
    ISBN 0-471-98573-2
    1. Sand dunes Congresses. 2. Aeolian processes Congresses.
I. Goudie, Andrew. II. Livingstone, Ian. III. Stokes, S.
IV. International Conference on Aeolian Research (4th : 1998 :
Oxford, U.K.) V. Series.
    GB611.A356   1999
    551.3'75—dc21   99–32324
                                                                        CIP

British Library Cataloguing in Publication Data

A catalogue record for this book is available from the British Library
ISBN 0 471 98573 2

Typeset in 10/12pt Times by Mayhew Typesetting, Rhayader, Powys
Printed and bound in Great Britain by Biddles Ltd, Guildford and King's Lynn
This book is printed on acid-free paper responsibly manufactured from sustainable forestry, in which at least two trees are planted for each one used for paper production.

Contents

List of Contributors

B. O. Bauer Department of Geography, University of Southern California, Los Angeles, California 90089-0255, USA

D. A. Gillette Atmospheric Sciences Modeling Division, Air Resources Laboratory, National Oceanic and Atmospheric Administration, Research Triangle Park, North Carolina 27711, USA

A. S. Goudie School of Geography, University of Oxford, Mansfield Road, Oxford OX1 3TB, UK

G. Kocurek Department of Geological Sciences, University of Texas, Austin Texas, 78712, USA

N. Lancaster Desert Research Institute, 2215 Raggio Parkway, Reno, Nevada, 89512-1095, USA

J. Leys Centre for Natural Resources, Department of Land and Water Conservation, Gunnedah Research Centre, Gunnedah, PO Box 462, NSW 2380, Australia

I. Livingstone School of Environmental Science, Nene – University College Northampton, Northampton, NN2 7AL, UK

C. McKenna Neuman Department of Geography, Trent University, Peterborough, Ontario A91 7B8, Canada

G. McTainsh Faculty of Environmental Sciences, Griffith University, Northern Campus, Brisbane, Queensland 4111, Australia

W. G. Nickling Department of Geography, University of Guelph, Guelph, Ontario N1G 2W1, Canada

K. Pye Department of Geology, Royal Holloway, University of London, Egham, Surrey TW20 0EX, UK

D. J. Sherman Department of Geography, University of Southern California, Los Angeles, California 90089-0255, USA

D. Sherwin Department of Geography and Geology, City College Norwich, Ipswich Road, Norwich NR2 2LJ, UK

A. K. Singhvi Physical Research Laboratory, Earth Science Division, Navrangapura, Ahmedabad, Gujarat 380 009, India

S. Stokes School of Geography, University of Oxford, Mansfield Road, Oxford OX1 3TB, UK

V. P. Tchakerian   Department of Geography, College of Geosciences, Texas A&M University, College Station, Texas 77843-3147, USA

D. S. G. Thomas   Sheffield Centre for International Dryland Research, Department of Geography, University of Sheffield, Winter Street, Sheffield S10 2TN, UK

A. G. Wintle   Institute of Geography and Earth Sciences, University of Wales, Penglais, Aberystwyth, Dyfed SY23 3DB, UK

Preface

In July 1998 the Fourth International Conference on Aeolian Research (ICAR 4) was held in the School of Geography and St Catherine's College, University of Oxford. Previous conferences have been held in Denmark (1985 and 1990), and Zyzxx, California, USA (1994). The next conference (ICAR 5) will be held in Australia in 2002. The Oxford conference was attended by over 100 delegates from 23 different countries. In all some 80 papers were presented over four days, and some of these were published in a special issue of the *Journal of Arid Environments* (1998, vol. 39, pp. 343–547). Others have been published in special issues of *Earth Surface Processes and Landforms* (1999, vol. 24, pp. 381–479) and *Zeitschrift für Geomorphologie* (supplement b page 116).

One of the features of the conference was the presentation of a series of keynote overview papers by some of the leading researchers in a range of subdisciplines. It is these that form the bulk of the present volume. The only exceptions to this are Chapter 1, which provides an historical introduction and context to aeolian research studies, Chapter 8 and Chapter 14 which was written after the conference and uses some of the information and ideas that the conference generated to give a view of major future trends in aeolian research.

The Conference was held in association with the British Geomorphological Research Group, which provided financial assistance to enable post-graduates to attend, and with IGCP No. 413.

A. S. Goudie
I. Livingstone
S. Stokes
February 1999

1 The History of Desert Dune Studies over the Last 100 Years

ANDREW S. GOUDIE

School of Geography, University of Oxford, UK

The study of dunes between the 1880s and 1980s has been characterised by certain major trends and contributions. These include the work of French scientists in the Western Sahara, the pioneer studies of Vaughan Cornish, the researches of British explorers in the Eastern Sahara, the classic work of R.A. Bagnold, and the increasing use of air photos and satellite images in recent years. Recent advances include a greater concern with the physics of aeolian processes, the emergence of planetary studies, the development of new field technology, a concern with human-related issues, and the utilisation of new dating techniques.

INTRODUCTION

Although aeolian processes and forms are widespread and have been the subject of a large amount of literature over a long period of time (see the bibliographies of Warren, 1969; Lancaster, 1988; and Busche *et al.*, 1984), most standard histories of geomorphology (e.g. Chorley *et al.*, 1964; Tinkler, 1985) have no consideration of dunes and other aeolian features. In this paper some of the major developments that have taken place in the study of desert dunes between the 1880s and the 1980s are reviewed but detailed consideration of the very latest developments is left to other contributors to this volume.

THE FRENCH IN THE WESTERN SAHARA

In the early years of the twentieth century, France gradually extended its control over North Africa and the Sahara. Morocco became a French protectorate in 1912, while Algeria had largely been brought under military control by 1870 and formally became part of metropolitan France in 1882. Tunisia became a French protectorate in 1881. Niger was brought under French control between 1896 and 1900, Chad became part of French Equatorial Africa in 1910, while Mali was made a French Colony in 1880. Thus it was that French studies of Sahara dunes developed in the late nineteenth and early twentieth centuries.

Aeolian Environments, Sediments and Landforms. Edited by A.S. Goudie, I. Livingstone and S. Stokes.
© 1999 John Wiley & Sons, Ltd.

A notable pioneer attempt to understand the Saharan dunes was that of Rolland (1881), a mining engineer, who produced five propositions which are summarised in Rolland (1890, pp. 158–9):

1.   Les dunes sont de formation contemporaine et leurs éléments proviennent de la désagrégation des roches sous les influences atmosphériques et sous les effets du climat saharien (dans le Saharien algérien, elles proviennent surtout des terrains d'atterrisement; dans le désert libyque, des grès de Nubie etc.).
2.   L'amoncellement des sables en grandes dunes est dû entièrement au vent (dont le rôle prédominant était contesté par la plupart des géologues s'étant occupés du Sahara).
3.   Il y a relation, directe ou indirecte, entre les chaînes de grandes dunes et le relief du sol (la direction des chaînes étant indépendante de l'orientation des dunes élémentaires), et c'est le relief qui est la cause première de l'amoncellement des sables en certains entroits déterminés.
4.   Le va-et-vient des sables sous l'action alternative des courants atmosphériques se traduit finalement par un transport suivant la direction de la résultante mécanique des vents et cette direction est indiquée par les emplacements des grandes dunes par rapport aux régions qui les alimentent (dans l'ensemble du Sahara algérien, et sauf exceptions locales, elle est dirigée du N.-O. au S.-E.).
5.   Les grandes dunes ne sont pas, à proprement parler, mobiles, mais présentent une progression lente suivant la résultante mécanique des vents.

These propositions were debated at length by the Geographical Society of France (see *Compte Rendu des Séances de la Société de Géographie*, Année 1890, pp. 114–19, 158–64, 256–61, 305–6, 320–8 and 363–72), but are remarkably advanced for their time, dealing as they do with the source of sand, the distribution of ergs, the relationship of dunes to wind direction, and the nature of dune mobility. His need to assert that dunes were due entirely to the work of wind is interesting in that some geologists thought they had a rock core while others invented somewhat implausible mechanisms involving ground vibrations by earthquakes (e.g. Frere, 1870).

Flammand (1899) described the dunes of the Grand Erg Occidental in very limited detail following his traverse of that sand sea, while Foureau (1905) gave a rather fuller description of the dunes and other aeolian phenomena of the Grand Erg Oriental. Chudeau (1909) studied the dunes on the south side of the Sahara, and in particular drew attention to, and mapped the fossil dunes (*ergs morts*) that stretched in a zone between Dakar and Lake Chad. In 1920 Chudeau gave a more general account of Saharan dunes and discussed the nature of the sand, its possible sources, the localisation of dunes in great fluvial basins, and the main dune forms (those parallel to the wind, those transverse to the wind, barchans, and small obstacle dunes of various types, including nebkas and zemlat).

Perhaps the most productive of the French dune workers between the two world wars was Aufrère (1928, 1931, 1932, 1934), though notable studies were also produced, inter alia, by Monod (1928), Dufour (1936) and Urvoy (1933). Aufrère (1931 and 1934) sought to relate dune trends and forms to wind characteristics, using the sparse meteorological data then available, and although there were problems with his model (see Capot-Rey, 1945), it was a very major step forward. In addition, he applied the Davisian cyclic scheme to dune morphology, distinguishing young ergs (e.g. the Erg Occidental), with narrow, sandy, inter-dune furrows, and old ergs where the breadth of the sand-free surfaces exceeds that of the sand ridges (e.g. the Erg Oriental and the Erg Chech).

Aufrère also provides a tentative map of the main ergs and dune forms of the Sahara, together with accurate diagrams of dune forms based on air photo analysis. Indeed, Aufrère (1932) presented a paper to the International Geographical Congress in Paris on the use of vertical air photography for dune mapping, and believed that the mapping of dune trends would help one to understand the nature of wind flow over the Sahara. As he remarked (p. 161): 'La carte des chaînes de sable du Sahara et de l'Arabie est une carte des vents, et la photographie de la mer de sable est quelque chose comme une photographie du vent'. Aufrère's use of air photography and his concern to relate dune trends to wind directions was extremely influential, not least in terms of Madigan's pioneering studies on the dunes of Australia (Madigan, 1936).

After the second world war, French researches continued apace. Alimen (1953) concentrated on the granulometry of dune sands and Bellair (1953) on their mineralogy, but in the immediate post-war years, the most important figure in the study of Saharan dunes was probably Capot-Rey, who published a number of major papers, particularly in the *Travaux Institute de Recherches Sahariennes* (Algiers), including one that brought the work of Bagnold to French readers (Capot-Rey and Capot-Rey, 1948). However, another major contributor was Dubief (1953), who undertook extensive statistical analysis of wind data in time and space and produced wind vectors for sand movement for the whole of the Sahara and identified lines of convergence and divergence in sand flow. He also produced a map of the numbers of days upon which sand movement occurred.

THE CONTRIBUTION OF VAUGHAN CORNISH

Vaughan Cornish (1862–1948) was the first British scientist to study dunes in detail, and in Europe only the work of Rolland, and above all, Sokolow, had appeared by around 1900 (Warren, 1969). In America, George Perkins Marsh (1864) made some useful observations on dunes, but little or no subsequent work was done there. Cornish was an amateur British geographer who made remarkable contributions to the study of wave forms in nature (Goudie, 1972). He appreciated that these were found in seas, deserts, snow, clouds and earthquake activity, and he tried to elucidate what common factors produced such an important and widespread morphological type (Cornish, 1914). He termed the general study of wave-forms 'kumatology', but the word has not gained wide acceptance (Cornish, 1899).

His interest in waves seems to have started with the study of sand waves, ranging in size from mere ripples to large desert sand-ridges, though early on he saw that waves in sand were analogous to waves in snow. In Egypt, Cornish (1900) made some simple morphometric measurements, and established the existence of certain constant relationships between the height of a dune or ripple, and the length between one dune or ripple and the next. He also became interested in the spacing and heights of peaks along a dune ridge and tried to relate them to the dynamics of dune formation. Such measurements now appear both simple and obvious but, given the background of geomorphological ideas at the same time, his originality of thought is clear.

Cornish's observational data and measurements on dunes were undoubtedly of considerable importance. They were frequently referred to by overseas scientists such as Sven Hedin and Johannes Walther, and their publication in French in 1900 made them available to a wide European readership. Some of the conclusions he reached in attempting to explain the shapes and patterns of sand-wave forms now appear unsatisfactory. Bagnold (1941) refuted Cornish's hypothesis, which he shared with workers such as King, that ripples are embryonic dunes. The famous and at first widely adopted 'lee eddy hypothesis' of barchan formation was also questioned by Bagnold (1941), though severe doubts had already been expressed by three experienced field men, Beadnell (1910, p. 387), King (1918, pp. 21–3) and Hume (1925, p. 48). Cornish obtained this idea from Darwin (1884, p. 42), who noted a fixed eddy in the lee of current ripples in water, and applied it to wind ripples and transverse sand dunes. This 'unfortunate fallacy' (Cooper, 1958, p. 42) became deeply entrenched in the literature as 'a striking demonstration of the way in which an unproved concept becomes embedded in a body of scientific thought and how it is transmitted from author to author until it finally reaches the textbooks'. The views of Cooper on the lee-eddy hypothesis, which were shared by Sharp (1966), and which were based on experimental observations, have in part themselves been questioned by Inman *et al.* (1966) and by Hoyt (1966). They have found lee eddies of low velocity behind barchans but only ascribe a minor role to them in shaping the dunes. Thirdly, not surprisingly, Cornish lacked the physical and mathematical approach of later workers. Vaughan Cornish himself, expressed the hope that a younger generation of mathematicians, might subdue theoretically his own refractory observational data.

BRITISH EXPLORERS IN THE EASTERN SAHARA

Given that the British Empire at its zenith was so large, it is remarkable how little of it was hyper-arid. The Germans, albeit briefly, administered most of the Namib. As for the Sahara, this was essentially French territory and as Lord Salisbury put it, 'we have given the Gallic cockerel an enormous amount of sand. Let him scratch it as he pleases' (Porch, 1985). However, the British did administer Egypt, occupying it in 1892 and declaring it a protectorate in 1914, and it was here that the most important British work on dunes was undertaken.

Apart from Cornish, who did not travel far from the Nile, the first major figure was Beadnell (1910), who argued that on mineralogical grounds Kharga dunes were not necessarily derived from the denudation of the Nubian Sandstone. More importantly he made detailed observations of wind directions and velocity and of rates of barchan movement. He argued that smaller barchans moved more speedily than large, but that an average annual rate of movement was 15 to 16 metres. He observed that the threshold wind velocity for their movement was 13 mph (21 km h^{-1}). He also measured the angle of barchan lee slopes which he noted was close to the angle of rest of dry blown sand.

King (1916) argued that dunes are not merely overgrown ripples as Cornish had maintained, and he also suggested, *contra* Cornish, that the resemblance between dune form and dune movement and waves was only a very superficial one. He also did ripple experiments using a trough that could be turned to meet any serious change in the direction of wind during the experiments. This was acknowledged by Bagnold (1941, p. xxii) as the stimulus to his wind tunnel studies. As Ball was to do some years later, King also carried out some field studies of the electrical charging of dune particles.

King (1918) mapped the 350 km long Abu Moharik barchan dune belt, produced some surveyed profiles and used smoke to test Cornish's lee eddy hypothesis. This was possibly the earliest example of the use of flow visualisation techniques. He also poured sand onto a tray in sandstorms and noted that either a cigar-shaped or a D-shaped minor dune formed.

Ball (1927), Director of Desert Surveys, noted that as early as the Senussi Campaign in 1915, there were concerns that dunes could inhibit motor transport, but he also demonstrated that much valuable information on dunes could be derived from motor journeys into sand seas. He discussed the distribution of the dunes, and used rates of barchan movement to argue that the Abu Moharik belt might have taken 35 000 years to form. His most original argument, however, was that 'the clean-cut character' and narrow width of the Kharga dune belts was 'the consequence of lateral attraction by the dunes themselves on the flying electrified particles of sand'.

It is evident, therefore, that by the late 1920s Ball, Beadnell and King had already undertaken some very significant and original observations on dune form, movement and process.

However, in 1929 and 1930 Major R.A. Bagnold (1931) undertook some motorised journeys in the Western (Libyan) Desert, and this heralds another era in dune research. However, Bagnold acknowledged his indebtedness to Ball whom he regarded as the father of Egyptian exploration, and who had also pioneered desert travel with the Light Car Patrols in the First World War (Bagnold, 1990, p. 70). A biography and collection of some of Bagnold's major papers is provided by Thorne *et al.* (1988).

On his travels in the Western Desert in 1929 and 1930, Bagnold became fascinated by the vast scale of organisation of the dunes. He was invalided out of the army in 1934 (56 years before he died) and decided that the way to sort out some of the fundamental questions relating to dunes was to build a wind tunnel. He embarked between 1935 and 1939 on his study of the physics of blown sand at Imperial College, London, undertaking a field expedition in 1938 to check the wind tunnel results in the true waterless desert conditions of the Egyptian–Libyan Desert. One of his travelling companions was the desert geomorphologist, R.F. Peel. (Bagnold, 1990, pp. 115–18). His first thoughts on the interactions between the wind and individual flying sand grains were presented to the Royal Geographical Society in 1935 (Bagnold, 1935), but were not at this stage based on wind tunnel experimentation. He did, however, use horizontal rotating discs to investigate the reality of bouncing grains, and saw the movement and deposition of bouncing grains as the explanation for why sand accumulated in sharply defined dunes – a problem that had intrigued King (1916).

The wind tunnel work was written up in the late 1930s (e.g. Bagnold 1936, 1937) and brought together in *The Physics of Blown Sand and Desert Dunes* (1941), the greatest classic in the study of aeolian phenomena. The physics of blown sand make up the first eight chapters whereas the second half of the book deals with grain size distributions, ripples, the main types of dune, the internal structure of sand deposits and singing sands. It contains his famous model of the transition from barchan to seif. In retrospect Bagnold may have been over-concerned with barchans, but given their ubiquity and perfection of development in the Western Desert this is scarcely surprising. Besides the Western Desert, Bagnold was much influenced by a spate of work in fluid mechanics developed in the 1930s by von Karman (1934, 1935), Prandtl (1935) and others.

In 1951, CNRS, with the help of the Rockefeller Foundation, held a colloquium in Algiers on 'Actions Eoliennes, Phénomènes d'Evaporation et d'Hydrologie super-ficielle dans les Régions arides' to which Bagnold contributed (Bagnold, 1953). There were a number of significant papers that looked at the physics of sand movement (e.g. by Kawamura, Knapp and Zingg) and which built on the ideas of Bagnold.

A review of the influence and limitations of *The Physics of Blown Sand and Desert Dunes* in the light of more recent advances is provided by Tsoar (1994), who though he points to areas where subsequent research has necessitated a revision of Bagnold's ideas, recognises that (p. 91) it 'is the first textbook in dynamic geomor-phology to deal with the process of fluid (wind) action on a sediment as a problem of fluid dynamics'. He also acknowledged that it was 40 years ahead of its time, and that his work was (p. 91) 'rediscovered at the end of the 1970s by a group of scholars in Denmark, USA and UK, as a consequence of significant international increase in interest in aeolian studies, as well as the result of development of boundary-layer wind tunnels, improved high-precision instruments and applications of modern mathematical and statistical techniques'.

Although when he returned to science after the Second World War Bagnold devoted much of his time to the study of the dynamics of water and sediment flow, rather than to aeolian processes, he did produce a paper (Bagnold, 1953), in which he speculated, quite diffidently, that linear sand dunes might be caused by roll vortices (Livingstone, 1988). This model was widely supported by both meteoro-logists and geomorphologists (see, for example Folk, 1971), but more recent small-scale single dune studies suggest that the key to linear dune origin and maintenance may lie less with thermally driven vortices than with the dune's own modification of the pattern of air flow by its intrusion into the atmospheric boundary layer (e.g. Tsoar, 1978, and Livingstone, 1986).

THE EXPANSION OF DUNE STUDIES

In the post-war era, the geographical spread of dune studies developed substantially. This was partly because of the wider availability of good maps and air photographs

Table 1.1 Major regional dune studies accomplished between 1945 and 1980

Region	Author
Wyoming	Ahlbrandt (1974)
Thar (India)	Allchin *et al.* (1978); Singh (1977); Vestappen (1970)
Brazil	Bigarella (1975)
Australia	Brookfield (1970); Folk (1971); Jennings (1975); King (1960); Mabbutt (1968); Twidale (1972)
Egypt	Embabi (1971)
Peru	Finkel (1959); Hastenrath (1967); Lettau and Lettau (1978)
Pakistan (Thal)	Higgins *et al.* (1974)
Saudi Arabia	Holm (1960)
Namib	Lancaster (1980); Besler (1980)
New Mexico	McKee and Moriola (1975)
Sahara	Mainguet (1975); Monod (1957); Mainguet and Callot (1978)
California	Norris and Norris (1961); Sharp (1966); Long and Sharp (1964)
Kansas	Simonett (1960)
Nebraska	Smith (1965); Warren (1976)
Mauretania	Tricart and Brochu (1955); Monod (1958)
Sinai	Tsoar (1974)
Ténéré	Warren (1971)
Kalahari	Grove (1969)

of dune areas, partly because of resource evaluation surveys by bodies like CSIRO in Australia, and partly because of the establishment of arid zone research centres such as those of Jodphur in India, Lanzhou in China and Gobabeb in the Namib. Table 1.1 lists some of the more important regional studies that were accomplished between 1945 and 1980.

In addition, however, some global views of ergs began to appear, and the work of Wilson, tragically foreshortened, is of particular significance. His work on ergs (e.g. Wilson, 1971, 1973), on dune hierarchies and the role of grain size characteristics in determining dune form (Wilson, 1972) was both innovative and influential (see, for example, Wasson and Hyde, 1983). Global studies of dunes were to develop further as satellite images became more widely available.

THE AIRPHOTO AND SATELLITE ERAS

Although an airphoto mosaic of dunes at El Arish in Sinai was flown in 1916 and published by Hume in 1925 (Hume, 1925, Fig. 9A) and subsequent use of air photos was made by Aufrère (1931) and Madigan (1936), it was not until after the Second World War that there was widespread use of air photography to map and describe dune forms. Smith (1963), for example, used Second World War Tri-Metrogon photography of the Sahara and Clos-Arceduc (1969) and Mainguet and Callot (1978) made extensive use of vertical air photographs. Air photography also proved very useful in the recognition and mapping of relict dune fields in Nigeria (Grove 1958), the Sudan (Grove and Warren, 1968) and in the Kalahari (Grove, 1969).

In the 1970s the availability of satellite images enabled a continental and global scale picture to emerge of the diversity of dune forms, their colour, and their orientation with respect to formative winds. In particular Fryberger (1979) undertook a global study of the relationship between dune forms and their resultant drift potentials. It was possible to compare dune morphologies between different deserts, and to have a picture of patterns in areas where little access had previously been possible (e.g. the Taklamakan). Satellite-based studies were brought together in *A Study of Global Sand Seas* (McKee, 1979), while Mainguet was a major figure in the mapping of aeolian land forms in relation to wind flow in the Sahara (e.g. Mainguet, 1975). Satellite images were also important in identifying ancient dune fields, including those of the Pantanal and Llanos in South America (reviewed by Clapperton, 1993).

DUNE RESEARCH IN THE 1980S AND 1990S

Dune research in the last two decades has proliferated as never before and a series of major syntheses of aeolian geomorphology have appeared (e.g. Greeley and Iverson, 1985; Pye and Tsoar, 1990; Cooke *et al.*, 1993; Livingstone and Warren, 1996; Lancaster, 1995). There have also been regular conferences (e.g. Brookfield and Ahlbrandt, 1983; Nickling, 1986; Pye, 1993; Pye and Lancaster, 1993; Barndorff-Nielsen and Willetts, 1991).

Tchakerian (1995, pp. 3–6) recognises no less than eight key tendencies over this period.

1. An attempt to refine the theoretical foundations of aeolian geomorphology developed by Bagnold, focusing on aeolian grain mechanics and the basic physics of aeolian sediment transport. Associated with this, and facilitated by the developments in computing technology have been the refinement of models to simulate entrainment and sediment transport. In Chapter 6, Gillette discusses the physics of aeolian movement of saltating grains.
2. There has been the development of remote sensing technology to which reference has already been made. This has enabled us to gain a much clearer view of the geomorphology of sand seas, a topic discussed by Lancaster in Chapter 4.
3. There have been major advances created by the emergence of planetary geomorphology. A major stimulus to aeolian studies developed as a result of the American space programme and the recognition of dune forms and wind streaks on Mariner and Viking images of Mars (Greeley and Iverson, 1985; Wells and Zimbelman, 1997). Wind tunnels were developed that could simulate sand movement in planetary atmospheres.
4. Field technology has changed. The development of various flow visualisation techniques and the data logger revolution of the 1970s and 1980s enabled monitoring to be undertaken in transects across dunes of both sand and wind flow. Early examples of detailed single dune monitoring (Livingstone, 1990) included the studies of Tsoar in the Negev (Tsoar, 1978) and of Livingstone in the Namib (Livingstone, 1986). New surveying techniques (such as differential

GPS) also have great potential for monitoring dune activity. Nickling and McKenna Neuman review airflow and sediment transport in Chapter 2.

5. There has been an increasing awareness of human-related issues such as desertification and the need for management of sand and dune movement (e.g. Watson, 1990), and the possible impacts of global change on aeolian systems (e.g. Muhs and Maat, 1993). Thomas reviews management issues in Chapter 5.

6. Aeolian sedimentary environments have been seen not only as useful analogues for studying hydrocarbon reservoirs in the stratigraphic record, especially in Mesozoic sandstones of the Western USA and North-West Europe, but can also contribute to the exploration for, and recovery of oil and natural gas. Considerable attention has been paid to identifying and explaining sand surface textures employing the SEM and to recognising aeolian dune structures (e.g. Hunter, 1980), the nature of interdune deposits (and of bounding surfaces (e.g. Kocurek, 1981). This is a theme developed in Chapter 10 of this volume by Kocurek.

7. Dune studies in general have benefited from studies of dune dynamics and sediment transport in coastal environments (e.g. Sarre, 1988; Hesp and Fryberger, 1988). Bauer and Sherman review developments in this area in Chapter 4.

8. There has been a development in techniques for dating dunes. Of particular significance has been the development of thermoluminescence and optical dating techniques. These enable the ages and histories of dunes to be directly determined, which not only has palaeoclimatic significance (e.g. Stokes *et al.*, 1997) but also permit estimates of some dynamic characteristics of dunes, including the speed at which they can form or degrade. Luminescence dating is discussed by Singhvi and Wintle in Chapter 12, while Tchakerian discusses aeolian palaeoenvironments in Chapter 11.

REFERENCES

Ahlbrandt, T.S., 1974. The source of sand for Killpecker sand-dune field, south-western Wyoming. *Sedimentary Geology*, **11**, 39–57.

Alimen, H., 1953. Variations granulométriques et morphoscopiques du sable le long de profils dunaires au Sahara occidental. *Colloques Internationaux CNRS*, **35**, 219–235.

Allchin, B., Goudie, A. and Hegde, K.T.M., 1978. *The Prehistory and Palaeogeography of the Great Indian Desert*. Academic Press, London.

Aufrère, L., 1928. L'orientations des dunes et la direction des vents. *Comptes Rendus Académie des Sciences*, **187**, 833–835.

Aufrère, L., 1931. Le cycle morphologique des dunes. *Annales de Géographie*, **40**, 362–385.

Aufrère, L., 1932. Utilisation de la photographie zénithale dans l'étude morphologique et dans la cartographie des dunes. *Comptes Rendus du Congrés International de Géographie Paris 1931*, **1**, 155–162.

Aufrère, L., 1934. Les dunes du Sahara algérien (notes de morphologie dynamique). *Bulletin de l'Association de Géographies Français*, **83**, 130–142.

Bagnold, R.A., 1931. Journeys in the Libyan Desert. *Geographical Journal*, **78**, 13–39, 524–535.

Bagnold, R.A., 1935. The movement of desert sand. *Geographical Journal*, **85**, 343–369.

Bagnold, R.A., 1936. The movement of desert sand. *Proceedings of the Royal Society of London, Series A*, **157**, 594–620.

Bagnold, R.A., 1937. The size grading of sand by wind. *Proceedings of the Royal Society of London, Series A*, **163**, 250–264.

Bagnold, R.A., 1941. *The Physics of Blown Sand and Desert Dunes*. Methuen, London.

Bagnold, R.A., 1953. The surface movement of blown sand in relation to meteorology. *Research Council of Israel Special Publication*, **2**, 89–96.

Bagnold, R.A., 1953. The surface movement of sand in relation to meteorology. In *Desert Research. International Symposium Held in Jerusalem 1952*, pp. 89–96.

Bagnold, R.A., 1990. *Sand, Wind and War*. University of Arizona Press, Tucson.

Ball, J., 1927. Problems of the Libyan Desert. *Geographical Journal*, **70**, 21–38, 105–128, 209–224.

Barndorff-Nielsen, O.E. and Willetts, B.B. (Eds), 1991. *Aeolian Grain Transport*, Vols 1 and 2. Springer Verlag, Vienna.

Beadnell, H.J.L., 1910. The sand dunes of the Libyan Desert. *Geographical Journal*, **35**, 379–395.

Bellair, P., 1953. Sables désertiques et morphologie éolienne. *Nineteenth International Geological Congress, Algiers, 1952*, **7**, 113–118.

Besler, H., 1980. Die Dünen-Namib: Entstehung und Dynamik eines Erg. *Stuttgarter Geographische Studien*, **96**, 208 pp.

Bigarella, J.J., 1975. Lagoa dunefield, state of Santa Catarina, Brazil: a model of eolian and pluvial activity. *Boletim Paranaense de Geociências*, **33**, 133–167.

Brookfield, M., 1970. Dune trends and wind regime in central Australia. *Zeitschrift für Geomorphologie NF Suppl.*, **10**, 121–153.

Brookfield, M.E. and Ahlbrandt, T.S., 1983. *Eolian Sediments and Processes*. Elsevier, Amsterdam.

Busche, D., Draga, M. and Hagedorn, H., 1984. *Les Sables Éoliens Modelés et Dynamique. La Menace Éolienne et son Controle. Bibliographie Annotée*. Eschborn, GTF.

Capot-Rey, R., 1945. Dry and humid morphology of the Western Erg. *Geographical Review*, **35**, 391–407.

Capot-Rey, R. and Capot-Rey, G., 1948. Le déplacement des sables éoliens et la formation des dunes desértiques, d'aprés R.A. Bagnold. *Travaux, Institut de Recherches Sahariennes*, **5**, 47–80.

Chorley, R.J., Dunn, A.J. and Beckinsale, R.P., 1964. *The History of the Study of Landforms*, Vol. 1. Methuen, London.

Chudeau, R., 1909. *Sahara Soudanais*. Amand Colin, Paris, 322 pp.

Chudeau, R., 1920. Etude sur les dunes sahariennes. *Annales de Géographie*, **29**, 334–351.

Clapperton, C.M., 1993. *Quaternary Geology and Geomorphology of South America*. Elsevier, Amsterdam.

Clos-Arceduc, A., 1969. Essai d'explication des formes dunaires sahariennes. *Etudes de Photo-Interpretation*, 4, Institut Géographique National, 66 pp.

Cooke, R.U., Warren, A. and Goudie, A.S., 1993. *Desert Geomorphology*. UCL Press, London.

Cooper, W.S., 1958. Coastal sand dunes of Oregon and Washington. *Geological Society of America, Memoir*, **72**.

Cornish, V., 1899. On Kumatology. *Geographical Journal*, **13**, 618–624.

Cornish, V., 1900. On desert sand-dunes bordering the Nile Delta. *Geographical Journal*, **15**, 1–30.

Cornish, V., 1914. *The Waves of Sand and Snow and the Eddies Which Make Them*. Fisher Unwin, London.

Darwin, G.H., 1884. On the formation of ripple marks in sand. *Proceedings Royal Society of London*, **36**, 18–43.

Dubief, J., 1951. Les vents de sable au Sahara Français. *Colloques Internationaux de CNRS*, **35**, 47–68.

Dubief, J., 1953. Les vents de sable dans le Sahara Français. *Colloques Internationaux CNRS*, **35**, 45–70.

Dufour, L., 1936. Observations sur les dunes du Sahara méridional. *Annales de Géographie*, **45**, 276–285.

Embabi, N.S., 1971. Structures of barchan dunes at the Kharga oases depression, the western Desert, Egypt. *Bulletin de la Société de Géographie d'Egypte*, **43/44**, 53–71.

Finkel, H.J., 1959. The barchans of southern Peru. *Journal of Geology*, **67**, 614–647.

Flamand, G.B.M., 1899. La traversée de l'erg occidental. *Annales de Géographie*, **8**, 231–241.

Folk, R.L., 1971. Genesis of longitudinal and oghurd dunes elucidated by rolling upon grease. *Bulletin Geological Society of America*, **82**, 3461–3468.

Foureau, F., 1905. *Documents scientifiques de la Mission Saharienne*. Masson, Paris.

Frere, H.B.E., 1870. Notes on the Runn of Cutch and neighbouring region. *Journal of The Royal Geographical Society*, **40**, 181–207.

Fryberger, S.G., 1979. Dune form and wind regime. *US Geological Survey Professional Paper*, **1052**, 137–169.

Goudie, A.S., 1972. Vaughan Cormish: Geographer. *Transactions of the Institute of British Geographers*, **55**, 1–16.

Greeley, R. and Iversen, J.D., 1985. *Wind as a Geological Process*. Cambridge University Press, Cambridge.

Grove, A., 1958. The ancient erg of Hausaland and similar formations on the south side of the Sahara. *Geographical Journal*, **124**, 528–533.

Grove, A.T., 1969. Landforms and climatic change in the Kalahari and Ngamiland. *Geographical Journal*, **135**, 191–212.

Grove, A.T. and Warren, A., 1968. Quaternary landforms and climate on the south side of the Sahara. *Geographical Journal*, **134**, 194–208.

Hastenrath, S.L., 1967. The barchans of the Arequipa region, southern Peru. *Zeitschrift für Geomorphologie NF*, **11**, 300–331.

Hesp, P.A. and Fryberger, S., 1988. Special issue on eolian sediments. *Sedimentary Geology*, **55**.

Higgins, G.M., Ahmad, M. and Brinkmann, R., 1974. The Thal Interfluve, Pakistan, Geomorphology and depositional history. *Geologie en Mijnbouw*, **52**, 147–155.

Holm, D.A., 1960. Desert morphology on the Arabian peninsula. *Science*, **132**, 1369–1379.

Hoyt, J.H., 1966. Air and sand movements in the lee of dunes. *Sedimentology*, **7**, 137–144.

Hume, W.F., 1925. *The Geology of Egypt*. Ministry of Finance, Cairo.

Hunter, R.E., 1980. Quasi-planar adhesion stratification – an eolian structure formed in wet sand. *Journal of Sedimentary Petrology*, **50**, 263–266.

Inman, D.L., Ewing, G.C. and Corliss, J.D., 1966. Coastal sand dunes of Guerero Negro, Baja California, Mexico. *Bulletin of the Geological Society of America*, **77**, 787–802.

Jennings, J.N., 1975. Desert dunes and estuarine fill in the Fitzroy Estuary (North-western Australia). *Catena*, **2**, 215–258.

Kármán, T. von, 1934. Some aspects of the turbulence problem. *Proceedings of the Fourth International Congress on Applied Mechanics*, Cambridge, 54–91.

Kármán, T. von, 1934. Turbulence and skin friction. *Journal of Aeronautical Sciences*, **1**, 1–20.

King, D., 1960. The sand ridge deserts of South Australia and related aeolian landforms of the Quaternary arid cycles. *Transactions of the Royal Society of South Australia*, **83**, 99–109.

King, W.H.J., 1916. The nature and formation of sand ripples and dunes. *Geographical Journal*, **47**, 189–209.

King, W.H.J., 1918. Study of a dune belt. *Geographical Journal*, **51**, 16–33.

Kocurek, G., 1981. Significance of interdune deposits and bounding surfaces in aeolian dune sands. *Sedimentology*, **28**, 753–780.

Lancaster, N., 1980. The formation of seif dunes from barchans – supporting evidence for Bagnold's model from the Namib Desert. *Zeitschrift für Geomorphologie NF*, **24**, 160–167.

Lancaster, N., 1988. *A Bibliography of Dunes: Earth, Mars and Venus*. NASA Contractor Report 4149.

Lancaster, N., 1995. *Geomorphology of Desert Dunes*. Routledge, London.

Lettau, L. and Lettau, H., 1978. Bulk transport of sand by the barchans of the Pampa de la Joya in southern Peru. *Zeitschrift für Geomorphologie NF*, **13**, 182–195.

Livingstone, I., 1986. Geomorphological significance of wind flow patterns over a Namib linear dune. In *Aeolian Geomorphology*, W.G. Nickling (Ed.). Allen and Unwin, Boston, pp. 87–112.

Livingstone, I., 1988. New models for the formation of linear sand dunes. *Geography*, **73**, 105–115.

Livingstone, I., 1990. Desert sand dune dynamics: review and prospect. In *Namib Ecology: 25 Years of Namib Research*, M.K. Seely (Ed.). Transvaal Museum Monograph No. 7, Pretoria, pp. 47–53.

Livingstone, I. and Warren, A., 1996. *Aeolian Geomorphology*. Longman, Harlow.

Long, J.T. and Sharp, R.P., 1964. Barchan dune movement in Imperial Valley, California. *Bulletin of the Geological Society of America*, **75**, 149–156.

Mabbutt, J.A., 1968. Aeolian landforms in central Australia. *Australian Geographical Studies*, **6**, 139–150.

Madigan, C.T., 1936. The Australian sand-ridge deserts. *Geographical Review*, **26**, 205–227.

Mainguet, M., 1975. Etude comparée des ergs à l'échelle continentale (Sahara et déserts d'Australie). *Bulletin de l'Association Géographie Français*, **25**, 135–140.

Mainguet, M. and Callot, Y., 1978 L'erg de Fachi-Bilma (Tchad-Niger) *CNRS Mémoires et Documents*, NS **18**, 184 pp.

Marsh, G.P., 1864. *Man and Nature*. Scriber, New York.

McKee, E. (Ed.), 1979. A study of global sand seas. *US Geological Survey Professional Paper* 1052.

McKee, E. and Moiola, R.J., 1975. Geometry and growth of the White Sands dune field, New Mexico. *US Geological Survey Journal of Research*, **3**, 59–66.

Monod, R., 1957. Majâbat al-Khoubrâ: contribution à l'étude de l' 'Empty Quarter' ouest-saharien. *Mémoires IFAN*, **52**, 406 pp.

Monod, T., 1928. Une traversée de la Mauritanie occidentale. *Revue de Géographie Physique et de Géologie Dynamique*, **1**, 3–25, 88–106.

Muhs, D.R. and Maat, P.B., 1993. The potential response of aeolian sands to greenhouse warming and precipitation reduction on the Great Plains of the United States. *Journal of Arid Environments*, **25**, 351–361.

Nickling, W.G. (Ed.), 1986. *Aeolian Geomorphology*. Allen and Unwin, Boston.

Norris, R.M. and Norris, K.S., 1961. Algodones dunes of south-eastern California. *Bulletin Geological Society of America*, **72**, 605–620.

Porch, D., 1985. *The Conquest of the Sahara*. Cape, London.

Prandtl, L., 1935. The mechanics of viscous fluids. In *Aerodynamic Theory*, Vol. III, Durand, W.F. (Ed.). Springer, Berlin, pp. 304–208.

Pye, K. (Ed.), 1993. *The Dynamics and Environmental Context of Aeolian Sedimentary Systems*. The Geological Society, London.

Pye, K. and Lancaster, N. (Eds.), 1993. Aeolian sediments ancient and modern. In *International Association of Sedimentologists Special Publication 16*, Blackwell Science, Oxford.

Pye, K. and Tsoar, H., 1990. *Aeolian Sand and Sand Dunes*. Unwin Hyman, London.

Rolland, G., 1881. Sur les grands dunes de sable du Sahara. *Bulletin de la Société Géologique de France*, Sér. 3, **10**, 30–47.

Rolland, G., 1890. Les grandes dunes de sable du Sahara. *Compte Rendu des Séances de la Société de Géographie*, Année 1890, pp. 158–165.

Sarre, R.D., 1988. Evaluation of aeolian sand transport using intertidal zone measurements. Saunton Sands, England. *Sedimentology*, **35**, 671–679.

Sharp, R.P., 1966. Kelso dunes, Mohave Desert, California. *Bulletin of the Geological Society of America*, **77**, 1045–1074.

Simonett, D.S., 1960. Development and grading of dunes in western Kansas. *Annals of the Association of American Geographers*, **50**, 216–241.

Singh, S., 1977. Sand dunes and palaeoclimate in Jodhpur district, western Rajasthan (India). *Man and Environment*, **1**, 7–15.

Smith, H.T.U., 1963. *Eolian geomorphology, wind direction, and climatic change in North Africa*. Final Report AF19 (628) 298 Air Force Cambridge Research Laboratories, Bedford, MA.

Smith, H.T.U., 1965. Dune morphology and chronology in central and western Nebraska. *Journal of Geology*, **73**, 557–578.

Stokes, S., Thomas, D.S.G. and Washington, R., 1997. Multiple episodes of aridity in southern Africa since the last interglacial period. *Nature*, **388**, 154–158.

Tchakerian, V. (Ed.), 1995. *Desert Aeolian Processes*. Chapman and Hall, London.

Thorne, C.R., MacArthur, R.C. and Bradley, J.B. (Eds), 1988. *The Physics of Sediment Transport by Wind and Water*. American Society of Civil Engineers, New York.

Tinkler, K.J., 1985. *A Short History of Geomorphology*. Croom Helm, London.

Tricart, J. and Brochu, M., 1955. Le grand erg ancien du Trarza et du Cayor (sud-ouest de la Mauritanie et nord du Sénégal). *Revue de Géomorphologie Dynamique*, **4**, 145–176.

Tsoar, H., 1974. Desert dunes morphology and dynamics, El Arish (Northern Sinai). *Zeitschrift für Geomorphologie Suppl.*, **20**, 41–61.

Tsoar, H., 1978. *The Dynamics of Longitudinal Dunes. Final Technical Report*. European Research Office, US Army, London.

Tsoar, H., 1994. Classics in physical geography revisited. Bagnold R.A. The physics of blown sand and desert dunes. *Progress in Physical Geography*, **18**, 91–96.

Twidale, C.R., 1972. Evolution of sand dunes in the Simpson Desert, central Australia. *Transactions of the Institute of British Geographers*, **56**, 97–109.

Urvoy, Y., 1933. Les formes dunaires a l'ouest du Tchad. *Annales de Géographie*, **42**, 506–515.

Verstappen, H.T., 1970. Aeolian geomorphology of the Thar Desert and palaeoclimates. *Zeitschrift für Geomorphologie*, NF Suppl. 10, 1-4-20.

Warren, A., 1969. A bibliography of desert dunes and associated phenomena. In *Arid Lands in Perspective*, W.G. McGinnies and P. Paylore (Eds). University of Arizona Press, Tucson.

Warren, A., 1971. Dunes in the Ténéré Desert. *Geographical Journal*, **137**, 458–461.

Warren, A., 1976. Morphology and sediments of the Nebraska Sand Hills in relation to Pleistocene winds and development of eolian bedforms. *Journal of Geology*, **84**, 685–700.

Wasson, R.J. and Hyde, R., 1983. A test of granulometric control of desert dune geometry. *Earth Surface Processes and Landforms*, **8**, 301–312.

Watson, A., 1990. The control of blowing sand and mobile desert dunes. In *Techniques for Desert Reclamation*, A.S. Goudie (Ed.). Wiley, London, pp. 35–86.

Wells, G.L. and Zimbelman, J.R., 1997. Extraterrestrial arid surface processes. In *Arid Zone Geomorphology*, 2nd edn, D.S.G. Thomas (Ed.). Wiley, Chichester, pp. 659–690.

Wilson, I.G., 1971. Desert sandflow basins and a model for the development of ergs. *Geographical Journal*, **137**, 180–199.

Wilson, I.G., 1972. Universal discontinuities in bedforms produced by the wind. *Journal of Sedimentary Petrology*, **42**, 667–669.

Wilson, I.G., 1973. Ergs. *Sedimentary Geology*, **10**, 77–106.

2 Recent Investigations of Airflow and Sediment Transport Over Desert Dunes

WILLIAM G. NICKLING[1] and CHERYL MCKENNA NEUMAN[2]
[1] Department of Geography, University of Guelph, Canada
[2] Department of Geography, Trent University, Canada

During the past two decades, desert dune research has changed significantly from descriptive studies of dune morphology and the textural characteristics of their sediments, to those more concerned with the processes of sediment transport and deposition as they relate to the initiation, development and maintenance of dune forms. This work has been carried out at a range of spatial and temporal scales and includes both field and laboratory wind tunnel measurements of wind flow and sediment transport dynamics in relation to dune development and morphology. Significant advances have also been made in the development of numerical models that can aid in the description and explanation of dune initiation and migration of dunes as well as the simulation of air flow patterns and sediment transport over them. Investigations of dune morphology based on traditional geomorphological techniques have been thoroughly addressed in numerous recent texts and research publications. Consequently, this review concentrates on recent research on the modelling and field verification of airflow and sediment transport over free form dunes. In doing so we define the present understanding of these processes, identify those factors which have hindered research progress, and offer some suggestions for future research. We begin with a summary of the state of knowledge regarding numerical models of airflow and dune dynamics, include a discussion on the utility and limitations of hardware simulation, and progress to a review of recent contributions from field measurement programs. In the paper we also attempt to evaluate the wide range of instrumentation that has been used in the laboratory and field to investigate airflow and sediment transport over dunes and make suggestions on instrument needs for the future.

INTRODUCTION

To many individuals, dunes and extensive sand seas (ergs) are synonymous with deserts, despite the fact that major sand deposits occupy only approximately 25% of the world's deserts. Major sand seas are most extensive in the old world deserts of the Sahara, Arabia, central Asia, Australia and southern Africa where they occupy between 20 and 45% of the land area that is classified as arid (Lancaster, 1994). However, smaller sand accumulations and dune fields, of much more limited areal extent, can be found throughout the world in almost all arid or semi-arid regions. Although desert dunes cover a relatively small percentage of the earth's surface, they represent an important geomorphological form that has been studied

Aeolian Environments, Sediments and Landforms. Edited by A.S. Goudie, I. Livingstone and S. Stokes.
© 1999 John Wiley & Sons, Ltd.

extensively, resulting in a voluminous literature that extends back more than a century (see Chapter 1).

Over the past decade this literature has been thoroughly reviewed and augmented in books on dunes by Pye and Tsoar (1990), Lancaster (1995), in more general desert geomorphology texts by Thomas (1989), Cooke *et al.* (1993), Livingstone and Warren (1996) and in review articles by Nickling (1994) and Lancaster (1994). More specific reviews of linear dunes have been published by Tsoar (1989) and Lancaster (1982) and star dunes by Lancaster (1989).

Loose sand deposits in ergs and more localised sand accumulations are frequently characterised by the development of bedforms which vary in size, form and pattern. Bedforms frequently develop as regular repeating patterns resulting from the complex interaction between the shearing force of the wind and the sediment on the surface. As the bedforms begin to grow they alter the airflow above them, often developing a quasi-equilibrium between the fluid flow and the emerging bedform morphology. This quasi-equilibrium can be further complicated by the temporal and spatial variability in sand supply, wind direction, wind speed and in some cases, the presence of vegetation.

Many researchers have argued that aeolian bedforms develop as an hierarchical arrangement of superimposed forms (e.g. Allen, 1968; Wilson, 1972; Lancaster, 1995). Lancaster (1995) suggests that three orders of aeolian bedforms can be identified with only the first two occurring in all sand seas: (a) wind ripples (0.1 to 1 m spacing); (b) individual simple dunes or superimposed dunes on compound and complex dunes (50 to 500 m spacing); and (c) compound and complex dunes or draa (spacing >500 m). Each of these hierarchical orders operates at differing temporal and spatial scales. Ripples, for example, develop on very short time scales (minutes to hours) in response to changes in wind speed and direction. In contrast draa, because of their large areal extent and sand volume, may take millennia to approach some form of equilibrium with the wind regime and available sand supply. However, in that wind regimes and sediment supply may well change over these extended time scales, it is possible that the higher order bedforms never reach equilibrium with prevailing wind conditions.

A wide range of transverse and longitudinal dune forms have been described in the literature with a variety of local names often being used to describe the same forms. In an attempt to bring some order to the variety and complexity of desert dune forms, numerous classification systems have been developed (e.g. Melton, 1940; Hack, 1941; McKee, 1979; Greeley and Iversen, 1985; Thomas, 1989; Pye and Tsoar, 1990; Livingstone and Warren, 1996). These classifications take many forms and are based on numerous criteria including: formative processes, orientation, general morphology, and the presence or absence of vegetation. A recent comprehensive classification that focuses on dune formation has been presented by Livingstone and Warren (1996) (Figure 2.1). This system divides dunes into free and anchored forms with further subdivision based on morphology/orientation, vegetation and topography.

During the past two decades, desert dune research has changed significantly from descriptive studies of dune morphology and the textural characteristics of their sediments, to those more concerned with the processes of sediment transport and deposition as they relate to the initiation, development and maintenance of dune

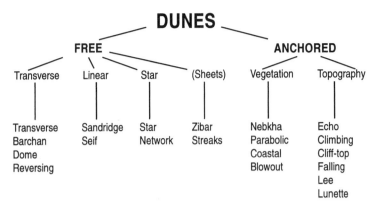

Figure 2.1   A framework for dune classification (after Livingstone and Warren, 1996)

forms. This work has been carried out at a range of spatial and temporal scales and includes both field and laboratory wind tunnel measurements of wind flow (e.g. Lancaster, 1985; Tsoar *et al.*, 1985; Livingstone, 1986; Mulligan, 1988; Frank and Kocurek, 1996a,b; White, 1996; Wiggs *et al.*, 1996a,b; Hesp and Hastings, 1998) and sediment transport dynamics (e.g. Tsoar, 1983, 1993; Lancaster *et al.*, 1996; Wiggs *et al.*, 1996a; McKenna Neuman *et al.*, 1997) as they relate to dune development and morphology. Significant advances have also been made in the development of numerical models that can aid in the description and explanation of dune initiation and migration of dunes as well as the simulation of airflow patterns and sediment transport over them (e.g. Howard *et al.*, 1978; Hunt *et al.*, 1988a,b; Howard and Walmsley, 1985; Werner, 1995).

As previously noted, investigations of dune morphology based on traditional geomorphological techniques (e.g. topographic survey, ripple orientation analysis, sedimentological and stratigraphic analyses, regional wind data collection, erosion pin survey, etc.) have been thoroughly addressed in numerous texts and research publications. Consequently, the following review concentrates on recent research on the modelling and field verification of airflow and sediment transport over free form dunes. In doing so we will attempt to define our present understanding of these processes, identify those factors which have hindered research progress, and finally, offer some suggestions for future research. We begin with a summary of the state of knowledge regarding numerical models of airflow and dune dynamics, include a discussion on the utility and limitations of hardware simulation, and progress to a review of recent contributions from field measurement programmes.

NUMERICAL MODELS

Topographic Effects on Flow

Numerical modelling of the effects of topography on airflow has been carried out in the context of both meteorology and fluid mechanics. There are innumerable applications for these models, many involving the siting of wind turbines, and the diffusion of air pollution. In rare instances, theorists have looked at the flow over aeolian sand dunes. The literature is dominated by the early analysis of Jackson and Hunt (1975), with many modifications to and extensions of this work published over the last two decades. Owing to the comprehensive scope of this paper, we address below only those aspects which are pertinent to the understanding of dune dynamics.

The 1975 analysis of Jackson and Hunt divides the flow over a hill into two regions which include an inner layer which is assumed to follow the contour of the hill, and an outer inviscid region. Central to this theory is the concept of a perturbation velocity. If $u_0(z)$ is the reference upwind velocity at height z, and the distance above the surface of the hill is referred to as the displacement Δz, then in the region close to the surface of the hill the velocity u is approximated by the upstream velocity at the same displacement, $u(x,z) = u_0(\Delta z\ (x,z))$. Continuity implies the existence of a vertical velocity so that in the outer region $u\ (x,z) = u_0(z) + \Delta u\ (x,z)$, though as $z \to \infty$, Δu and $v \to 0$. In the outer region, the perturbation velocity Δu induces a perturbation pressure p. Since pressure is continuous, a pressure gradient must exist in the inner region which locally affects the velocity. Therefore within this inner region, the horizontal velocity cannot simply equal the displaced upwind velocity $u_0(\Delta z)$, but has a perturbation $\Delta u'$ so that $u = u_0(\Delta z) + \Delta u'$. Central to the Jackson and Hunt (JH) model is estimation of the perturbation velocities Δu, $\Delta u'$ and v.

The JH analysis was later revisited and extended by Hunt et al. (1988a,b), owing to deficiencies highlighted in other theoretical and field based work (i.e. Sykes, 1980; Walmsley et al., 1982; Bradley, 1983; Jensen and Zeman, 1985). The revised theory further divides the outer region into an outer and middle layer; and the inner region into a shear stress layer and an inner surface layer where shear stress is effectively constant with height (Figure 2.2). The model still applies only to very low hills where $h \gg L$, and h is the height of the hill at its crest. L is a characteristic length scale whereby $2L$ is taken as the distance between points corresponding to the half height of the hill. The wind speed at the top of the middle layer (h_m) is taken as the upwind reference speed which determines the pressure perturbations in the flow. In the original version of the JH model, the reference wind speed was taken as that at the top of the boundary layer. However, it is now recognised that the disturbance only reaches the top of the boundary layer in the case of hills which are very long in comparison to the boundary layer depth. For an adiabatic atmosphere with ln $(L/z_0) \gg 1$ and $L < h_{bl}$, the height of the middle layer can be approximated from $h_m = L/(\ln(L/z_0))^{0.5}$. The thickness of the inner region (l), is given by l ln $(l/z_0) = 2k^2L$. Within this inner region, the maximum of the perturbation velocity is located at height h_{max}, with Hunt et al. (1988) suggesting $h_{max} \sim l/3$, though this approximation is somewhat affected by the roughness length (z_0).

The depth of the inner surface layer can be estimated from $l_s \sim (lz_0)^{0.5}$. Within this inner surface layer, the perturbation velocity decreases to zero near the surface, while the gradient of shear stress is also zero. In the context of sediment transport on dunes, wind speed measurement within this strata is required for evaluation of the surface shearing stress (τ_0). As an example of its typical depth for desert dunes, McKenna Neuman et al. (1997) estimate that l and l_s are in the order of 1 m and

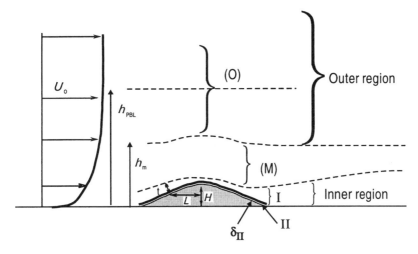

Figure 2.2   Definition sketch for airflow over a low hill (after Hunt *et al.*, 1988b)

1 cm, respectively, for two transverse reversing dunes of variable size at Silver Peak, Nevada (e.g. h = 6 m and L= 20 m; h = 13 m and L = 28 m). This illustrates a significant limitation for field studies in that it is commonly impossible to place a rotating cup anemometer in the constant stress region of most dunes. This problem was first identified, and the implications addressed, in work by Rasmussen (1988) and Jensen and Zeman (1985).

At the core of the JH analysis is the computation of the fractional speed-up ratio. This ratio is expressed by $\Delta S \, (x, \, \Delta z) = \Delta u \, (x, \Delta z)/u_0(\Delta z)$, where $\Delta u(x,\Delta z)$ is the increase in wind speed at a given point above the hill relative to the undisturbed speed at the same displacement above the surface. At the ridge crest, ΔS is approximated by $2h/L$, so that the fractional speed up ratio is very sensitive to the slope of the hill (h/L), whereas the roughness length plays a relatively minor role.

Another way of describing the wind over a hill is by means of the speed-up factor, given as $S = u(z)/u_0(z)$. It is commonly used in the context of field studies of dune development and maintenance (e.g. Lancaster, 1985; Lancaster *et al.*, 1996; McKenna Neuman *et al.*, 1997). Jackson and Hunt suggest that S is much more sensitive to displacement (Δz) above the surface than is ΔS, and so, in the absence of a standard instrument height ΔS is the preferred parameter.

An example of the normalised variation in shear stress along a hillslope is shown in Figure 2.3 (a) and (b), as computed by Jensen and Zeman (1985). The general trend which appears in Figure 2.3 (a) is characteristic of all numeric models of topographic modification of airflow, though in detail and magnitude substantive differences between model outcomes do arise. Shear stress is moderately reduced on the lower flank of relatively steep hills where stagnation can be significant. At some point midway up the slope, shear stress increases rapidly in response to the speed up in airflow, and may double that for the upwind reference surface. However, slightly before the hill crest, the numerical model indicates that shear stress drops. As explained by Jensen and Zeman (1985), this forward phase shift is inherent in the

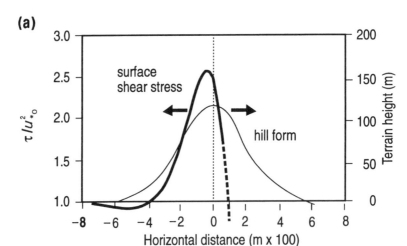

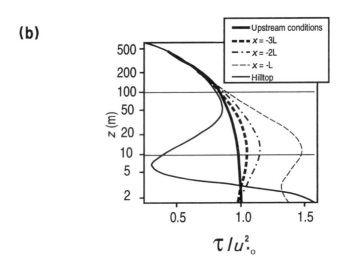

Figure 2.3 (a) A model based prediction of the variation in surface shear stress along a hillslope. (b) Normalised stress profiles at various locations on the hill (after Jensen and Zeman, 1985)

equations of motion, even in their most rudimentary form. The precise location of the stress maximum depends on the turbulence closure scheme, though it is usually somewhere between -0.2 and $-0.5\ L$. Assuming that on a sand dune the mass transport rate (q) is approximately a function of u_*^3, the implication is that when erosion occurs on the stoss side of a dune, increasing in magnitude as elevation increases, the potential exists for some deposition beyond the shear stress maximum (e.g. within the dune crest).

Most of the early numerical models assume that the incident airflow is normal to a two-dimensional ridge. In situations where the flow is oblique to the ridge, it can

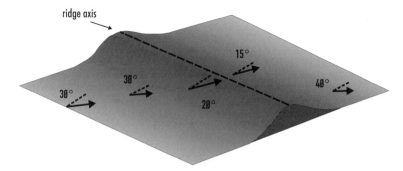

Figure 2.4   Orientation of the velocity vector at 1 m height across a two-dimensional ridge with lee side separated flow. Upstream wind approach angle is 30° from the normal to the ridge axis (after Rasmussen, 1989, based on Mikkelsen, 1989)

be divided up into its vector components relative to the dune axis. As explained by Jackson (1977), the perturbation pressure set up by the component normal to the hill causes a directional change in the wind, with the angle to the ridge axis increasing as the flow passes over the top. Figure 2.4 illustrates this effect from the work of Mikkelsen (1989) on a dune at Ferring on the Danish North Sea coast.

The extension of the JH model to three dimensions followed quickly on the heels of its introduction in the work of Mason and Sykes (1979). Walmsley *et al.* (1982) later adopted the Mason and Sykes extension to produce a very sophisticated three-dimensional, terrain-following, numerical scheme.

Paralleling the JH analysis is the considerable contribution of Taylor and his colleagues. In 1974, Taylor and Gent introduced two separate approaches to modelling airflow over low hills; one based on the mixing length hypothesis (similar to JH), and the other based on closure of the turbulent energy equation. However, the limitations of mixing length theory and the assumption of constant advection velocities soon became apparent in studies by Walmsley *et al.* (1982), Taylor *et al.* (1983) and Walmsley *et al.* (1986). Subsequently, Beljaars *et al.* (1987) founded a linear mixed spectral finite difference model (MSFD) on the analysis of turbulent energy. Though considerably more sophisticated, this model was still not found to perform well in predicting airflow on the lee side of hills. Similarly, a non-linear extension of the MSFD model, developed by Xu and Taylor (1992) to simulate airflow over an infinite series of two-dimensional wave-like topographic perturbations, still has problems with model closure in the case of flow separation. Various other turbulence closure schemes are explored further in Ayotte *et al.* (1994).

Numerical Models of Dune Dynamics

The first comprehensive numerical model of dune maintenance was proposed by Howard *et al.* (1978). This model takes a finite difference approach to solving by iteration the equation of sediment continuity for interdependent sections of a barchan dune surface. The rectilinear section boundaries are defined in part by the streamlines of flow. Velocity and streamline data are provided for the model through field measurement and hardware simulation. Sediment movement through the individual sections of dune surface is based on computation from Bagnold's mass transport model. The effect of this sediment motion on momentum loss from the airstream is addressed through an adjustment of the roughness term in Bagnold's (1941) velocity profile equation. The model appears to offer a reasonable first approximation to the dynamics of a barchan dune. No improvement in model performance was observed with adjustment for the effect of slope angle on the local mass transport rate (q), and the inclusion of a lag between the inception of a wind event and sediment transport. The authors conclude their work with the following statement regarding goals for future research: 'If the interactions between sand transport, dune form and fluid flow could be completely modelled, the limits of stability of the dune and the factors controlling the form and size of the dune could be readily determined. Ideally, such a model should first predict the air flow over any three-dimensional sand deposit . . . Then, using the appropriate bedload transport equations, the obstacle would be modified by erosion and deposition in a simulation model, proceeding by steps to an equilibrium form (if, indeed, one is established).' Further discussion points to the difficulty of working with such models of fluid flow and the need for field measurement of sand transport.

Shortly thereafter, Howard and Walmsley (1985) used this very approach to produce a simulation model of the development of a barchanoid form from a symmetrical 'cosine-squared' hill. The main limitations of this model are that: (1) the form must be isolated on a flat plate; and (2) model instability develops after 20 to 30 iterations, so that the surface becomes 'rumpled'. This instability is believed to be a response to the acute sensitivity of q to small variations in u, and to small-scale topographic perturbations. The high computational costs of running the model are also noted.

In the most recent numerical model of aeolian dunes, Werner (1995) takes a completely fresh, and alternative approach to modelling not just a single dune form, but an entire field of dunes of varied form. Werner argues that because of the complicated nature of the relationship between air flow and transport/depositional processes over dunes, a reductionist path from the physics of sand transport to the evolution of a dune field is currently not feasible. Moreover, he contends that most available models of sand transport and dune development suffer from an empirical nature that precludes wider application and the development of testable hypotheses at the dune field scale. Werner's model is based on the premise that dune-forming systems are complex systems subject to strong non-linear dissipative processes that evolve to a finite number of steady states (*attractors*). Systems of this nature demonstrate emergent behaviour that is self organising on macroscopic space and time scales. In the computer simulation, dunes are built from slabs of sand, the positions of which are constrained to lie on a square lattice. Sand slabs are moved individually in a manner that simulates the movement of sand grains by saltation in air. The movement and ultimate placement of the individual slabs is a function of the specified wind regime, and follows a set of relatively simple rules and probabilities that are based on established theory and empirical observation (Figure 2.5). In accord with its apparent simplicity, this model requires only a few input parameters to simulate the development of barchans and transverse ridges, as well

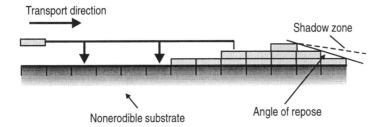

Figure 2.5 Conceptual basis of Werner's dune simulation transport algorithm (after Werner, 1995)

as linear and star dunes (Figure 2.6). This modelling approach has the potential to be a very powerful tool in that it can make testable predictions for complex three-dimensional morphologies within complicated wind regimes. To paraphrase Hallet (1990), 'the systematic study of geomorphic self-organisation is just starting. It offers difficult challenges but also exceptional opportunities.'

Modelling of Dune Initiation

Numerous conceptual models have been put forward for the initiation of various dune forms. These models, however, are most frequently based on inference from observations of present dune forms and limited direct field measurements (e.g. Bagnold, 1941; Cooper, 1958; Hanna, 1969; Folk, 1971, 1976; Tseo, 1993). As a result of the long time normally associated with the initiation and development of aeolian dunes, the evolutionary sequence has been documented in only a few cases for very simple systems (e.g. Kocurek *et al.*, 1992).

Numerical modelling of dune initiation and development is still in its infancy and in some ways has lagged behind research in related fields. It is therefore a major challenge for future research. There is no single reason for the relatively slow advancement in numerical modelling, but it may in part be related to the fact that no fundamental unifying theories of dune development have emerged, and no path to an all encompassing framework has been identified (Werner, 1995).

WIND TUNNEL MODELLING

Preliminary Scaling Considerations

The central involvement of wind tunnel experimentation in the study of airflow over hills and dunes has been the production of empirical data for numerical model extension and calibration, as well as, the simulation of settings for which an analytical solution is currently impossible. For example, in wind tunnel simulation of steep hills, the effects of flow separation can be examined while this situation is not well addressed by current numerical models. The flow can also be fully con-

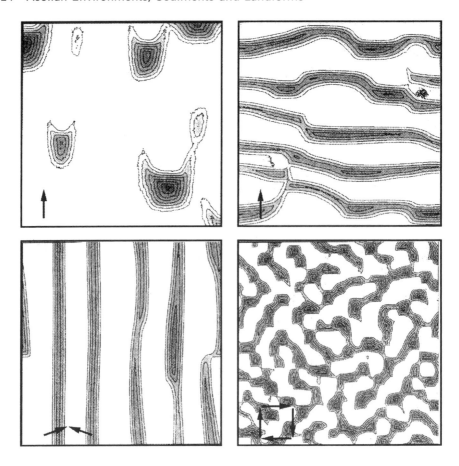

Figure 2.6 Simulated dune fields (after Werner, 1995).

trolled over a broad range of wind speed, and because of the small scale, more detailed measurements can be made than in the case of field work.

Still, physical modelling of dunes and associated surface features has been rare because of the difficulties in achieving correct simulation. In his comprehensive review of the scaling requirements, White (1996) suggests that 'Great insight and understanding of the physical flow of the atmosphere can be obtained if correct similitude principles are obeyed'. Most boundary layer wind tunnels reasonably simulate the surface layer flow in the lowest 10–100 m of the atmospheric boundary layer. Fortunately, this scale of flow is directly relevant to the problem of sand dune formation and sediment movement by aeolian processes. However, large scale features such as entire dune fields (e.g. greater than a few km in length or higher than 100 m) cannot be studied in wind tunnels because of the inability to simulate the three-dimensional Ekman spiral which lies above the surface layer.

From the equations of motion, White (1996) indicates that matching three non-dimensional coefficients (Rossby number, bulk Richardson number, Reynolds number) will simulate exactly in a wind tunnel, the physical processes of the full scale flow. The Rossby and Richardson numbers can usually be ignored in the limited case of the lowest 10–15% of an adiabatic atmosphere. The Reynolds number ($Re = uL/\nu$, where ν is the kinematic viscosity of the fluid) is the most significant ratio. However, if strict adherence to the Reynolds number criterion were to be upheld, almost no field cases could be modelled, as the model Re is usually several orders of magnitude less than that of the field topography. For example, in their simulation of a varying angle of incident flow over a two-dimensional ridge, Tsoar et al. (1985) report that even with the highest freestream velocity attainable, Re was two orders of magnitude lower than the field prototype. Fortunately, there is a route around this dilemma. Provided the Re number exceeds the minimum Reynolds independence number (Townsend, 1956), the model flows will be dynamically similar to the full scale prototype. White (1996) suggests that this independence number is between 100 000 and 500 000 for most aeolian dunes. In any event, wind tunnel simulation generally requires a relatively large scale reproduction of the dune, and so, it is not possible to model very large features such as multiple dunes and dune fields.

Additional requirements for adequate wind tunnel simulation include: (1) a geometrically similar model; (2) a Reynolds roughness number (u_*z_0/ν, where z_0 is the roughness length) in excess of 2.5 for aerodynamically rough flow (White, 1996), and (3) a z_0/h value equivalent to the full scale prototype so that correct simulation of the pressure distribution is obtained across the dune (Jensen, 1958). The third criterion is often distorted to uphold the requirement of rough flow, as z_0/h is usually very small in the field. A related conflict arises in the correct simulation of the inner layer in the sense of the Jackson and Hunt two layer model for flow over a hill. For this inner layer to be deep enough to accommodate measurement in wind tunnel simulation, the full scale feature must be either very large or very smooth (Finnigan et al., 1990). If the feature does not fit either of these criteria, then the modeller must choose between development of a thick inner layer in which instruments can be placed, and correct simulation of the turbulence. This is especially a dilemma for modellers of grassy or forested slopes.

Similarly, the atmospheric boundary layer flow has a negligible longitudinal pressure gradient. The wind tunnel gradient can be set to zero by adjusting the ceiling height along the working section. The model cross-section should be sufficiently small that local flow accelerations caused by flow blockage do not affect the longitudinal pressure gradient, and thereby, distort the flow. As one example of this potential effect, Finnigan et al. (1990) attribute a small increase in the fractional speed-up ratio (e.g. 0.06 of a total of 0.54 for ΔS) to blockage in their wind tunnel study of flow over a two-dimensional ridge. Unfortunately, little can be done to avoid the influence of the tunnel walls which becomes particularly severe in the case of a two-dimensional model ridge stretching across the entire tunnel floor.

The fetch length of the tunnel has an important bearing on the boundary layer height and the depth of the inner constant stress layer, as well as the longitudinal velocity spectrum. White (1996) states that as a general principle, the longer the flow development section, the better is the match of the normalised turbulence intensity profile. As a rule of thumb, he suggests that 10–25 boundary layer height (BLH) lengths are needed to match the mean dimensional velocity profile, 50 BLH lengths will give a match to the turbulence intensity profile, and 100–500 BLH lengths are required for simulations of the normalised turbulent energy spectra. These distances can be shortened appreciably by the introduction of flow spires or similar flow tripping devices at the tunnel entrance (Counihan, 1969).

Finally, a rather disappointing outcome of White's examination of scaling criteria for wind tunnel simulation is the realisation that it is generally not possible correctly to model wind flow past a large dune with saltation present. The central problem stems from the need to satisfy two dimensionless parameters simultaneously: (1) $\rho D_d / \rho_p\, D_p$ where D_d is dune length and D_p is particle diameter, ρ is atmospheric density and ρ_p is particle density; and (2) the Froude number u^2/gL. The reader is referred to White's meticulous analysis which demonstrates that the first criterion is virtually impossible to match. Froude number matching requires a very low wind tunnel speed, so that the modelling material must also have a very low threshold for motion. Unfortunately, meeting this requirement will result in suspension of the material once threshold is reached, as opposed to the desired saltation. Also, *Re* number matching for correct simulation of fully turbulent flow generally requires very high wind speeds. Not surprisingly, wind tunnel simulation usually leads to geometric distortion of erosion and deposition patterns around dunes.

Simulation of Topographically Altered Boundary Layer Flows

Five primary objectives appear in the published work on wind tunnel simulation of topographically altered boundary layer flows. They include:

(i)   validation of numerical analysis for stoss flow on a low symmetrical, two-dimensional ridge
(ii) investigation of the effects of streamline curvature; detailed analysis of Reynolds stresses
(iii) detailed measurement of flow separation characteristics
(iv) extension of analysis to complex hill shapes, including three-dimensional forms
(v)   consideration of the effects of oblique flow

We briefly address here the current state of knowledge regarding each of these objectives, concentrating on a small selection of recent and benchmark papers.

Wind tunnel experiments on airflow over low hills (e.g. Tsoar *et al.*, 1985, Gong and Ibbetson, 1989; Finnigan *et al.*, 1990; Wiggs *et al.*, 1996) generally confirm the main qualitative features of the stoss side mean velocity field as predicted by the JH model, provided the hill shape is taken to be defined by the continuation of the upstream surface streamline. That is, if there is a steady separation bubble, it must be considered part of the hill. Beginning at the toe, an adverse pressure gradient leads to a reduction in velocity. This is followed by flow acceleration toward the crest, with peak speed up occurring just short of the brink. The degree of phase shift, however, does appear to drift depending on the specific experimental situation, with data from Finnigan *et al.* (1990) showing least departure of the maximum speed up from the crest. The considerable disagreement in the literature over the height of maximum velocity perturbation (i.e. ranging from l to $l/4$) does not appear to have been resolved at this point through physical simulation.

Apart from analysis of the mean flow, the emphasis in most wind tunnel simulation studies has been on detailed measurement of the turbulent stresses and the momentum budget using hot wire or pulse wire anemometers. The importance of streamline curvature clearly emerges in this simulation work (e.g. Finnigan *et al.*, 1990; Gong and Ibbetson, 1989; Wiggs *et al.*, 1996). The basis for understanding streamline curvature effects lies in the well developed analogy between the action of centrifugal forces on a curved shear flow and the effects of buoyancy (Bradshaw, 1973). Concave streamlines, for example, appear near the upwind and downwind edges of the hill. Here they are associated with instability and thereby promote increased turbulent mixing. Streamwise vortex generation is initiated, with the vortex cores either fixed in one position or migrating laterally. The possible association of these with sand 'streamers' remains unexplored. The reverse is true at the hill crest where convex streamlines damp vertical motion and suppress turbulence. Finnigan *et al.* suggest that concave streamline curvature augments the turbulent shearing stress caused by streamwise flow acceleration (relative to mean shear), while the opposite is again true of convex curvature and deceleration. These authors discovered through physical simulation that there are also important differences in the response of individual stresses to streamline curvature and acceleration. While vertical Reynolds stresses (w'^2) respond to curvature, longitudinal stresses (u'^2) respond to acceleration. Therefore, it is suggested that phase differences in the streamwise variation of the three Reynolds stresses (u'^2, w'^2, and $u'w'$) pose problems for simple models that aim to predict turbulence. The wind tunnel experiments of Gong and Ibbetson (1989) also confirm that the major increase in the vertical component of turbulence and shear stress on the upwind slope, which is partly attributed to streamline curvature, is not well predicted in numerical models. Similarly, Wiggs *et al.* (1996) suggest that underprediction of shear stress at the upwind toe of the dune can result from failure of rotating cup anemometry to register these additional Reynolds stresses.

In comparison to the satisfactory description of attached flow through numerical modelling and physical simulation, no absolute rules exist for predicting the onset of flow separation. As reviewed by Tsoar *et al.* (1985), there also is strong disagreement in the literature regarding the importance of the separation vortex in the determination of the morphology of sand dunes. Finnigan (1988) suggests that the critical slope angle for steady separation is around 20° for smooth, symmetrical two-dimensional ridges, but is reduced with increasing surface roughness.

Despite the lack of a coherent theory of rough wall turbulent separation (Smith, 1986), wake behaviour is extensively described in empirical terms throughout the literature. Methods of observation include flow visualisation using wool tufts and smoke to interpret the region of flow separation or flow reversal, as well as the use of detailed anemometry to characterise this region of momentum deficit and active turbulent mixing. The depth of the wake is suggested by Finnigan *et al.* (1990) to be roughly in the order of the height of the topographic feature. The simulation data presented by these authors for a two-dimensional ridge show that shear stress near the surface reaches a maximum in the area of the hill crest, but is reduced in the lee to levels below the upstream value. However, at progressively higher streamlines, this maximum migrates leeward. It is postulated that this lag may result from a delay in reaction to the imposition of stabilising curvature, whereby larger, longer lived eddies become progressively more important in transferring stress. As a result, an obvious kink appears in velocity profiles within the lee region, this kink aligning with the Reynolds stress maxima. Similarly in the lee, Gong and Ibbetson (1989) show that a speed reduction is observed only at low levels, whereas a positive speed perturbation persists at heights $\sim h/2$ above the surface. Turbulent diffusion thereby transports kinetic energy from the high speed upper part of the wake to the low speed wall region, so that eddy scale is more closely related to the depth of the wake region, or to hill height, than to distance from the ground (Finnigan et al., 1990).

While most physical simulations have concentrated on the simplest case of two-dimensional flow over a low symmetrical straight crested ridge, a few have considered more complex shapes. Gong and Ibbetson (1989) compare the flow characteristics of an undisturbed boundary layer, with those corresponding to a two-dimensional ridge and a circular hill, both with maximum slopes of 15°. They found that mean flow and turbulence characteristics were broadly similar over both the two- and three-dimensional features, though the latter was associated with reduced velocity perturbation. As expected, some degree of horizontal divergence was observed in the case of the three-dimensional flow, as well as low level flow convergence in the wake to the lee. In comparison, Tsoar et al. (1985) focus principally on two-dimensional flow, but investigate the effects of cross-section asymmetry and crest shape on flow acceleration and separation. Their flow simulations are replicated with three varied ridge models as characterised by: (1) a triangular cross-section, typical of a longitudinal or seif dune; (2) an asymmetrical triangular cross-section; and (3) a symmetrical cross-section with a rounded crest, typical of a vegetated dune. In the case of the third model, flow separation resulted from an adverse pressure gradient and the flow separation bubble was smaller than that for the sharp crested feature where separation ensued from the discontinuity of shape at the crestline. Similarly, the rounded crest was associated with lower velocities on the line of flow reattachment, and shorter distances between the lines of separation, convergence and reattachment.

In a further extension of the work on complex topographic forms, Wiggs et al. (1996) recently constructed a geometrically identical, 1:200 model of a barchan dune from the eastern part of the Sultanate of Oman. Detailed pulse wire anemometry was carried out along the centre line and the two flanks of this three-dimensional dune form in order to characterise the mean velocity field and turbulence structure. The authors found that 'the velocity, shear velocity, and Reynolds stress components broadly correspond to previously published work from dune studies and measurements over low hills'. Despite a striking disparity between similitude parameters reported for the model and the field prototype, and the failure to meet the minimum independence Re criteria (Re_{model} only 3200), the authors demonstrate reasonable agreement between anemometer data from the wind tunnel simulation and field measurement programmes. The main anomaly was found to rest in the depressed shear stress values measured at the toe of the field dune, while no such reduction was

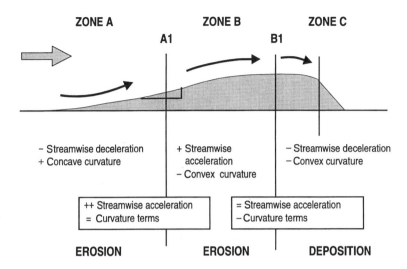

Figure 2.7 Model of dune dynamics based on the effects of streamwise acceleration and streamline curvature on shear stress production (after Wiggs *et al.*, 1996)

observed in the physical simulations. The authors attribute this anomaly to the effects of streamline divergence, and present a conceptual model of dune dynamics based on inference regarding the equilibrium between slope morphology, and the effects of streamwise acceleration and streamline curvature on shear stress production (Figure 2.7). Unlike the early finite difference model of Howard *et al.* (1978), which is founded upon on a barchan dune in the Salton Sea, California, the Wiggs *et al.* representation of barchan dune adjustment is strictly two-dimensional.

The final permutation considered in physical simulation addresses the deviation in wind approach from that which is typically assumed in numerical modelling (i.e. normal to the crestline). As far as we are aware, only one study so far has examined this effect in the context of desert dunes. This work by Tsoar *et al.* (1985) was introduced above with regard to the effects of ridge shape. The authors explore the influence of three different angles of incidence (15°, 25° and 35°) upon flow separation and vortex formation to the lee of the model ridge. These characteristics of the lee side flow are regarded as important in the development and maintenance of longitudinal dunes (Tsoar, 1983). The empirical findings of this study support the previous suggestion based on physical reasoning (see above) that there is a tendency of the wind on the lee slope to turn aside from its oblique approach towards a direction parallel to the crestline (Figure 2.8). This flow deflection is greatest on and near the dune surface within the separation bubble, whereas the flow becomes more aligned with the main flow as elevation increases. Velocity drops to its lowest magnitude close to the crest where the flow is deflected, while in the lower part of the lee slope the streamlines converge. This convergence brings about an increase in speed, which if in excess of threshold, may be associated with a corridor of sediment transport and erosion.

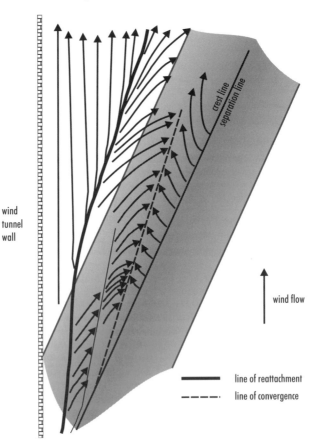

Figure 2.8 Streamline orientation from flow visualisation on a symmetrical, tri-angular ridge with a 25° angle of incidence in the approaching airflow (after Tsoar *et al.*, 1985)

Effects of Slope on Particle Entrainment and Transport

Understanding of the physics of dune initiation, growth and maintenance requires knowledge of not only how airflow and thereby shear stress varies along the topographic form, but also how sediment is transported and deposited in response to these along-slope gradients. As noted above, correct wind tunnel simulation of particle transport over a model dune is essentially impossible because of unavoidable violation of the rules of similitude. Such work therefore remains a priority for field study. This is not to say that with appropriate consideration of Froude number effects (Owen and Gillette, 1985) small pieces of the problem cannot be addressed in physical simulation. For example, Iversen and Rasmussen (1994) have used a sloping wind tunnel to assess the effects of bed angle on the threshold of motion, while saltation layer dynamics on sloping surfaces have been simulated by Rasmussen *et al.* (1996). As summarised by White (1996), 'the true forte of wind tunnel applications lies in the study of saltation. In effect, there is little compromise on the physical process of saltation occurring in the wind tunnel. This is due to the fact that saltation is not scaled but simply replicated in the tunnel.'

FIELD-BASED MODEL DEVELOPMENT AND VALIDATION

As a general rule, there have been many more field-based studies of the effects of topography upon airflow than those carried out in wind tunnels (reviewed above). Papers pertaining specifically to flow over low symmetrical hills (e.g. Kettle's Hill, Brent Knoll, and Askervein Hill) are numerous, and largely aimed at extension and calibration of the JH model. In this regard, we refer the reader to the work of Bradley (1983), Smedman and Bergstrom (1984), Walmsley (1989), Walmsley and Padro (1990) and Walmsley et al. (1982). Extensive reviews are also provided by Taylor and Teunissen (1987) and Finnigan (1988).

Partly in response to developments in the meteorological-based study of airflow over low hills, and in recognition of the significance of airflow speed-up in interpreting dune morphology (Lancaster, 1985; Tsoar, 1983), there has been a great deal of interest during the last decade in field-based measurement of wind speed and sediment transport along a variety of dune slopes at a range of scales (e.g. Arens et al., 1995; Burkinshaw et al., 1993; Burkinshaw and Rust, 1993; Frank and Kocurek, 1996 a,b; Lancaster, 1985; Lancaster et al., 1996; McKenna Neuman et al., 1997; Mulligan, 1988; Rasmussen, 1989). Complementing and reinforcing this effort has been the concurrent development of low cost data logging systems, data processing software and computing hardware. Anemometry remains the chief limitation in terms of cost and physical size.

Direct measurements of wind speed within flows which are transverse to the dune crest orientation generally are in good agreement with predictions of inner layer depth and fractional speed up based on the JH model (e.g. Burkinshaw et al., 1993; Rasmussen, 1989). At the same time, they have served to reinforce the important principle that in situations where the flow is non-uniform (i.e. accelerated or decelerated), the velocity profile throughout much of the boundary layer is non-logarithmic (Arens et al., 1995; Burkinshaw et al., 1993; Frank and Kocurek, 1996a; Lancaster et al., 1996; McKenna Neuman et al., 1997; Mulligan, 1988; Rasmussen, 1989; Wiggs et al., 1996). Therefore stress cannot be treated as constant in a vertical sense, except within a thin stratum directly adjacent to the surface. In the terminology of the revised JH model (Hunt et al., 1988), this stratum is referred to as the inner surface layer. As outlined above, surface shearing stress also varies along the profile of the dune, as a general rule increasing from the toe towards the crest. Unfortunately, the specific form of this relation and the absolute magnitude of τ_0 remain highly speculative, varying from dune to dune depending on the profile shape and the sediment transport situation. This is because at present there is no means of measuring surface shearing stress directly. The customary procedure is to compute its value from the law of the wall based on the logarithmic model. However, unless the dune is very large or very smooth, the inner surface or constant stress layer is far too thin (i.e. in the order of 1 cm) to place a conventional cup anemometer within it.

In limited circumstances, the use of considerably more compact, fast response hot wire anemometers may allow some provision for measurement either directly within or very near the constant stress region. These instruments have been introduced with encouraging success in field studies addressing aeolian transport (Butterfield, 1991) and airflow along dune surfaces (Frank and Kocurek, 1994).

So at this point in time, the best we can do in practice is to get as close to the surface as possible with an anemometer and estimate the friction velocity (u_*) from the log model, assuming an appropriate value for z_0. While the absolute magnitude of the estimate for u_* will certainly contain some unknown degree of error, the computed trend along the dune slope would appear to be reasonably correct (i.e. McKenna Neuman *et al.*, 1997). Using a similar approach, Howard *et al.* (1978) also obtained satisfactory model performance in simulating transport on a barchan dune. If sediment transport is occurring at the time of measurement, then it is necessary to correct for the apparent additional 'roughness' of saltation as well. Owen's (1964) formulation for the dependency of the roughness length (z_0) on the mass transport rate (i.e. $z_0 \sim u_*^2/2g$) may be used in this regard. The final potential source of error concerns the well known non-linearity of rotation anemometers, wherein they respond more quickly to an increase in wind speed than a decrease. In a turbulent airstream, the measured value of the mean longitudinal wind speed may exceed that of the true mean by an amount somewhere between 3 and 10%, depending on the anemometer height (Kaganov and Yaglom, 1976). Given the well acknowledged variation in Reynolds stress along a hillslope (see above), it is highly probable that this overspeeding error will not be constant.

From the pioneering work of Bagnold (1941), it is widely recognised that the mass transport rate essentially varies as the cube of the friction velocity, as determined through wind profile analysis. Since this time, many more sophisticated, semi-empirical and numerical models have been proposed, but this fundamental rule of thumb holds. Through the principle of sediment continuity, the local mass transport rate determines the rate of surface lowering, as can be computed from Rubin and Hunter's (1982) formulation, $dh/dt = - dQ/dx$. Q is the local volumetric transport rate, given as $1/\gamma(q/x)$ where γ is the bulk density of the sediment. From this simple relation, a variety of conceptual models have been proposed which link the dune shape in profile to the along slope variation in velocity and shearing stress (Lancaster, 1985; Burkinshaw and Rust, 1993; Wiggs *et al.*, 1996). The finite difference model of Howard *et al.* (1978) is based on a similar continuity equation, but includes an additional term for the streamline separation.

The crest shape appears in the literature as being a particularly important determinant of the longitudinal variation in stress (Burkinshaw *et al.*, 1993; Burkinshaw and Rust, 1993; Lancaster, 1985; Mulligan, 1988; Wiggs *et al.*, 1996). Rounded, convex crests with a distinct crest–brink separation are associated with streamline divergence, deceleration, and thereby stress reduction and sediment deposition. The maximum velocity perturbation occurs well in advance of the steepening crest, as also indicated in the numerical models. In comparison, triangular shaped crests with no crest–brink separation, as characteristic of reversing dunes, demonstrate large flow acceleration from midslope all the way up to the crest. The maximum velocity perturbation coincides with the point of highest elevation. The crest, therefore, is subject to very high rates of erosion and eventually flattens out. In the case of wind direction reversal, extraordinarily high shear stresses and mass transport rates are associated with the rapid reworking of a former slipface with slope angle in the order of 30°, as compared with the usual 10–15° for the stoss side (Burkinshaw *et al.*, 1993; McKenna Neuman *et al.*, 1997) The potential for self-regulation of the dune form, particularly in the area of the crest, is well illustrated in these conceptual models.

In response to the uncertainty regarding measurement of shear stress from velocity profiles, a number of researchers have undertaken to measure the mass transport rate directly using sediment traps and erosion pins (Burkinshaw and Rust, 1993; Lancaster *et al.*, 1996; McKenna Neuman *et al.*, 1997; Wiggs *et al.*, 1996). Despite a large degree of uncertainty regarding trap collection efficiency and the concerns listed above regarding velocity measurement and u_* estimation, McKenna Neuman *et al.* (1997) found a very high degree of correlation between direct measurements of the mass transport rate and values of q modelled from velocity data sampled at a standardised height ($z = 0.3$ m) along the stoss slope of a reversing dune at Silver Peak, Nevada (Figure 2.9). The power function $q \sim u_{0.3}^n$, where $n = 5.6$ and $r^2 = 0.91$, was also found to give an equally good estimation of q for velocities in excess of 6 ms^{-1} (i.e. above threshold) (Figure 2.10). This work would seem to suggest that with only minor modification, any of the recent models of airflow over hills could be used to give a reasonably good prediction of the variation in the mass transport rate along the stoss side of a transverse dune. Detailed accommodation for the vertical momentum loss associated with saltation (e.g. Anderson and Haff, 1988, 1991; Janin and Cermak, 1988; Spies *et al.*, 1995; McKenna Neuman and Maljaars, 1997) would not appear to be necessary, at least in practice, in the context of modelling dune dynamics on the scale of several 10s to 100s of metres.

Having dealt with non-uniform airflows, the next hurdle in understanding and modelling dune dynamics involves the absence of a steady flow regime, as is usually simulated in wind tunnels. While it would seem possible to model steady flows lasting in the order of seconds and perhaps minutes, the problem of dealing with more realistic, unsteady airflow is a daunting one. Under these conditions, the relative amounts of erosion and deposition can vary both temporally and spatially along the profile of the dune. For example, field observations by a variety of researchers indicate that dune crests are active on a more frequent basis than lower, windward areas where wind speeds are often below threshold (e.g. Lancaster, 1985; Lancaster *et al.*, 1996; McKenna Neuman *et al.*, 1997). Given that the mass transport rate can adjust itself to varying wind speed in a matter of seconds, while the profile of the dune is altered on the basis of hours, days and years, it might be further argued that the dune shape is likely rarely in equilibrium with the local wind on a short-term basis. In a preliminary examination of the temporal linking of airflow ($\Delta t = 10$ s) and sediment transport ($\Delta t = 2$ s) on a sharp crested, reversing dune, McKenna Neuman *et al.* (1999, in press) have found very close synchronisation between six anemometers along a transect running from the dune toe up to the crest for wind gusts lasting from 1 to 18 min (Figure 2.11(a) and (b)). Dominant frequencies include gusts lasting 3, 5 and 8 min. Gusts lasting in the order of

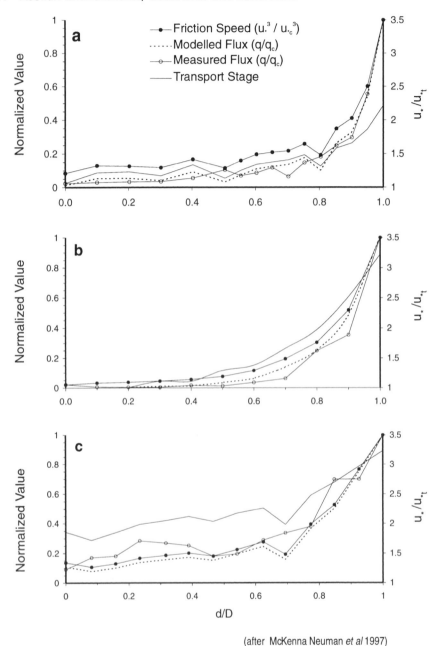

(after McKenna Neuman *et al* 1997)

Figure 2.9   Stoss slope variation in friction speed and the mass transport rate, normalised by that for the dune crest (i.e. $u*^3/u*_c^3$ and q/q_c). The model estimate for q/q_c, based on Owen's (1964) transport equation, is also shown as the dashed line. The solid line represents the influence of threshold ($u*/u*_t$, values on secondary vertical axis). Field measurements are based on two reversing dunes at Silver Peak, Nevada (h= 6 m, L = 22 m for (a) and (b); h = 13 m, L = 28 m for (c)). (after McKenna Neuman *et al.*, 1997)

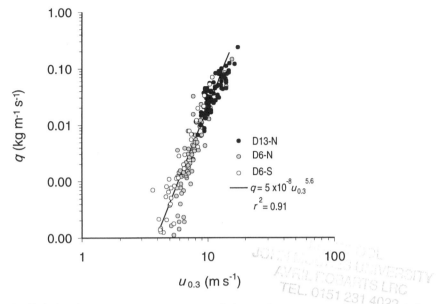

Figure 2.10 Relation between measurements of along slope mass transport rate (q) and wind speed at z = 0.3 m, Silver Peak, Nevada (after McKenna Neuman *et al.*, 1997)

seconds do not appear to be either spatially or temporally correlated, though this outcome may well be influenced by limitations of the instrumentation. While there is more noise in the mass transport data, owing to variable trap alignment, sand streamers, and scouring, the temporal agreement between q and u is quite striking in many cases, especially at higher elevations. Similar to the earlier finding of Howard *et al.* (1978) through a model calibration exercise, q does not appear to lag u. As a rule of thumb, as the crest is approached, low frequency transport events are observed to assume even greater importance. So while wind speed up is obviously paramount, there are other intermittence effects (Stout and Zobeck, 1997) which appear to reinforce the observation of a very large increase in mass transport towards the crest of this dune.

The apparent emphasis in this discussion upon stoss side flow arises from the amount of field work attending to this phenomena relative to what might be cited as a lack of attention to lee side, separated flow. This is somewhat surprising given that the structure of lee side flow is widely believed to be important in the organisation of dune fields, especially in regard to the spacing of dunes and the nature of interdune flats.

Within the last decade, one of the few studies to address this region of flow is that by Frank and Kocurek (1994, 1996a). Their conceptual model of this flow is presented in Figure 2.12. It is based on detailed anemometry of transverse flow over 15 dunes in the southwestern USA. The model bears a strong similarity to that for flow over subaqueous dunes (i.e. McLean and Smith, 1986; Nelson and Smith, 1989). In the case of flow separation, a back-flow eddy develops that extends about

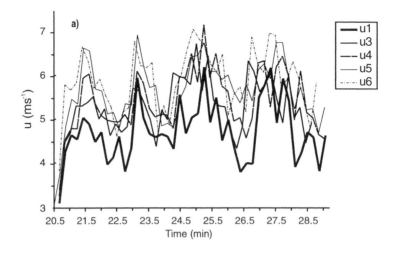

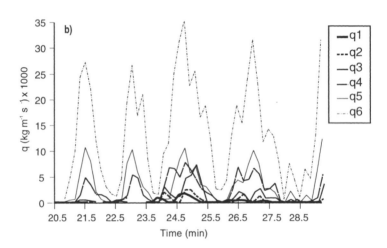

Figure 2.11   (a) Temporal synchronization (Δt = 10 s) between wind speed measurements at six locations (1= toe; 6=crest) along the stoss slope of a reversing dune at Silver Peak, Nevada. (b) Coincident measurements of the mass transport rate (Δt=2s) at the same six sites. (after McKenna Neuman *et al.*, 1999, in press)

four dune-brink heights downwind from the brink of the dune. Recognised kinks in the empirically derived velocity profiles divide the flow into four additional regions which include the interior, the upper and lower wake and the internal boundary layer. It is in this latter, very thin layer that the velocity gradient is particularly steep so that interdune flats are suggested to be partially erosional.

Similarly, the only substantive numerical study of grainfall deposition and grain flow initiation on the lee slope is that recently published by McDonald and Anderson (1995). In this work, they compare modelled relations between grainfall deposition and distance from the dune brink with field measurements from a low

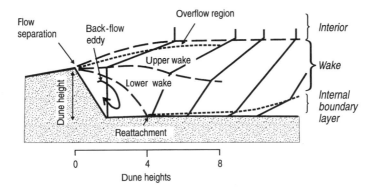

Figure 2.12   Conceptual model of lee side air flow (after Frank and Kocurek, 1996a)

transverse dune on the coast of California. Numerical experiments are replicated for varied forms of an assumed wind field with best results given by a wake without return flow. As a result of this grainfall deposition, a topographic bulge appears in the upper lee face which eventually fails as grainflow. McDonald and Anderson's work suggests that with increased wind speed, the frequency but not the size of grainflows increases, so that large thick grainflows require some other mechanism to explain their development.

DISCUSSION

During the past decade considerable strides have been made in the study of aeolian processes but much still remains to be done, particularly with regard to our understanding of the initiation and development of aeolian dunes. Recent research has led to the development of several numerical transport models (e.g Anderson and Hallett, 1986; Ungar and Haff, 1987; Werner, 1990; Sorensen, 1991; McEwan and Willetts, 1994) that have their roots in the work of Bagnold (1941) and Owen (1964) and draw on numerous complementary theoretical and empirical studies of grain dynamics (e.g White and Shultz, 1977; Rumpel, 1985; Nickling, 1988; Willetts and Rice, 1988, 1989). These models have led to a much clearer understanding of the physical processes controlling sediment movement but have not lead to more reliable predictions of the mass transport rate (Anderson *et al.*, 1991). Moreover, these models are normally based on assumptions which are not typical of actual field conditions (e.g. steady wind flow and a flat bed consisting of a uniform grain size). On dunes for example, winds are unsteady, surfaces are sloped and often irregular, and the sediments are usually comprised of mixed grain sizes. Also, the turbulence scales associated with field conditions vary spatially and temporally, and are not well addressed by current numerical models.

Anderson *et al.* (1991) have argued that theory has in many ways caught up to experiment and that new experimental work is required for the verification of postulated effects. The implementation and verification of these mathematical models in actual field conditions is a daunting task for a variety of reasons, the most significant encompassing problems relating to scale, instrumentation requirements, and interdisciplinary collaboration.

The Issue of Scale

The study of aeolian processes and landforms encompasses a wide array of topics, methodologies and disciplines that cover a range of temporal and spatial scales. These extend from detailed laboratory studies of grain mechanics, to field studies on airflow and sediment transport on dunes, as well as to the investigation of dune field development and the maintenance of sand seas. An important question that has not been adequately addressed is the nature of the linkages between these various types of empirical and theoretical studies which cover such a wide range of temporal and spatial scales. That is to say, what relevance do inter-grain collisions in aeolian sediment transport ultimately have to do with the development of sand seas? Intuitively there must be a link but at present, these links are poorly defined at best. Such perplexing scale issues, however, are not confined to the study of aeolian processes and landforms. As Kirkby (1989, p. 256) points out, '. . . the difficulty of spanning scales or of aggregation and disaggregation is just as severe a problem in physics and other fundamental sciences. Reconciliation between atomic physics and Newtonian physics is not a trivial exercise and we must expect difficulties in our own, less developed, field'.

Schumm and Lichty (1965) in their classic paper address the issue of temporal and spatial scales in relation to geomorphic systems. They indicate that many variables responsible for the development of landforms and landscape are time and space dependent and therefore, may either change from independent to dependent or become irrelevant as the time scale increases in length. Warren and Knott (1983) suggest that dune-forming processes operate at three main spatial and temporal scales which approximately correspond to the steady, graded and cyclic time scales outlined by Schumm and Lichty. At the steady scale, aeolian processes operate over very short time periods (instantaneous to days) and distances, and are associated with individual grain transport and the formation of ripples. Dune dynamics and morphology most likely are associated with the somewhat longer graded time scale (1 to 100 years). Over this time scale, dunes may tend towards some form of equilibrium in response to rates and directions of sand movements generated by surface winds. Equilibrium over this time scale, however, is in part controlled by the areal extent of the dune/dune field and the volume of sand contained within it. Form flow interactions and feedbacks are important at this scale and are critical in defining the morphology and transport dynamics of the dunes (Lancaster, 1994). Sand sea development is associated with cyclic time scales (10^3 to 10^6 years). These scales incorporate the long-term physiographic, tectonic and climatic changes that are required for the massive accumulation of sand associated with these sedimentary deposits. Thus, while inter-grain collisions lie at the core of transport mechanics and the development of ripples, they become much less important in determining the overall morphology of dunes and dune fields, and they most likely are irrelevant in the development of sand seas. These arguments do not negate the importance of process studies, but they do point to the need for careful selection of the most important or relevant variables when modelling, especially when addressing research questions which cover varying temporal and spatial scales.

Instrumentation

As outlined above, simultaneous measurements of wind flow and sediment transport are required at appropriate temporal and spatial scales in order to validate present numerical models and to extend understanding of the controls on sediment transport in complex natural settings. While measurements of wind flow in natural settings are usually made using sensitive cup anemometry, these instruments are not suitable for short temporal scales (seconds) that are required for validation of turbulence-based models.

Butterfield (1991) was one of the first to use tower-mounted fast response, rugged hot wire anemometers (four heights: 2 to 10 cm) in a natural dune setting during active saltation in conjunction with more conventional cup anemometry. At each instrument site, detailed velocity profile data (hot wire) were compared with instantaneous sediment flux measurements using a modified form of the Rasmussen and Mikkelsen (1991) trap design fitted with a sensitive load cell. Complementary studies using hot wire anemometry on sloping dune surfaces have also been carried out by Frank and Kocurek (1994) who found that velocity profiles showed up to three distinct segments joined by curving links, a pattern similar to that observed by Butterfield (1991). Typically a thin low shear saltation layer exists near the surface (up to 3 cm), increasing in thickness with wind shear. Above this lies a high shear internal boundary layer, associated with up-slope flow acceleration, which increases in thickness up slope with the corresponding increase in u_*.

The problem of determining the actual near surface shear stresses associated with sediment transport has been addressed more thoroughly in the fluvial literature (e.g McLean and Smith, 1986; Bennett and Best, 1995; Kostaschuk and Villard, 1996). In this literature, there has been a growing trend to move away from the law of the wall paradigm in favour of the Reynolds stress approach which describes the turbulent structures and intensities of the flow based on measurements of the instantaneous horizontal and vertical components of velocity. This new emphasis in empirical studies has been supported by rapid, coincident advances in measurement technology, including the refinement of laser velocimeters which can be used in both laboratory and field testing. As reviewed above, Wiggs et al. (1996) were the first to use cross-wire anemometry and the Reynolds stress approach in studying airflow on a barchan dune.

Although considerable advances have been made through limited use of hot wire or hot film anemometry, these instruments are not without significant problems in field situations. Most ruggedised probes are single element and therefore, are normally designed for measurement of the horizontal component of velocity. This design precludes their use for full determination of the Reynolds stresses, though measurement of the turbulence intensity remains possible. Cross-wire probes are normally required for Reynolds stress determination, giving simultaneous measurement of both the horizontal and vertical fluctuating components of the wind velocity. However, the wires in these instruments are extremely fragile. Personal experience has shown that they may last only a matter of a few seconds when placed into the saltation cloud, which can be a very expensive learning experience. Correct alignment of these instruments over irregular or sloping surfaces is also problematic. In an inhospitable aeolian environment, the use of very sophisticated and sensitive electronic equipment, primarily developed for clean air laboratory use, can be a serious logistical problem. The very high cost of these instruments must also be considered, particularly if multiple arrays are used, as in cup anemometry, to investigate the flow patterns over even very small dunes.

In recent studies of flow over sub-aqueous bedforms, laser and acoustic velocimetry has proven to be a very useful measurement technique which for the most part, has replaced hot-wire/hot-film anemometry in both field and laboratory investigations. At present this technique is not amenable to aeolian systems dealing with saltation in that it requires a tracer (approximately neutrally buoyant particle or bubble) moving with the same approximate velocity as the horizontal and vertical components of the flow. While suspended sediment particles in water match these criteria, saltating sand in air does not. Acoustic doppler anemometers have also been used extensively in meteorology and the study of dust transport (Moosemüller et al., 1998), but these have limitations because their very large size does not allow measurement closer than 20–30 cm above the bed. As well, the emitting and receiving transducers are prone to abrasion and therefore damage, from saltating grains.

Based on this discussion, there is a critical research need to develop rugged, relatively low cost, fast response instrumentation to measure near surface flow fields and Reynolds stresses on desert dunes and other complex natural surfaces.

Accurate measurement of the mass transport rate fluxes at small temporal scales is also an important issue that has seen significant progress during the past few years. Several innovative, near isokinetic traps of differing designs for field use have been reported in the literature (Butterfield, 1991; Rasmussen and Mikkelsen, 1991; Nickling and McKenna Neuman, 1997; Bauer and Namikas, 1998). These trap designs often incorporate sensitive load cells or balances continuously to weigh trapped sediment down to one second intervals providing excellent temporal resolution for the evaluation and development of sediment transport models. These traps, however, are frequently not designed to orientate themselves into the wind, and still have problems with scouring at the base when either used in high wind speeds or for extended periods of time. There is also the critical problem of mounting these instruments on steep dune slopes, particularly if they are self-orientating or have load cells that must be kept level. Further design modifications and field test programmes are required to overcome these limiting issues. Alternative instrumentation for the indirect measurement of aeolian saltation includes the saltiphone, developed in the Netherlands (Spaan and Van den Abeele, 1991), and the commercially available Sensit (Gillette and Stockton, 1986). Both of these instruments record particle impacts at a point elevation with either a microphone (saltiphone) or a piezoelectric crystal (Sensit). These instruments have the advantage of being omnidirectional. However, they must be calibrated in either a wind tunnel or the field against a traditional sediment collection trap, using the same sediment as transported at the field site. These instruments are extraordinarily robust and play a very useful role in sediment transport studies, especially when used in determination of the relative mass transport rate at a given site.

Interdisciplinary Collaboration

A characteristic trend that has developed over the past decade in the study of desert dunes has been the incorporation of theory, modelling techniques, instrumentation and methodologies from a wide range of related disciplines (e.g. engineering, fluvial geomorphology, meteorology) that has added greatly to our understanding of dune dynamics.

Of particular importance has been the parallel work on bedform development in fluvial systems. This work, which for the most part has been flume based (e.g. Bennett and Best, 1995; McLean and Smith, 1986; McLean et al., 1994; Nelson and Smith, 1989; Nelson et al., 1993), has provided great insight and a modelling framework in regard to our understanding of turbulent structures and flow fields over dunes as they relate to bedform initiation, dune morphology and sediment transport. These flume studies have been complemented by field measurements over large bedforms with associated high flows in the Fraser River by Kostaschuk and Church (1993) and Kostaschuk and Villard (1996). Concepts and models developed in these field and laboratory studies have been successfully used in the modelling of flow and sediment transport over aeolian dunes by Frank and Kocurek (1996a,b) and McKenna Neuman et al. (1997).

Similarly, the concepts and numerical models of airflow over hills (e.g. Taylor and Gent, 1974; Jackson and Hunt, 1975; Hunt et al., 1988) have been applied, at varying levels of sophistication, to flow over aeolian dunes by Howard and Walmsley (1985), Rasmussen (1989), McKenna Neuman et al. (1996) and Wiggs et al. (1996a,b). The complexity of these numerical models, especially in the case of three-dimensional flow, as well as their present limitations in simulating flow separation and re-attachment, has restricted their general application.

Finally, the recent detailed, process-based studies on beaches and coastal dunes have made important contributions to our understanding of aeolian processes that have wide application. For example, work by Bauer et al. (1992, 1996), Davidson Arnott et al. (1997a, b), Gares et al. (1996), Sherman (1992), and Sherman and Hotta (1990) has provided important new insights into wind profile development over complex surfaces that directly affect determination of the surface shear stress and the modelling of sediment transport rates.

Owing to the demonstrated interdisciplinary nature of research on bedform dynamics, full progress continues to be limited to some degree by the lack of coordination in these independent efforts, and the tendency to publish in discipline specific journals. A good example of this is the excellent work carried out on coastal dune processes and morphology which is infrequently used or cited in the desert dune literature though it has direct applicability. Aeolian researchers have also been relatively slow in venturing outside their discipline boundaries to make use of instrumentation, methodologies and models that are used in other disciplines. For example, it has taken some time to assimilate either the models describing the flow over sub-aqueous dunes, developed in engineering and fluvial geomorphology, or the meteorological models of airflow over low hills. It is an unfortunate trait (albeit improving) that many geomorphologists do not have the numerical and computing skills to make full use of these models and to extend their application to the study of desert dunes. In this regard, there is a critical need for more cross-disciplinary research that brings investigators having strong numerical and modelling skills together with those researchers having extensive empirical field and laboratory experience.

CONCLUSION

Research during the past decade has been very productive, showing major advances in our understanding of dune dynamics, and has been bolstered by the significant and innovative contributions of numerous young scientists who have recently begun to publish. However, as stated by Anderson and Willetts (1991) in their review of aeolian sediment transport processes at the Second International Aeolian Research Conference, much remains to be done if the aeolian community is to tackle the realistic problems of significant geological and environmental interest. This will require a greater degree of integration between researchers working in different disciplines and sub fields, and will necessitate closer co-operation among theoreticians, numerical modellers, and empirically based field and laboratory scientists.

As a final outcome of the present review of progress in the understanding of dune dynamics during the last decade, we suggest the following shortlist of specific research needs:

1. Development of a rugged, relatively low cost, fast response instrument to measure near surface flow fields and Reynolds stresses on desert dunes and other complex natural surfaces.
2. Field-based measurement of flow separation, grain fallout and sediment transport in the interdune corridor.
3. Improved accessibility of numerical airflow models for use by non-specialist earth scientists.
4. Field-based model extension and evaluation of airflow models addressing: (a) flow separation; and (b) rhythmic topography.
5. Field-based measurement and analysis of the spatial implications of transport event intermittence in unsteady flows along individual stoss slopes.
6. Increased attention to the relations existing between large scale turbulence in the upper atmosphere, event intermittence, and dune dynamics.
7. Renewed interest in the mechanisms underlying dune initiation.
8. Continued effort in the development of non-linear, numerical models which emulate the spatial self-organization which is characteristic of aeolian dunes on a large scale.

REFERENCES

Allen, J.R.L., 1968. The nature and origin of bedform hierarchies. *Sedimentology*, **10**, 161–182.

Anderson, R.S. and Haff, P.K., 1988. Simulation of eolian saltation. *Science*, **241**, 820–823.

Anderson, R.S. and Haff, P.K., 1991. Wind modification and bed response during saltation of sand in air. *Acta Mechanica (Suppl)*, **1**, 21–52.

Anderson, R.S. and Hallet, B., 1986. Sediment transport by wind: Toward a general model. *Geological Society of America Bulletin*, **97**, 523–535.

Anderson, R.S., Sorensøn, M. and Willetts, B.B., 1991. A review of recent progress in our understanding of aeolian sediment transport. *Acta Mechanica (Suppl)*, **1**, 1–19.

Arens, S.M., van Kaam, P. and van Boxel, J.H., 1995. Airflow over foredunes and implications for sand transport. *Earth Surface Processes and Landforms*, **20**, 315–315.

Ayotte, K.W., Xu, D. and Taylor, P.A., 1994. The impact of turbulence closure schemes on predictions of the mixed spectral finite-difference model for flow over topography. *Boundary-Layer Meteorology*, **68**, 1–33.

Bagnold, R.A., 1941. *The Physics of Blown Sand and Desert Dunes*. Methuen, London.

Bauer, B.O. and Namikas, S.L., 1998. Design and field test of a continuously weighing tipping-bucket assembly for aeolian sand traps. *Earth Surface Processes and Landforms*, **23**, 1171–1183.

Bauer, B.O., Sherman, D.J. and Wolcott, J.F., 1992. Sources of uncertainty in shear stress and roughness length estimates derived from velocity profiles. *Professional Geographer*, **44**, 453–464.

Beljaars, A.C.M., Walmsley, J.L. and Taylor, P.A., 1987. A mixed spectral finite-difference model for neutrally stratified boundary-layer flow over roughness changes and topography. *Boundary-Layer Meteorology*, **38**, 273–303.

Bennett, S.J. and Best, J.L., 1995. Mean flow and turbulence structure over fixed, two-dimensional dunes: Implications for sediment transport and bedform stability. *Sedimentology*, **42**, 491–513.

Bradley, E.F. (1983) The influence of thermal stability and angle of incidence on the acceleration of wind up a slope. *Journal of Wind Engineering and Industrial Aerodynamics*, **15**, 231–242.

Bradshaw, P. (1973) *Effect of Streamline Curvature on Turbulent Flow*. AGARDograph No. 169 Paris, AGARD.

Burkinshaw, I.K. and Rust, I.C. (1993) Aeolian dynamics on the windward slope of a reversing dune, Alexandria coastal dunefield, South Africa. *Special Publication International Association of Sedimentologists*, **16**, 13–21.

Burkinshaw, J.R., Illenberger, W.K. and Rust, I.C., 1993. Wind-speed profiles over a reversing transverse dune. In *The Dynamics and Environmental Context of Aeolian Sedimentary Systems*, K. Pye (Ed.). Geological Society Special Publication No. 72, pp. 25–36.

Butterfield, G.R., 1991. Grain transport rates in steady and unsteady turbulent airflows. *Acta Mechanica (Suppl)*, **1**, 97–122.

Cooke, R.U., Warren, A. and Goudie, A., 1993. *Desert Geomorphology*. UCL Press, London.

Cooper, W.S., 1958. Coastal sand dunes of Oregon and Washington. *Geological Society of America Memorandum*, **72**. GSA, New York.

Counihan, J., 1969. An improved method of simulating a neutral atmospheric boundary layer in a wind tunnel. *Atmospheric Environment*, **3**, 197–214.

Davidson-Arnott, R.G.D., White, D.C. and Ollerhead, J., 1997a. The effects of artificial pebble concentration on aeolian sand transport on a beach. *Canadian Journal of Earth Sciences*, **34**, 1499–1508.

Davidson-Arnott, R.G.D., Nielsen, N., Aagaard, T. and Greenwood, B., 1997b. Alongshore and onshore aeolian sediment transport, Skallingen, Denmark. *Proceedings of the Canadian Coastal Conference*, 463–477.

Finnigan, J.J., 1988. Air flow over complex terrain. In *Flow and Transport in the Natural*

Environment: Advances and Applications, W.L. Steffan and O.T. Denmead (Eds). Springer-Verlag, Heidelberg, pp. 183–229.

Finnigan, J.J., Raupach, M.R., Bradley, E.F. and Aldis, G.K., 1990. A wind tunnel study of turbulent flow over a two-dimensional ridge. *Boundary-Layer Meteorology*, **50**, 277–317.

Folk, R.L., 1971. Genesis of longitudinal and oghurd dunes elucidated by rolling upon grease. *Bulletin of Geological Society of America*, **82**, 3461–3468.

Folk, R.L., 1976. Rollers and ripples in sand, streams and sky: Rhythmic alteration of transverse and longitudinal vortices in three orders. *Sedimentology*, **23**, 649–669.

Frank, A. and Kocurek, G., 1994. Models for airflow velocity profiles in natural settings: accounting for atmospheric convection, and secondary flow over dunes (on both the windward and lee slopes), *Desert Research Institute* (1994), 37–38.

Frank, A. and Kocurek, G., 1996a. Toward a model of airflow on the lee side of aeolian dunes. *Sedimentology*, **43**, 451–458.

Frank, A.J. and Kocurek, B., 1996b. Airflow up the stoss slope of sand dunes: Limitations of current understanding. *Geomorphology*, **17**, 47–54.

Gares, P., Davidson-Arnott, R.G.D., Bauer, B.O., Sherman, D.J., Carter, R.W.G., Jackson, D.W.T. and Nordstrom, K.F., 1996. Alongshore variations in aeolian sediment transport, Carrick Finn Strand, Ireland. *Journal of Coastal Research*, **12**, 673–682.

Gillette, D.A. and Stockton, P.H., 1986. Mass momentum and kinetic energy of saltating particles. In *Aeolian Geomorphology, Proceedings of the 17th Binghamton Symposia*, Nickling, W.G. (Ed.), pp. 35–56.

Gong, W. and Ibbetson, A., 1989. A wind tunnel study of turbulent flow over model hills. *Boundary-Layer Meteorology*, **49**, 113–146.

Greeley, R. and Iversen, J.D., 1985. *Wind as a Geological Process on Earth, Mars, Venus and Titan*. Cambridge University Press, London.

Hack, J.T., 1941. Dunes of the western Navajo Country. *Geographical Review*, **31**, 240–263.

Hallett, B., 1990. Spatial self-organization in geomorphology: from periodic bedforms and patterned ground to scale invariant topography. *Earth Science Reviews*, **29**, 57–76.

Hanna, S.R., 1969. The formation of longitudinal sand dunes by large helical eddies in the atmosphere. *Journal of Applied Meteorology*, **8**, 874–883.

Hesp, P.A. and Hastings, K., 1998. Width, height and slope relationships and aerodynamic maintenance of barchans. *Geomorphology*, **22**, 193–204.

Howard, A.D., Morton, J.B., Gad-El-Hak, M. and Pierce, D.B., 1978. Sand transport model of barchan dune equilibrium. *Sedimentology*, **25**, 307–338.

Howard, A.D. and Walmsley, J.L., 1985. Simulation model of isolated dune sculpture by wind. In *Proceedings of International Workshop on the Physics of Blown Sand*, O.W. Barndorff-Nielsen, J.T. Moller, K. Rasmussen and B.B. Willetts (Eds). University of Aarhus, Aarhus, pp. 377–392.

Hunt, J.C.R., Richards, K.J. and Brighton, P.M.W., 1988a. Stable stratified flow over low hills. *Quarterly Journal of the Royal Meteorological Society*, **114**, 859–886.

Hunt, J.C.R., Leibovich S. and Richards, K.J., 1988b. Turbulent shear flows over low hills. *Quarterly Journal of the Royal Meteorological Society*, **114**, 1435–1470.

Iversen, J.D. and Rasmussen, K.R., 1994. The effect of surface slope on saltation threshold. *Sedimentology*, **41**, 721–728.

Jackson, P.S., 1977. Aspects of surface wind behaviour. *Wind Engineering*, **1**, 1–14.

Jackson, P.S. and Hunt, J.C.R., 1975. Turbulent wind flow over a low hill. *Quarterly Journal of the Royal Meteorological Society*, **101**, 929–955.

Janin, L.G. and Cermak, J.E., 1988. Sediment-laden velocity profiles developed in a long boundary-layer wind tunnel. *Journal of Wind Engineering and Industrial Aerodynamics*, **28**, 159–168.

Jensen, J., 1958. The model-law for phenomena in natural wind. *Ingeniøren*, **2**, 121–128.

Jensen, N.O. and Zeman, O., 1985. Perturbations to mean wind and turbulence in flow over topographic forms. In *Proceedings of the International Workshop on the Physics of Blown Sand*, O.E. Barndorff-Nielsen, J.T. Moller, K. Rasmussen and B.B. Willetts (Eds). University of Aarhus, Aarhus, pp. 351–368.

Kaganov, E.J. and Yaglom, A.M., 1976. Errors in wind speed measurements by rotation anemometers. *Boundary-Layer Meteorology*, **10**, 229–244.

Kirkby, M., 1989. The future of modelling in physical geography. In *Remodelling Geography*, B. Macmillan (Ed.). Blackwell Science, Cambridge, pp. 255–272.

Kocurek, G., Townsley, M., Yeh, E. and Havholm, K., 1992. Dune and dunefield development stages on Padre Island, Texas: effects of lee flow and sand saturation levels and implications for interdune deposition. *Journal of Sedimentary Petrology*, **62**, 622–635.

Kostaschuck, R.A. and Church, M.A., 1993. Macroturbulence generated by dunes: Fraser River, Canada. *Sedimentary Geology*, **85**, 25–37.

Kostaschuk, R. and Villard, R., 1996. Flow and sediment transport over large subaqueous dunes: Fraser River, Canada. *Sedimentology*, **43**, 849–863.

Lancaster, N., 1982. Dunes on the Skeleton Coast, Namibia (South West Africa): Geomorphology and grain size relationships. *Earth Surface Processes and Landforms*, **7**, 575–587.

Lancaster, N., 1985. Variations in wind velocity and sand transport on the windward flanks of desert sand dunes. *Sedimentology*, **32**, 581–593.

Lancaster, N., 1989. The dynamics of star dunes: an example from the Gran Desierto, Mexico. *Sedimentology*, **36**, 273–289.

Lancaster, N., 1994. Dune morphology and dynamics. In *Geomorphology of Desert Environments*, A.D. Abrahams and A.J. Parsons (Eds). Chapman and Hall, London.

Lancaster, N., 1995. *Geomorphology of Desert Dunes*. Routledge, London.

Lancaster, N., Nickling, W.G., McKenna Neuman, C. and Wyatt, V.E., 1996. Sediment flux and airflow on the stoss slope of a barchan dune. *Geomorphology*, **17**, 55–62.

Livingstone, I., 1986. Geomorphological significance of wind flow patterns over a Namib linear dune. In *Aeolian Geomorphology, Proceedings of the 17th Binghamton Symposia*, W.G. Nickling (Ed.), pp. 97–112.

Livingstone, I. and Warren, A., 1996. *Aeolian Geomorphology: An Introduction*. Longman, Singapore.

Mason, P.J. and Sykes, R.I., 1979. Flow over an isolated hill of moderate slope. *Quarterly Journal of the Royal Meteorological Society*, **105**, 383–395.

McDonald, R.R. and Anderson, R.S., 1995. Experimental verification of aeolian saltation and leeside deposition models. *Sedimentology*, **42**, 39–56.

McEwan, I.K. and Willetts, B.B., 1994. On the prediction of the bedload transport rate in air. *Sedimentology*, **41**, 1241–1251.

McKee, E.D., 1979. An introduction to a study of global sand seas. In *A Study of Global Sand Seas*, E.D. McKee (Ed.). United States Geological Survey, Professional Paper 1052.

McKenna Neuman, C. and Maljaars, M., 1997. Wind tunnel measurement of boundary-layer response to sediment transport. *Boundary-Layer Meteorology*, **84**, 67–83.

McKenna Neuman, C., Lancaster, N. and Nickling, W.G., 1997. Relations between dune morphology, air flow, and sediment flux on reversing dunes, Silver Peak, Nevada. *Sedimentology*, **44**, 1103–1113.

McKenna Neuman, C., Lancaster N. and Nickling, W.G., 1999 in press. The effect of unsteady winds on sediment transport on the stoss slope of a transverse dune, Silver Peak, Navada. *Sedimentology*.

McLean, S.R., Nelson, J.M. and Wolfe, S.R., 1994. Turbulence structure over two-dimensional bedforms: Implications for sediment transport. *Journal of Geophysical Research*, **99**, 12729–12747.

McLean, S.R. and Smith, J.D., 1986. A model for flow over two-dimensional bed forms. *Journal of Hydraulic Engineering*, **112**, 300–317.

Melton, F.A., 1940. A tentative classification of sand dunes: its implication to dune history in the southern high plains. *Journal of Geology*, **48**, 113–174.

Moosemüller, H., Gillies, J.A., Rogers, C.F., DuBois, D.W., Chow, J.C., Watson, J.G. and Langston, R., 1998. Particle emission rates for unpaved shoulders along a paved road. *Journal of Air and Waste Management*, **48**, 174–185.

Mikkelsen, H.E., 1989. Wind flow and sediment transport over a low coastal dune. *Geoskrifter nr.* Institute of Geology, University of Aarhus, Denmark.

Mulligan, K.R., 1988. Velocity profiles measured on the windward slope of a transverse dune. *Earth Surface Processes and Landforms*, **13**, 573–582.

Nelson, J.M., McLean, S.R. and Wolfe, S.R., 1993. Mean flow and turbulence fields over two-dimensional forms. *Water Resources Research*, **29**, 3935–3953.

Nelson, J.M. and Smith, J.D., 1989. Mechanics of flow over ripples and dunes. *Journal of Geophysical Research*, **94**, 8146–8162.

Nickling, W.G., 1988. The initiation of particle movement by wind. *Sedimentology*, **35**, 499–511.

Nickling, W.G. and McKenna Neuman, C., 1997. Wind tunnel evaluation of a wedge-shaped aeolian transport trap. *Geomorphology*, **18**, 333–345.

Owen, P. and Gillette, D., 1985. Wind tunnel constraint on saltation. In *Proceedings of the International Workshop on the Physics of Blown Sand*, O.E. Barndorff-Nielsen, J.T. Moller, K.R. Rasmussen and B.B. Willetts (Eds). University of Aarhus, Denmark, pp. 253–269.

Owen, P.R., 1964. Saltation of uniform grains in air. *Journal of Fluid Mechanics*, **20**, 225–242.

Pye, K. and Tsoar, H., 1990. *Aeolian Sand and Sand Dunes*. Unwin Hyman, London.

Rasmussen, K.R., 1989. Some aspects of flow over coastal dunes. *Proceedings of the Royal Society of Edinburgh*, **96B**, 129–147.

Rasmussen, K. and Mikkelsen, H.E., 1991. Wind tunnel observations of aeolian transport rates. *Acta Mechanica (Suppl)*, **1**, 135–144.

Rasmussen, K.R., Iversen, J.D. and Rautahemio, P., 1996. Saltation and wind-flow interaction in a variable slope wind tunnel. *Geomorphology*, **17**, 19–28.

Rubin, D. and Hunter, R.E., 1982. Bedform climbing in theory and nature. *Sedimentology*, **29**, 129–138.

Rumpel, D.S., 1985. Successive aeolian saltation: Studies of idealized collisions. *Sedimentology*, **32**, 267–280.

Schumm, S.A. and Lichty, R.W., 1965. Time, space and causality in geomorphology. *American Journal of Science*, **263**, 110–119.

Sherman, D.J., 1992. An equilibrium relationship for shear velocity and apparent roughness length in aeolian saltation. *Geomorphology*, **5**, 419–431.

Sherman, D.J. and Hotta, S., 1990. Eolian sediment transport: theory and measurement. In *Coastal Dunes: Process and Morphology*. K. Nordstrom, K. Psuty and R. Carter (Eds). John Wiley, New York, pp. 17–37.

Smedman, A.-F. and Bergstrom, H., 1984. Flow characteristics above a very low and gently sloping hill. *Boundary-Layer Meteorology*, **29**, 21–37.

Smith, F.T., 1986. Steady and unsteady boundary layer separation. *Ann. Rev. Fluid Mechanics*, **18**, 197–220.

Sorensen, M., 1991. Estimation of some aeolian saltation transport parameters from transport rate profiles. *Acta Mechanica (Suppl)*, **1**, 141–190.

Spaan, W.P. and van den Abeele, G.D., 1991. Wind borne particle measurements with acoustic sensors. *Soil Technology*, **4**, 51–63.

Spies, P.-J., McEwan, I.K. and Butterfield, G.R., 1995. On wind velocity profile measurements taken in wind tunnels with saltating grains. *Sedimentology*, **42**, 515–521.

Stout, J.E. and Zobeck, T.M. (1997) Intermittent saltation. *Sedimentology*, **44**, 959–970.

Sykes, R.I., 1980. An asymptotic theory of incompressible turbulent boundary-layer flow over a small hump. *Journal of Fluid Mechanics*, **101**, 647–670.

Taylor, P.A. and Gent, P.R., 1974. A model of atmospheric boundary-layer flow above an isolated two-dimensional 'hill': An example of flow above 'gentle topography'. *Boundary-Layer Meteorology*, **7**, 349–362.

Taylor, P.A., Walmsley, J.L. and Salmon, J.R., 1983. A simple model of neutrally stratified boundary-layer flow over real terrain incorporating wavenumber-dependent scaling. *Boundary-Layer Meteorology*, **26**, 169–189.

Taylor, P.A. and Teunissen, H.W., 1987. The Askervein Hill Project: Overview and background data. *Boundary-Layer Meteorology*, **39**, 15–39.

Thomas, D.S.G., (Ed.), 1989. *Arid Zone Geomorphology*. Belhaven Press, London.

Townsend, A.A., 1956. *The Structure of Turbulent Shear Flow*. Cambridge University Press, Cambridge.

Tseo, G., 1993. Two types of longitudinal dune fields and possible mechanisms for their formation. *Earth Surface Processes and Landforms*, **18**, 627–643.

Tsoar, H., 1983. Dynamic processes operating on a longitudinal (seif) dune. *Sedimentology*, **30**, 567–578.

Tsoar, H., 1989. Linear dunes-forms and formation. *Progress in Physical Geography*, **13**, 507–528.

Tsoar, H., Rasmussen, K.R., Sorensen, M. and Willetts, B.B., 1985. Laboratory study of flow over dunes. In *Proceedings of the International Workshop on Physics of Blown Sand*, Memoirs 8. Department of Theoretical Statistics, Aarhus University.

Ungar, J.E. and Haff, P.K., 1987. Steady state saltation in air. *Sedimentology*, **34**, 289–299.

Walmsley, J.I., Salmon, J.R. and Taylor, P.A., 1982. On the application of a model of boundary-layer flow over low hills to real terrain. *Boundary-Layer Meteorology*, **23**, 17–46.

Walmsley, J.L., 1989. Internal boundary-layer height formulae – a comparison with atmospheric data. *Boundary-Layer Meteorology*, **47**, 251–262.

Walmsley, J.L. and Padro, J., 1990. Shear stress results from a mixed spectral finite-difference model: Application to the Askervein hill project data. *Boundary-Layer Meteorology*, **51**, 169–177.

Walmsley, J.L., Taylor, P.A. and Keith, T., 1986. A simple model of neutrally stratified boundary-layer flow over complex terrain with surface roughness modifications (MS3DJH/3R). *Boundary-Layer Meteorology*, **36**, 157–186.

Warren, A. and Knott, P., 1983. Desert dunes: A short review of needs in desert dune research and a recent study of micro-meteorological dune initiation mechanisms. In *Eolian Sediments and Processes*, M.E. Brookfield and T.S. Ahlbrandt (Eds). Elsevier, Amsterdam, pp. 343–352.

Werner, B.T., 1990. A steady-state model of wind-blown sand transport. *Journal of Geology*, **98**, 1–17.

Werner, B.T., 1995. Eolian dunes: Computer simulations and attractor interpretation. *Geology*, 1107–1110.

White, B.R., 1996. Laboratory simulation of aeolian sand transport and physical modelling. *Annals of Arid Zone*, **35**, 187–213.

White, B.R. and Schultz, J.C., 1977. Magnus effect in saltation. *Journal of Fluid Mechanics*, **81**, 497–512.

Wiggs, G.F.S., Livingstone, I. and Warren, A., 1996. The role of streamline curvature in sand dune dynamics: Evidence from field and wind tunnel measurements. *Geomorphology*, **17**, 29–46.

Willetts, B.B. and Rice, M.A., 1988. Particle dislodgement from a flat sand surface by wind. *Earth Surface Processes and Landforms*, **13**, 717–728.

Willetts, B.B. and Rice, M.A., 1989. Collisions of quartz grains with a sand bed: The influence of incident angle. *Earth Surface Processes and Landforms*, **14**, 719–730.

Wilson, I.G., 1972. Aeolian bedforms – their developments and origins. *Sedimentology*, **19**, 173–210.

Xu, D. and Taylor, P.A., 1992. A non-linear extension of the mixed spectral finite difference model for neutrally stratified turbulent flow over topography. *Boundary-Layer Meteorology*, **59**, 177–186.

3 Geomorphology of Desert Sand Seas

NICHOLAS LANCASTER
Desert Research Institute, University and Community College System of Nevada, USA

Desert sand seas are major depositional landforms that comprise dunes of varying size, spacing and morphological type. Dune patterns in sand seas are the product of regional changes in wind regimes that promote the formation of dunes of different morphological type; and temporal changes in sand supply, availability, and mobility that give rise to the formation of multiple generations of dunes. Recent work demonstrates the importance of changes in sea level and climate that affect sediment supply, availability and wind energy. Understanding of the present form of sand seas therefore requires knowledge of the variation in time and space in the rates of sediment supply, availability and mobility, the spatial distribution and age of different dune generations, and the age of episodes of sediment input and stability.

INTRODUCTION

Desert sand seas are major depositional landforms that comprise dunes of varying size, spacing, and morphological type; areas of sand sheets; as well as interdune deposits (e.g. playas, lacustrine deposits) and extradune fluvial, lacustrine, and marine sediments. Sand seas form part of local to regional scale sediment transport systems in which sand is moved by wind from source areas to depositional sinks. The operation of these systems is governed by the supply of sediment for transport, the transport capacity of the wind, and the availability of this sediment for transport. Together, these factors determine the state of the system (Kocurek and Lancaster, 1999). Each of these variables is ultimately controlled by regional tectonics, climate and sea level.

Typically, sand seas cover an area of more than 100 km^2, with a sand cover greater than 20 per cent (Wilson, 1973). They contain many cubic kilometers of sediment (Table 3.1) and have accumulated over periods of time ranging from thousands to perhaps millions of years, during which Quaternary climatic cycles have likely played a major role in determining spatial and temporal patterns of sediment supply, availability, and mobility.

Many sand seas show a clear spatial patterning of dune types, dune size and spacing, sediment thickness, and sand particle size and composition (Breed and Grow, 1979; Lancaster, 1995a; Wilson, 1973). The pattern of dunes of different types and the spatial variation in their composition, size, spacing, and alignment in a sand sea are the surface expression of the factors that control its present day dynamics and history of accumulation. This chapter evaluates some of the characteristics of these

Aeolian Environments, Sediments and Landforms. Edited by A.S. Goudie, I. Livingstone and S. Stokes.
© 1999 John Wiley & Sons, Ltd.

Table 3.1   Estimates of sediment volume contained in some sand seas

Sand sea	Area (km^2)	Average thickness (m)	Sediment volume (km^3)
Erg Oriental	192 000	26	4992
Issaouane-n-Irarraren[a]	38 500	43	1655
Erg Occidental[a]	103 000	21	2163
Simpson Desert[a]	300 000	1	300
Namib Sand Sea[b]	34 000	20	680

[a] Data in Wilson (1973)
[b] Data from Lancaster (1989)

patterns and puts forward a conceptual model for sand sea accumulation and dune patterns in which the history of changes in sediment supply, availability, and mobility plays a major role in determining the present form of sand seas.

DUNE PATTERNS IN SAND SEAS

Dune patterns in sand seas can be analysed in terms of the morphological types of dunes that occur, their size and spacing, and their alignment. Many early workers (e.g. Capot-Rey, 1970; Holm, 1960; Mainguet and Callot, 1978) documented the variety of dune types in desert sand seas, but it was not until the synoptic regional view provided by satellite images that the true nature of spatial variations in dune morphology and morphometry was recognised and documented (Breed *et al.*, 1979a). Parallel studies of the occurrence of dune types in relation to regional variations in wind regimes provided information on the wind regime or process domains occupied by different dune types (Fryberger, 1979; Wasson and Hyde, 1983). Subsequent field studies of dune processes and dynamics provided explanations of how dune morphology is controlled by wind regime (e.g. Lancaster, 1989a; Sweet and Kocurek, 1990; Tsoar, 1983).

The paradigm that evolved from these studies was that wind regime complexity is the major determinant of dune type and that spatial variations in dune types within a sand sea could be explained by regional changes in wind regimes (e.g. Lancaster, 1983, 1989b), with the sand sea being in dynamic equilibrium with present-day wind and sand supply conditions.

While it appears generally true that the morphological type of dunes is primarily a response to wind regime characteristics, with sand supply or sediment budget, particle size and vegetation cover as secondary variables, the pattern of dunes in many sand seas is much more complex, and requires a more elaborate explanation.

DUNE GENERATIONS

Many sand seas are characterised by sharp transitions between dunes of different morphological type, size, alignment and composition. Such patterns can be recognised at all spatial scales from the small Silver Peak dunefield in central Nevada and the Algodones and Kelso dunefields in California (Lancaster, 1993) to large sand seas in Namibia (Lancaster, 1989b), the Sahara and the Arabian Peninsula (Holm, 1960). Wilson (1972, 1973) suggested that some of such patterns might be the result of changes in particle size that determined the effective wind regime for the area. A close examination of remote sensing images, together with data on dune composition and degree of post-depositional modification, suggests, however, that many sand seas are made up from a mosaic of different generations of dunes.

Different generations of dunes may be identified on remote sensing images by any one of the following, or a combination thereof:

1. Patterns of dune morphology (e.g. transverse *vs* linear dunes) and morphometry (small closely spaced dunes *vs* large dunes).
2. Variations in dune sediment composition, colour, and particle size.
3. Differences in dune trends and alignments.
4. Variations in vegetation cover and dune activity.
5. Geomorphic relations between dunes of different types (e.g. crossing patterns, superposition of dunes).

Field investigations provide further information on the degree of post-depositional modification of dune sediments, relative and absolute age of dunes (e.g. via luminescence dating); existence of bounding surfaces (sequence boundaries) separating dune generations, and the nature of geomorphic relations between dune generations. Below, some examples of sand seas comprised of a number of dune generations are reviewed.

The Gran Desierto Sand Sea

A very characteristic feature of Gran Desierto sand sea of northern Mexico (Figure 3.1) is the juxtaposition of dunes of contrasting composition, morphology, and degree of activity (Lancaster, 1992; Lancaster, 1995b). In many areas, active star dunes occur next to stabilised crescentic dunes. Elsewhere, there are several sets of crescentic dunes, each with a different morphology and alignment that coalesce with or are superimposed on each other. The pattern of dune morphology in this sand sea suggests that there are two scales of variability in dune morphology in this sand sea. At the regional scale, there is a progression from sand sheets in the northwest through a zone of crescentic and reversing dunes to the central star dune area. The eastern part of the sand sea is characterised by large compound/complex crescentic dunes, as well as areas of vegetated sand sheets and linear and parabolic dunes. At the local scale, however, different dune generations are juxtaposed or superposed on each other (Figure 3.2).

Spatial changes in dune morphology at the first scale are characterised by transitions from one dune type to another (e.g. from crescentic through reversing to star dunes) (Lancaster, 1989a). They are the result of regional changes in the wind regime from one dominated by northerly and westerly winds to the west and north of the sand sea (sand sheets and crescentic dunes migrating toward the southeast)

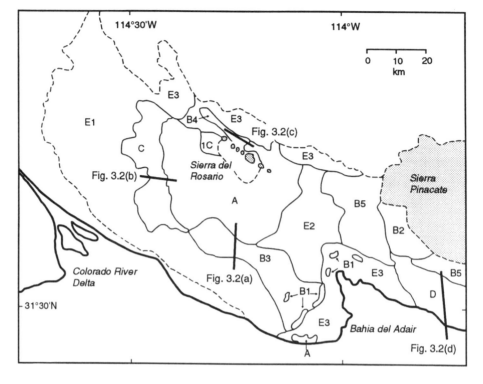

Figure 3.1 Areas of different dune morphology in the Gran Desierto sand sea. A: Chains and clusters of star dunes. B: Crescentic dunes. 1, 2: Active simple crescentic dunes; 3: large relict crescentic dunes, stabilised by vegetation; 4: coalescing multiple generations of largely relict crescentic dunes; 5: compound/complex crescentic dunes. C: Reversing dunes with active crescentic dunes on margins of the area, with stabilised and relict crescentic dunes in topographically lower areas. D: Linear and parabolic dunes, largely vegetated. E: Sand sheets. 1: Sparsely vegetated, low relief; 2: Moderately vegetated, 2–3 m local relief; 3: Undifferentiated. Reproduced by permission from Kluwer Academic Publishers

and southerly winds to the east and south (crescentic dunes migrating to the north and northwest). The star dunes occur in the zone where these two wind regimes interact. Other changes are partly due to the effect of increasing dune size on bedform reconstitution time in a seasonally varying wind regime: small dunes can be reformed in a single wind season, whereas larger dunes exhibit a morphology that is controlled by several wind directions. The extensive sand sheets probably represent a response to limited sediment supply and sediment bypassing.

The juxtaposition and/or superposition of dunes (Figure 3.2) with different morphologic types, alignments, grain-size composition and degree of post-depositional modification of sedimentary structures (e.g. cementation of laminae, carbonate accumulation, bioturbation) cannot be explained by regional changes in wind regimes. Detailed investigations of the geomorphic relations between adjacent areas of dunes of varying characteristics show that they represent genetically distinct

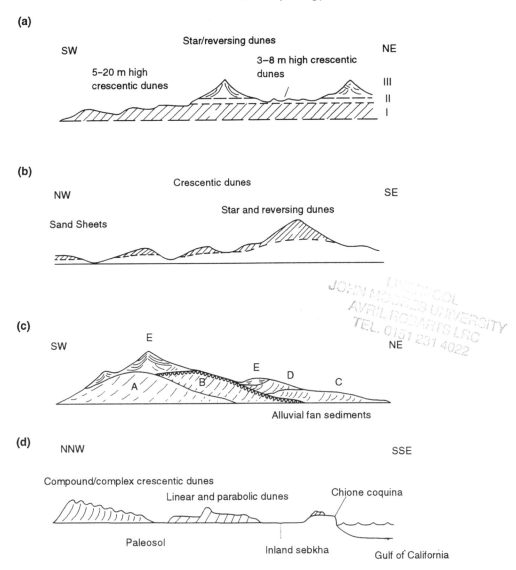

(a)

SW

Star/reversing dunes

NE

5–20 m high
crescentic dunes

3–8 m high crescentic
dunes

III

II

I

(b)

Crescentic dunes

NW

SE

Star and reversing dunes

Sand Sheets

(c)

E

SW

NE

E   D

A   B   C

Alluvial fan sediments

(d)

NNW

SSE

Compound/complex crescentic dunes

Linear and parabolic dunes

Chione coquina

Paleosol

Inland sebkha

Gulf of California

Figure 3.2 Dune generations in the Gran Deseierto Sand Sea. For locations see Figure 3.1. Reproduced by permission from Kluwer Academic Publishers

episodes of dune formation and/or modification of existing bedforms, each of which forms a separate generation of dunes. Up to five generations of dunes and three periods of sand sheet formation can be recognised in the Gran Desierto sand sea. Dune and sand sheet generations are separated by laterally-extensive bounding surfaces characterised by weakly developed soils, erosional unconformities and deflation lag surfaces. In the southeastern star dune area, the star dunes are super-imposed on two older generations of crescentic dunes (Figure 3.2(a)). Elsewhere in

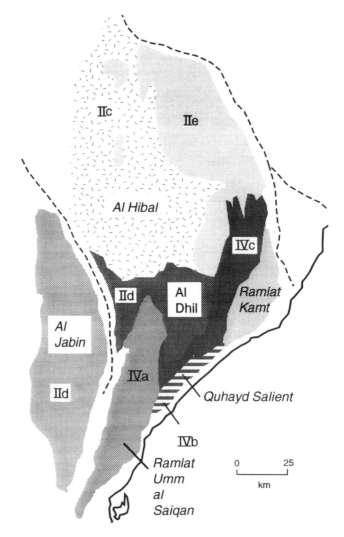

Figure 3.3   Dune generations in the Wahiba Sands (redrawn from figures in Warren, 1988)

the star dune area, the active star and reversing dunes overlie one main generation of crescentic dunes (Figure 3.2(b)), while north of the Rosario Mountains four super-posed and coalesced generations of crescentic dunes capped by star and reversing dunes can be recognised (Figure 3.2(c)). In the eastern part of the sand sea, the compound/complex crescentic dunes appear older than the vegetated linear and parabolic dunes closer to the coast. In turn, these dunes are overlain by active simple crescentic dunes near the coast of the Bahia del Adair (Figure 3.2(d)).

Some of these generations of dunes are composed of sediments of different origins, as identified via analysis of remote sensing images (Blout and Lancaster, 1990) although considerable mixing of sediments from different sources has occurred. For

Table 3.2 Sequence of dune-forming periods in the Wahiba Sand Sea (from Warren, 1988). Refer to Figure 3.3 for locations

Phase IV b	Ramlat Kamt and Ramlat Umn Al Siqan dune salients
Phase IV a	Quhayd salient dunes
Phase III	Mid Holocene (?8–6 ka) wet phase with interdune lacustrine and pond deposits
Phase II	
E	Compound crescentic dunes – Al Hibal
D	Smaller linear ridges – Al Hibal, Al Dhil, proto mega ridges of Al Jabin
C	Complex linear mega ridges of Al Hibal
B	Younger coastal aeolianites
A	Erosion of older coastal aeolianites
Phase I	Older aeolianites of Al Jabin

example, crescentic dunes north of the Bahia del Adair (B5 on Figure 3.2) are derived from a shallow marine source (Population 3), whereas crescentic dunes of B2 and B5 are composed of some of the oldest sands in the area (Population 1), as are the central star dunes and crescentic dunes of B3. Sand sheets and crescentic dunes in the northwestern sand sea are derived from a Colorado River source (Population 2).

The Gran Desierto sand sea therefore represents an amalgamation of multiple generations of dunes that were deposited adjacent to each other or superposed on one another. Each generation of dunes was the product of distinct episodes of dune formation and/or reworking of older dunes separated by periods of varying duration when active dune migration was replaced by geomorphic stability and incipient soil formation.

Dune Generations in Other Sand Seas

Similar patterns of multiple dune generations occur in many other sand seas. In the Wahiba Sand Sea of Oman (Figure 3.3), a large inland part of the sand sea is characterised by 50–100 m high S–N trending complex linear dunes, with a spacing of 1–2 km (Warren, 1988). In places, the crestal areas of these dunes are overlain by crescentic and network dunes. To the south of the large linear ridges are smaller (10–30 m high) vegetated linear dunes with a 300–500 m spacing. There are three groups of simple and compound crescentic and network dunes near the coast. These groups of dunes overlie cemented carbonate sands (aeolianites). Warren (1988) distinguished four main phases of dune development in this complex sand sea (Table 3.2).

In the United Arab Emirates, (Figure 3.4) there are two main generations of linear dunes: (1) an older generation of large complex linear dunes on WSW or WNW to ENE or ESE trends; and (2) N–S trending small linear dunes east of Sebkha Matti; as well as two generations of crescentic dunes: (1) small crescentic dunes that partly overlie dunes of group (1) above in the area north of Liwa Oasis; and (2) large compound crescentic dunes in the area of Liwa and to the south. The linear dunes of group (1) also overlie yet older cemented carbonate sands (aeolianites) (Glennie, 1998).

Linear dunes in the Simpson–Strzelecki Desert are of two generations formed in different climatic and hydrological conditions: reddened quartz rich linear dunes

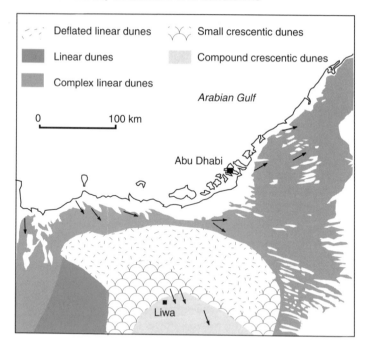

Figure 3.4 Dune generations in the United Arab Emirates (modified from Glennie, 1998). Reproduced by permission of A.A. Balkema

formed prior to 20 ka, and younger pale linear dunes. Three dune generations occur adjacent to the Finke River (Nanson *et al.*, 1995): (1) regional linear dunes largely reworked during the last Glacial (30–12 ka); (2) red coloured source-bordering dunes deposited at *ca* 100 ka; (3) overlying pale coloured dunes that consist of two units formed 17–9 and 5–0 ka.

At Kelso Dunes in the Mojave Desert, four main dune generations can be recognised on the basis of composition, morphology and alignment patterns (Lancaster, 1993). Geomorphic relationships between the different dune morphological units suggest that Kelso Dunes represent in part a stacked sequence of dunes of different generations. This has occurred because the expansion of the dune field in response to sediment inputs is restricted by its location on the piedmont of the Granite Mountains.

The Namib Sand Sea

Application of the paradigm of dune generations to the Namib Sand Sea suggests that the model proposed by Lancaster (1983, 1989b) should be revised. In the central and northwest parts of the sand sea, pale crescentic dunes can be clearly seen to overlie S–N orientated brown linear dunes. The close juxtaposition of large compound crescentic and large linear dunes is difficult to explain in terms of regional changes in wind regime and these dunes must be considered to be of different ages. Star dunes and reticulate linear dunes in the eastern sand sea represent further generations of dunes, perhaps with a different sediment source, as there is a sharp change in dune sand colour and mineral composition east of Gobabeb, on the boundary between the main linear dunes and chains of star dunes (Walden *et al.*, 1996).

The southern area of the sand sea exhibits several examples of dunes that overlie each other. Northeast of Sylvia Hill, small crescentic dunes bury low linear dunes. South of the Uri Hauchab, low compound linear dunes overlie the coarse sands and zibars of the trailing margins of the main S–N linear dunes. North of Lüderitz, similar compound linear dunes appear to be buried by crescentic dunes migrating in from the coast. The small areas of star dunes in the southern sand sea also appear to represent a further generation of dunes. A revised map of dune morphological variations in the Namib (Figure 3.5) suggests that there could be as many as nine different generations of dunes in this sand sea.

DUNES AS AMALGAMATIONS OF PERIODS OF AEOLIAN ACTIVITY

Many sand seas are characterised by large dunes (complex and compound dunes, mega dunes, draas). Where they have been investigated, such dunes are almost always an amalgamation of several periods of dune construction, stabilisation and reworking. In Mauritania, detailed studies of linear dune and interdune sediments in the Akchar Sand Sea of Mauritania show that they represent the amalgamation of Late Pleistocene and Holocene deposits (Kocurek *et al.*, 1991).

The prominent large complex linear dunes (Figure 3.6) are composite features. Their core consists of sand deposited during the period 20–13 ka. These linear dunes were stabilised by vegetation during a period of increased rainfall 11 to 4.5 ka, when soil formation altered dune sediments and lakes formed in interdune areas. Further periods of dune formation after 4 ka cannibalised existing aeolian deposits on the upwind margin of the sand sea. The currently active 'cap' of crescentic dunes superimposed on the linear dunes dates to the last 30 years. Similar sequences of dune deposits have been recognised in the Sahel (Talbot, 1985) and southern Sahara (Vökel and Grunert, 1990).

Large complex linear dunes in the Wahiba Sands also provide evidence of multiple periods in their formation. The main ridges are 50–10 m high and 1000–1800 m wide (Warren, 1988). Superimposed on their flanks are linear ridges 20 m high and 500 m apart (Glennie, 1970), which run oblique to the main dune trend. The crests of the complex ridges are characterised by small crescentic, linear and network dunes (Warren and Kay, 1987). Complex linear dunes in the Rub al Khali are composed of a core of sand that accumulated between 20 and 9 ka (McClure, 1978; Warren, 1988; Whitney *et al.*, 1983) and was stabilised by vegetation with weak pedogenesis and lakes in interdune areas 9 to 6 ka. The morphology of the modern dune system may be less than 2 ka old (Figure 3.7).

Simple linear dunes in the Kalahari and Australia are similarly composed of sand deposited and/or reworked in several generations of aeolian activity. The last major period of linear dune development in the partly vegetated southern Kalahari dune

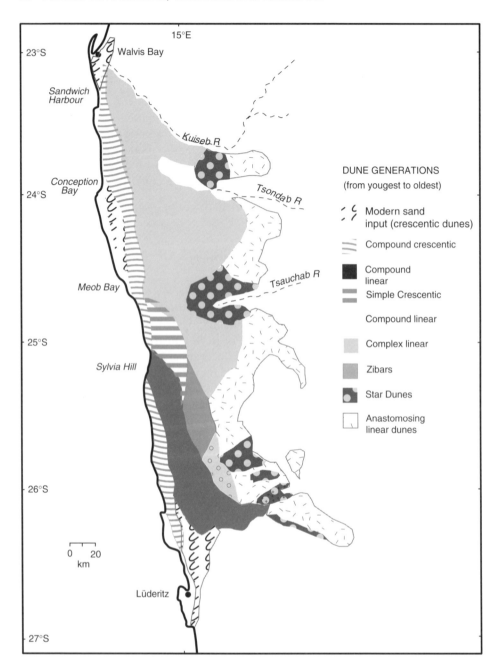

Figure 3.5 Dune generations in the Namib Sand Sea

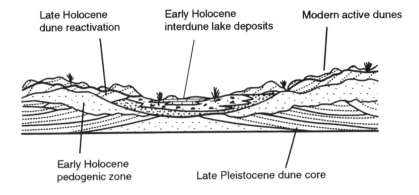

Late Holocene dune reactivation

Early Holocene interdune lake deposits

Modern active dunes

Early Holocene pedogenic zone

Late Pleistocene dune core

Figure 3.6 Amalgamated deposits of linear dunes in the Akchar Sand Sea, Mauritania (after Kocurek *et al.*, 1991)

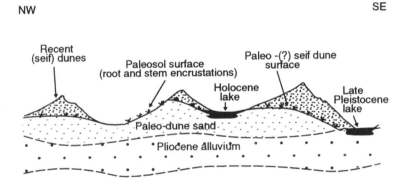

NW

SE

Recent (seif) dunes

Paleosol surface (root and stem encrustations)

Paleo -(?) seif dune surface

Holocene lake

Late Pleistocene lake

Paleo-dune sand

Pliocene alluvium

Figure 3.7 Inferred stratigraphic relations between periods of dune formation in the Rub' al Khali sand sea (after McClure, 1978). Reproduced by permission of Springer-Verlag

system occurred between 17 and 10 ka and probably involved reworking of sediment deposited in a previous period of dune building from 28 to 23 ka that may represent the initial deposition of sand in this region (Stokes *et al.*, 1997b). Holocene dune activity in the region was localised in extent and occurred at 6 ka and 1–2 ka (Thomas *et al.*, 1997). Large relict linear ridges in the northeastern Kalahari provide further evidence for multiple periods of dune building, each spanning 5–20 ka, that occurred during the Late Pleistocene (Stokes *et al.*, 1998). OSL ages for these periods are 95–115, 41–46 and 20–26 ka. Late Glacial and Holocene aeolian activity was restricted to reworking of the crestal areas of the linear dunes in the period 10–16 ka (Stokes *et al.*, 1998). In the Simpson Desert of Australia, extensive reworking of linear dune sand took place near Birdsville during the Holocene. These sands cover a dune core that was formed 80 ka or earlier (Nanson *et al.*, 1992).

ORIGIN AND DEVELOPMENT OF PATTERNS OF DUNE MORPHOLOGY

Based on evidence from the above examples, there are therefore two scales of variability in patterns of dune morphology in sand seas: (1) regional changes related to spatial variations in wind regime and sediment budget that determine the type and size of dune that will form. This scale is characterised in part by transitions between dune morphological type; and (2) juxtapositions of dunes that form a mosaic of different generations. In addition, many large dunes are composite forms.

Multiple dune generations represent the response of a sand sea to climatic, tectonic, or sea level changes that have affected sediment supply and/or dune mobility. Formation of areas of different dunes and their subsequent preservation also implies that later periods of sediment input and dune formation occurred in a manner that did not result in complete reworking of the older dunes.

Preservation of older dune generations is favoured by climatic changes that result in surface stabilisation by vegetation and soil formation that may result in bounding surfaces of regional extent (e.g. Kocurek et al., 1991; Talbot, 1985). In some areas, different dune generations are superposed on one another. This takes place where the accumulation is restricted by topographic obstacles, e.g. north of the Rosario mountains in the Gran Desierto (Figure 3.2(c)), at Kelso Dunes, California (Lancaster, 1993), and Great Sand Dunes, Colorado (Andrews, 1981).

Different dune types appear to respond in a different way to periods of dune construction, stability, and destruction. Dunes can change form or orientation only at defects or terminations, e.g. the tips of linear dunes or the ends of crescentic ridges and star dune arms (Werner, 1995; Werner and Kocurek, 1997). Linear dunes typically have lengths of tens of kilometres and can therefore persist without change for long periods of time. Werner and Kocurek (1997) estimate a 10 kyr time period to reorientate Simpson Desert linear dunes. Linear dunes also extend, rather than migrate, so they tend to be stable forms, although there may be considerable movement of sand along and/or across them (Thomas, 1992).

In many sand seas, there is a hierarchical system of dunes which consists of: (1) individual simple dunes or superimposed dunes on compound and complex form; and (2) compound or complex dunes (mega dunes or draas) (Warren and Allison, 1998; Wilson, 1972). There is a characteristic time period, termed the relaxation or reconstitution time (Allen, 1974), over which each element of the hierarchy will adjust to changed conditions. Because change in dunes involves movement of sediment, an increasing spatial scale is therefore involved at each level of the hierarchy.

The reconstitution time can be represented by the time taken for the dune to migrate one wavelength in the direction of net transport. This can vary from as little as 10–100 years for simple crescentic dunes (Havholm and Kocurek, 1988; Lancaster, 1988) to 100–1000 years for compound crescentic dunes (Stokes et al., 1997a) and 1000 to 10 000 years for large complex linear dunes (Lancaster, 1988; Warren, 1988; Wilson, 1972). Reconstitution time therefore increases by several orders of magnitude from simple through compound to complex dunes. In general, crescentic dunes have much shorter reconstitution times than linear dunes because their rate of migration is much more rapid.

The above implies that the morphology of simple dunes and superimposed dunes is governed principally by annual or seasonal patterns of wind speed and direction and by spatial changes in wind speed over the primary dune. Compound crescentic dunes are not sensitive to seasonal changes and their characteristics likely reflect Late Holocene conditions. Linear dunes are, however, relatively insensitive to short period changes and may persist for 10 to 100 ka, and many linear dunes appear to reflect conditions during the last Glacial maximum (Warren and Allison, 1998). Dune size and morphological type are therefore factors determining the persistence of dunes that are not in equilibrium with contemporary sand supply and wind regime characteristics, as recognised by Wilson (1971).

FORMATION OF SAND SEAS

The accumulation of sand seas is governed by principles of sediment mass conservation (Middleton and Southard, 1984; Rubin and Hunter, 1982). Accumulation of sand is the product of spatial changes in transport rates and temporal changes in sediment concentration such that:

$$\mathrm{d}h/\mathrm{d}t = -(\mathrm{d}q_s/\mathrm{d}x + \mathrm{d}C/\mathrm{d}t) \tag{3.1}$$

where h is the elevation of the deposition surface, t is time, q_s is the spatially averaged bulk volume sediment transport rate, x is the distance along the transport pathway, and C is the concentration of sediment in transport. The transport rate (q_s) consists of two components: that due to bedform migration (bedform transport, q_b) and that which is throughgoing (q_t) as a result of saltation transport. The sediment concentration (C) is a measure of the total amount of sediment in transport, and can be approximated by the average height of the dunes (Kocurek and Havholm, 1993). Deposition of sand and the accumulation of sand seas will therefore occur downwind of source zones wherever sand transport rates are reduced as a result of changes in climate or topography, so that sand seas are located in areas where sand transport rates are lower and/or more variable in direction, compared with adjacent areas without sand sea development (Fryberger and Ahlbrandt, 1979; Lancaster, 1985; Wilson, 1971).

Although sediment transport rates over sand surfaces are almost always at the capacity of the wind to transport sediment (sediment saturated), those over other desert surfaces (exposed bedrock, alluvial fans, playas) are frequently below transport capacity (undersaturated or metasaturated (Wilson, 1971)), because the sediment in transport is limited by supply or availability. Thus the actual sand transport rate (q_a) may be less than the potential rate (q_p) in proportion to the ratio q_a/q_p, which ranges between zero for completely undersaturated flows to 1 for fully saturated conditions (Kocurek and Havholm, 1993). The wind is thus potentially erosional until its transport capacity is reached, regardless of whether the wind is accelerating, steady or decelerating. Deposition occurs wherever there are local decreases in transport capacity (e.g. deceleration in the lee of obstacles or changes in surface roughness). The wind may still be transporting sand, as deposition only occurs until the transport rate is in equilibrium with changed conditions.

Following principles of sediment mass conservation, if transport rates decrease in the direction of flow, deposition will occur and the accumulation will grow. If, however, sediment transport rates increase in the direction of flow (e.g. acceleration of the wind), then the accumulation will be eroded. No change in transport rates in space will give rise to net sediment bypassing (Kocurek and Havholm, 1993). The relations between the sediment saturation level of the input to the system and the change in transport capacity in time and space (defined as the ratio between the potential transport rate upwind and downwind of the area of concern) determine the domains of erosion, bypass and accumulation of sand (Kocurek and Havholm, 1993).

These concepts can be used to assess the nature of changes that may occur through time. The concentration of sediment in the system may decrease if surfaces are stabilised by vegetation, the water table rises, or the source of sand is depleted (Kocurek and Havholm, 1993). The spatial pattern of wind energy may change as wind patterns shift position, leading to changes in the spatial distribution of sediment transport. This could lead to erosion of previously deposited sediment and its accumulation in other areas. The supply and/or availability for sediment transport and thus the saturation level of the input therefore plays a major role in determining the behaviour of the system, because growth of the accumulation can only occur if there is a supply of sediment.

The state of the sand sea system over time can be evaluated in terms of the supply of sediment (the deposition of sediment that is a source for the aeolian system either contemporaneously or at some later time, its availability for transport by the wind (the probability of entrainment of sand for transport), and the transport capacity of the wind (the potential rate of sediment transport, q_p). Variations in the transport capacity of the wind are governed by temporal changes in wind energy. Such changes over glacial–interglacial cycles have been predicted by climate models and are documented by evidence of increased dust transport in ocean and ice cores (e.g. deMenocal, 1995; O'Brien et al., 1995; Rea, 1990; Sarnthein et al., 1982). Sub-Milankovitch scale periods of increased windiness are also indicated (Alley et al., 1997). Sediment availability is strongly influenced by climatic changes that result in variations in vegetation cover and soil moisture, which determine the actual transport rate (q_a) (Lancaster and Baas, 1998). Changes in sediment availability over time may result in episodes of reworking of existing aeolian sediment (e.g. Holocene reactivation of the crests of linear dunes in Mauritania and Australia). In addition, changes in sea level and regional hydrology may make sediment stored in shallow marine or lacustrine environments available for transport.

The source of sediment for sand sea accumulation is commonly external to the sand body, although internal sources, e.g. alluvial deposits in interdune areas, do occur. Sediment for transport by wind may be derived from deflation of bedrock (Besler, 1980), but the primary source of sand-sized grains is sediments deposited by fluvial/alluvial, coastal, or lacustrine systems (e.g. Allison, 1988; Corbett, 1993; Muhs et al., 1995). In those sand seas for which the source is known, fluvial systems provide the major input of sand for wind transport, either directly through deflation of alluvial sediments, or indirectly through transfer of sediment in coastal sediment transport systems, as in the Namib, Gran Desierto, Sinai, Atacama and Arabian sand seas.

Availability limited	Transport limited	Transport and availability limited	
CI_{AL}	CI_{TL}	CI_{TAL}	Contemporaneous input
LI_{AL}	LI_{TL}	LI_{TAL}	Lagged input
S_{AL}	S_{TL}	S_{TAL}	Stored sediment

Figure 3.8 Matrix of possible states of sand seas

The relations between sediment supply, availability, and mobility at any point in time, as well as their variation through time, define a series of states of the system (Kocurek, 1998; Kocurek and Lancaster, 1999), as summarised in Figure 3.8. The interaction between varying levels of sediment supply, availability and mobility result in episodic input of sediment to the system is demonstrated by examples from the Mojave Desert (Clarke *et al.*, 1996a; Rendell and Sheffer, 1996; Wintle *et al.*, 1994), Tunisia (Blum *et al.*, 1998), the Wahiba (Warren, 1988), and the Namib (Corbett, 1993), discussed in detail by Tchakerian (Chapter 12).

In arid regions, periods of increased precipitation tend to result in an increased sediment supply from fluvial sources, but the availability of this sediment for transport by the wind is limited by vegetation cover. Such sediment is therefore stored. During periods of desiccation, this sediment becomes available as lagged input, with the magnitude of the input being limited mainly by the transport capacity of the wind. This was the situation in the Mojave Desert during the Early Holocene (Kocurek and Lancaster, 1999). Periods of desiccation may also increase sediment supply from lacustrine sources as well as from destabilisation of hillslopes and aggradation of alluvial fan systems (Harvey and Wells, 1994; Wells and McFadden, 1987).

Given that climatic and sea level changes produce multiple pulses of sediment input to sand seas, why do these pulses often produce different generations of dunes rather than adding to pre-existing dunes? The impact of external changes on sand sea accumulation is manifested mainly by changes in sand supply and the availability of the sediment for transport. Changes in sediment supply transgress time and space, because a period of time is required for newly supplied sediment to be moved from source areas to depositional sinks. Changes in the transport capacity of the wind and sediment availability tend, however, to be geologically instantaneous.

Because they are time- and space-transgressive, the effects of changes in sediment supply tend to be more marked in areas that are proximal to sand sources and/or in sand seas where sediment supply is high and the system is transport limited. In this case, each period of increased sediment supply produces a new generation of dunes, which then migrate away from the source. If sediment supply declines over time as a result of sea level rise or the exhaustion of the volume of supplied sediment, then this may be manifested in undersaturation of the incoming wind and erosion of the upwind margin of the sand accumulation. Examples of dune generations that have moved away from their source can be found in the Gran Desierto and the Wahiba sand seas (Blount and Lancaster, 1990; Warren and Allison, 1998). In general, proximal areas of sand seas tend to have the potential for more dune generations than distal areas.

In distal areas of the sand sea, and/or in areas where the sediment supply is low, the effects of changes in sediment supply to the system are relatively small and the major control on sand sea and dune accumulation is likely to be sediment availability. The response of the system is determined in part by the antecedent conditions, so that a change from lower to higher sediment availability (e.g. desiccation leading to lower vegetation cover) will be manifested as a reactivation of previously stabilised dunes. The multiple dune generations of Kelso Dunes are one example of this, as are the multiple periods of reworking of large linear dunes in the Ackchar sand sea of Mauritania (Kocurek *et al.*, 1991).

In summary, the effect of climate change on sand sea and dune accumulation is determined by the fundamental controls on the state of each aeolian system. The relations among sediment supply, availability, and mobility can be evaluated in time to determine the response of the system to external forcing factors. The average state of each aeolian system also affects its response. In situations where the external sand supply is low, then the primary control on aeolian accumulation is sediment availability and such sand seas will be characterised by multiple episodes of reworking of dunes (e.g. Simpson Desert, Kalahari, Mauretania). Where sediment supply is high, then this is the primary control on accumulation and such sand seas will likely be characterised by multiple dune generations (e.g. Gran Desierto, Namib, Wahiba). A third scenario is where aeolian accumulation is limited by the transport capacity of the wind (Kalahari, Australia).

These principles lead to the paradigm in which the state of any sand sea and its response to climate change can be characterised by the relations among sediment supply, availability, and mobility, so that a sand sea or dunefield can be described as:

1. Transport limited, in which $Q_a = Q_p$ and the system is limited only by the capacity of the wind to move sediment from source zones. Examples are the Namib sand seas and many coastal dune fields.
2. Availability limited, in which $Q_a < Q_p$ and the system response is controlled by vegetation cover, for example as in the Kalahari and Australia.
3. Supply limited, in which $Q_a \ll Q_p$ and the system is starved of sediment, as in many central and northern sand Saharan sand seas, some of which are apparently eroding in present conditions.

In practice, these factors are not independent, and also vary in relative importance over time, leading to a complex response of most sand seas to external forcing factors.

CONCLUSIONS

Recent work on sand sea and dune accumulation demonstrates that changes in sea level and climate that affect sediment supply, availability, and wind energy have played a major role in determining the geomorphology of desert sand sea. Dune patterns in sand seas are the product of two scales of variability: (1) regional changes in wind regimes that promote the formation of dunes of different morphological type; and (2) temporal changes in sand supply, availability, and mobility that give rise to the formation of multiple generations of dunes.

Sand seas with many different dune generations (e.g. the Gran Desierto, the Wahiba) have experienced multiple episodes of dune formation and/or reworking of older dunes. Complex dune patterns are therefore indicative of areas that are sensitive to climatic and sea level changes that have affected sediment supply and mobility. The effects of these changes are propagated through the system from the source areas, so that sand seas (or areas of sand seas) that are close to their source will tend to exhibit more complex patterns than those that lie distant from their source.

Dune size appears to be a function of sediment supply and time, so that many large, complex dunes are comprised of the sediments generated by multiple generations of accumulation. They are tens of thousands of years old and are being modified by modern winds, as evidenced by superimposition of dunes of different morphological type. Stratigraphic and dating studies demonstrate the great age of the cores of large complex dunes in Arabia and Africa, as well as smaller linear dunes in Australia. Many dunes that are on trends that are not consistent with modern sand transporting winds and have ancient cores are of linear types, which have long reconstitution times and are therefore conservative of changed conditions.

Quaternary climatic and sea level changes that have affected sediment production, availability, and mobility have therefore played an important role in shaping the present morphology of many sand seas. Rather than assume that their morphology is a product of contemporary processes, we need to analyse sand seas in the same way as we study alluvial fans or river flood plains – by using detailed mapping of landforms, stratigraphy, sedimentology and relative and absolute dating techniques to reconstruct their history. Understanding of the present form of sand seas therefore requires knowledge of three main aspects of their history:

1. The variation in time and space in the rates of sediment supply, availability, and mobility.
2. The spatial distribution and age of different dune generations.
3. The age of episodes of sediment input and stability.

REFERENCES

Allen, J.R.L., 1974. Reaction, relaxation and lag in natural sedimentary systems: general principles, examples and lessons. *Earth Science Reviews*, **10**, 263–342.

Alley, R.B., Mayewski, P.A., Sowers, T., Stuiver, M. *et al.*, 1997. Holocene climatic instability: a prominent, widespread event 8200 yr ago. *Geology*, **26**, 483–486.

Allison, R., 1988. Sediment types and sources in the Wahiba Sands. In R.W. Dutton (Ed.), Scientific Results of the Royal Geographical Society's Oman Wahiba Sands Project 1985–1987. *Journal of Oman Studies, Special Report* 3, Muscat, Oman, pp. 161–168.

Andrews, S., 1981. Sedimentology of Great Sand Dunes, Colorado. In *Recent and Ancient Non-Marine Depositional Environments: Models for Exploration*, F.P. Ethridge and R.M. Flores (Eds). The Society of Economic Paleontologists and Mineralogists, Tulsa, OK, pp. 279–291.

Besler, H., 1980. Die Dunen-Namib: Enstehung und dynamik eines Ergs. *Stuttgarter Geographische Studien*, **96**, 241.

Blount, G. and Lancaster, N., 1990. Development of the Gran Desierto Sand Sea. *Geology*, **18**, 724–728.

Blum, M., Kocurek, G., Deynoux, H., Swezey, C. *et al.*, 1998. Quaternary wadi, lacustrine, aeolian depositional cycles and sequences, Chott Rharsa basin, southern Tunisia. In *Quaternary Deserts and Climatic Change*, A.S. Alsharan, K.W. Glennie, G.L. Whittle and C.G.S.C. Kendall (Eds). Balkema, Rotterdam, pp. 539–552.

Breed, C.S. and Grow, T., 1979. Morphology and distribution of dunes in sand seas observed by remote sensing. In *A Study of Global Sand Seas*, E.D. McKee (Ed.). Professional Paper, United States Geological Survey, pp. 253–304.

Breed, C.S., Grolier, M.J. and McCauley, J.F., 1979. Morphology and distribution of common 'sand' dunes on Mars: comparison with the Earth. *Journal of Geophysical Research*, **84**, 8183–8204.

Breed, C.S., Fryberger, S.G., Andrews, S., McCauley, C., Lennartz, F., Geber, D. and Horstman, K., 1979. Regional studies of sand seas using LANDSAT (ERTS) imagery. In E.D. McKee (Ed.), *A Study of Global Sand Seas: Professional Paper*, United States Geological Survey, pp. 305–398.

Capot-Rey, R., 1970. Remarques sur les ergs du Sahara. *Annales de Geographie*, **79**, 2–19.

Clarke, M.L., Richardson, C.A. and Rendell, H.M., 1996a. Luminescence dating of Mojave Desert sands. *Quaternary Science Reviews*, **14**, 783–790.

Clarke, M.L., Wintle, A.G. and Lancaster, N., 1996b. Infra-red stimulated luminescence dating of sands from the Cronese Basins, Mojave Desert. *Geomorphology*, **17**, 199–206.

Corbett, I., 1993. The modern and ancient pattern of sandflow through the southern Namib deflation basin. *International Association of Sedimentologists Special Publication*, **16**, 45–60.

deMenocal, P.B., 1995. Plio-Pleistocene African Climate. *Science*, **270**, 53–59.

Fryberger, S.G., 1979. Dune forms and wind regimes. In *A Study of Global Sand Seas*, E.D. McKee (Ed.). Professional Paper, United States Geological Survey, pp. 137–140.

Fryberger, S.G. and Ahlbrandt, T.S., 1979. Mechanisms for the formation of aeolian sand seas. *Zeitschrift für Geomorphologie*, **23**, 440–460.

Glennie, K.W., 1970. *Desert Sedimentary Environments*. Developments in Sedimentology, 14. Elsevier, Amsterdam.

Glennie, K.W., 1998. The desert of southeast Arabia: a product of Quaternary climatic change. In *Quaternary Deserts and Climatic Chaos*, A.S. Alsharan, K.W. Glennie, G.L. Whittle and C.G.S.C. Kendall (Eds). Balkema, Rotterdam, pp. 279–292.

Harvey, A.M. and Wells, S.G., 1994. Late Pleistocene and Holocene changes in hillslope sediment supply to alluvial fan systems: Zzyxx, California. In *Environmental Change in Drylands*, A.C. Millington and K. Pye (Eds). John Wiley, Chichester, pp. 66–84.

Havholm, K.G. and Kocurek, G., 1988. A preliminary study of the dynamics of a modern draa, Algodones, southeastern California, USA. *Sedimentology*, **35**, 649–669.

Holm, D.A., 1960. Desert geomorphology in the Arabian Peninsula. *Science*, **123**, 1369–1379.

Kocurek, G., 1998. Aeolian System Response to External Forcing Factors – A Sequence Stratigraphic View of the Saharan Region. In *Quaternary Deserts and Climatic Change*, A.S. Alsharan, K.W. Glennie, G.L. Whittle and C.G.St.C. Kendall (Eds). Balkema, Rotterdam, pp. 327–338.

Kocurek, G. and Havholm, K.G., 1993. Eolian sequence stratigraphy – a conceptual framework. In *Siliciclastic Sequence Stratigraphy*, P. Weimer and H. Posamentier (Eds). American Association of Petroleum Geologists, Tulsa, OK, pp. 393–409.

Kocurek, G., Havholm, K.G., Deynoux, M. and Blakey, R.C., 1991. Amalgamated accumulations resulting from climatic and eustatic changes, Akchar Erg, Mauritania. *Sedimentology*, **38**, 751–772.

Kocurek, G. and Lancaster, N., 1999. Aeolian Sediment States: Theory and Mojave Desert Kelso Dunefield example. *Sedimentology*, **46**, 505–516.

Lancaster, N., 1983. Controls of dune morphology in the Namib sand sea. In *Eolian Sediments and Processes*, Developments in Sedimentology, T.S. Ahlbrandt and M.E. Brookfield (Eds). Elsevier, Amsterdam, pp. 261–289.

Lancaster, N., 1985. Winds and sand movements in Namib sand sea. *Earth Surface Processes and Landforms*, **10**, 607–619.

Lancaster, N., 1988. Controls of eolian dune size and spacing. *Geology*, **16**, 972–975.

Lancaster, N., 1989a. The dynamics of star dunes: an example from the Gran Desierto, Mexico. *Sedimentology*, **36**, 273–289.

Lancaster, N., 1989b. *The Namib Sand Sea: Dune Forms, Processes, and Sediments*. Balkema, Rotterdam.

Lancaster, N., 1992. Relations between dune generations in the Gran Desierto, Mexico. *Sedimentology*, **39**, 631–644.

Lancaster, N., 1993. Development of Kelso Dunes, Mojave Desert, California. *National Geographic Research and Exploration*, **9**, 444–459.

Lancaster, N., 1995a. *Geomorphology of Desert Dunes*. Routledge, London.

Lancaster, N., 1995b. Origin of the Gran Desierto Sand Sea: Sonora, Mexico: Evidence from dune morphology and sediments. In *Desert Aeolian Processes*, V.P. Tchakerian (Ed.). Chapman and Hall, New York, pp. 11–36.

Lancaster, N. and Baas, A., 1998. Influence of vegetation cover on sand transport by wind: field studies at Owens Lake, California. *Earth Surface Processes and Landforms*, **23**, 69–82.

Mainguet, M. and Callot, Y., 1978. L'erg de Fachi-Bilma (Tchad-Niger). *Mémoires et Documents CNRS*, **18**, 178.

McClure, H.A., 1978. Ar Rub' al Khali. *Quaternary Period in Saudi Arabia. 1: Sedimentological, hydrochemical, geomorphological investigations in central and eastern Saudi Arabia*, S.S. Al-Sayari and J.G. Zötl (Eds). Springer-Verlag, Vienna, New York, pp. 252–263.

Middleton, G.V. and Southard, J.B., 1984. *Mechanics of Sediment Movement*. S.E.P.M., Tulsa, OK.

Muhs, D.R., Bush, C.A., Cowherd, S.D. and Mahan, S., 1995. Source of sand for the Algodones Dunes. In *Desert Aeolian Proceses*, V.P. Tchakerian (Ed.). Chapman and Hall, New York, pp. 37–74.

Nanson, G.C., Ghen, X.Y. and Price, D.M., 1992. Lateral migration, thermoluminescence chronology, and colour variation of longitudinal dunes near Birdsville in the Simpson Desert, Australia. *Earth Surface Processes and Landforms*, **17**, 807–820.

Nanson, G.C., Chen, X.Y. and Price, D.M., 1995. Aeolian and fluvial evidence of changing climate and wind patterns during the past 100 ka in the western Simpson Desert, Australia. *Palaeogeography, Palaeoclimatology, Palaeoecology*, **113**, 87–102.

O'Brien, S.R., Mayewski, P.A., Meeker, L.D., Meese, D.A. *et al.*, 1995. Complexity of Holocene climate as reconstructed from a Greenland ice core. *Science*, **270**, 1962–1964.

Pye, K. and Lancaster, N. (Eds), 1993. *Aeolian Sediments: Ancient and Modern*. IAS Special Publication, 16. Blackwell Science, Oxford.

Rea, D.K., 1990. Aspects of atmospheric circulation: the Late Pleistocene (0–950 000 yr) record of eolian deposition in the Pacific Ocean. *Palaeogeography, Palaeoclimatology, Palaecology*, **78**, 217–227.

Rendell, H.M. and Sheffer, N.L., 1996. Luminescence dating of sand ramps in the eastern Mojave Desert. *Geomorphology*, **17**, 187–198.

Rubin, D.M. and Hunter, R.E., 1982. Bedform climbing in theory and nature. *Sedimentology*, **29**, 121–138.

Sarnthein, M., Theide, J., Pflaumann, W., Erlenkeuser, H. *et al.*, 1982. Atmosoheric and oceanic circulation patterns off Northwest Africa during the last 25 million years. *Geology of the Northwest African Continental Margin*, U. von Rad, K. Hinz, M. Sarnthein and E. Seibold (Eds). Springer-Verlag, Berlin, pp. 547–604.

Stokes, S., Haynes, G., Thomas, D.S.G., Higginson, M. and Mallifa, M., 1998. Punctuated aridity in southern Africa during the last glacial cycle: the chronology of linear dune construction in the northeastern Kalahari. *Palaeogeography, Palaeoeclimatology, Palaeocology*, **137**, 305–332.

Stokes, S., Kocurek, G., Pye, K. and Winspear, N.R., 1997a. New evidence for the timing of aeolian sand supply to the Algodones dunefield and East Mesa area, southeastern California, USA. *Palaeogeography, Palaeoeclimatology, Palaeocology*, **128**, 63–75.

Stokes, S., Thomas, D.S.G. and Shaw, P.A., 1997b. New chronological evidence for the nature and timing of linear dune development in the southwest Kalahari Desert. *Geomorphology*, **20**, 81–94.

Sweet, M.L. and Kocurek, G., 1990. An empirical model of aeolian dune lee-face airflow. *Sedimentology*, **37**, 1023–1038.

Talbot, M.R., 1985. Major bounding surfaces in aeolian sandstones: a climatic model. *Sedimentology*, **32**, 257–266.

Thomas, D.S.G., 1992. Desert dune activity: concepts and significance. *Journal of Arid Environments*, **22**, 31–38.

Thomas, D.S.G., Stokes, S. and Shaw, P.A., 1997. Holocene aeolian activity in the southwestern Kalahari Desert, southern Africa: significance and relationships to late-Pleistocene dune-building events. *The Holocene*, **7**, 273–281.

Tsoar, H., 1983. Dynamic processes acting on a longitudinal (seif) dune. *Sedimentology*, **30**, 567–578.

Vökel, J. and Grunert, J., 1990. To the problem of dune formation and dune weathering during the Late Pleistocene and Holocene in the southern Sahara and Sahel. *Zeitschrift für Geomorphologie*, **34**, 1–17.

Walden, J., White, K. and Drake, N.A., 1996. Controls on dune colour in the Namib Sand Sea: preliminary results. *Journal of African Earth Sciences*, **22**, 349–353.

Warren, A., 1988. The dunes of the Wahiba Sands. In R.W. Dutton (Editor), Scientific Results of the Royal Geographical Society's Oman Wahiba Sands Project 1985–1987. *Journal of Oman Studies, Special Report* 3, Muscat, Oman, pp. 131–160.

Warren, A. and Allison, D., 1998. The palaeoenvironmental significance of dune size hierarchies. *Palaeogeography, Palaeoeclimatology, Palaeocology*, **137**, 289–303.

Warren, A. and Kay, S., 1987. Dune networks. In *Desert Sediments: Ancient and Modern*, L.E. Frostick and I. Reid (Eds). Blackwell Science, Oxford, pp. 205–212.

Wasson, R.J., 1983. Dune sediment types, sand colour, sediment provenance and hydrology in the Strzelecki-Simpson Dunefield, Australia. In *Eolian Sediments and Processes*. Developments in Sedimentology, M.E. Brookfield and T.S. Ahlbrandt (Eds). Elsevier, Amsterdam, pp. 165–195.

Wasson, R.J. and Hyde, R., 1983. Factors determining desert dune type. *Nature*, **304**, 337–339.

Wells, S.G. and McFadden, L.D., 1997. Influence of Late Quaternary climatic changes in geomorphic processes on a desert piedmont, eastern Mojave Desert, California. *Quaternary Research*, **27**, 130–146.

Werner, B.T., 1995. Eolian dunes: computer simulations and attractor interpretation. *Geology*, **23**, 1107–1110.

Werner, B.T. and Kocurek, G., 1997. Bed-form dynamics: Does the tail wag the dog. *Geology*, **25**, 771–774.

Whitney, J.W., Faulkender, D.J. and Rubin, M., 1983. The environmental history and present condition of the northern sand seas of Saudi Arabia. OF-03-95, United States Geological Survey.

Wilson, I.G., 1971. Desert sandflow basins and a model for the development of ergs. *Geographical Journal*, **137**, 180–199.

Wilson, I.G., 1972. Aeolian bedforms – their development and origins. *Sedimentology*, **19**, 173–210.

Wilson, I.G., 1973. Ergs. *Sedimentary Geology*, **10**, 77–106.

Wintle, A.G., Lancaster, N. and Edwards, S.R., 1994. Infrared stimulated luminescence (IRSL) dating of late-Holocene aeolian sands in the Mojave Desert, California, USA. *The Holocene*, **4**, 74–78.

4 Coastal Dune Dynamics: Problems and Prospects

BERNARD O. BAUER and DOUGLAS J. SHERMAN
Department of Geography, University of Southern California, USA

Efforts toward prediction of coastal dune dynamics are hampered by several non-trivial linkages between dunes and their surrounding environments. The most significant among these is the intimate inter-relationship between coastal dunes and the suite of nearshore processes fronting them (e.g. wave-related dune scarping, overwash of blowout throats, and swash-related sediment deposition on the foreshore and back beach). In this regard, the only distinctive coastal dune – the foredune – is very unlike other dunes commonly found in secondary dune fields or in dry-land environments that are dominantly wind sculpted. The effects of vegetation, moisture and topographic slope are additional factors that have a pronounced influence on the diverse geometry and dynamics of coastal dunes.

Empirical studies of coastal dune dynamics have generally followed one of two divergent strategies: the *mechanistic-reductionist* approach, which focuses on progressively smaller and smaller constituent parts of a complex system, versus the *holistic-constructivist* approach, which strives to understand fundamental essences and behaviours of complete systems. Examples of the mechanistic-reductionist approach include studies into the hydrodynamic processes by which sediments are exchanged between the foreshore and foredune, the trapping of sediments by vegetation covers, or the mechanics of aeolian saltation. At the smallest end of the reductionist spectrum, the results of such studies can be generalised to any aeolian system, coastal or otherwise. Examples of the holistic-constructivist approach include studies into the sequential development of foredunes on overwash barriers, the cyclical evolution of blowouts, and the equilibrium relationship manifested in specific dune assemblages in association with characteristic nearshore morphodynamic states. The results of these holistic studies are usually for a specific coastal environment and thus they are often idiographic. Both research strategies are important and necessary because they provide complementary insights. Nevertheless, there remains a large disjuncture in the scales at which problems are attacked and in the types of information that are generated. It is not obvious how this disparate knowledge might be combined to yield a grand, unifying theory of coastal dune dynamics.

INTRODUCTION

Dunes are fundamental components of the geomorphic systems of most sandy coastal environments. The dunes offer a protective buffer against coastal erosion and flooding, they provide a critical ecological habitat, and they are possessed of great beauty and recreational utility. Although sandy dunes may pose hazards to human settlement, the contemporary view is frequently of a resource to be managed if not

Aeolian Environments, Sediments and Landforms. Edited by A.S. Goudie, I. Livingstone and S. Stokes.
© 1999 John Wiley & Sons, Ltd.

exploited (e.g. Cooper, 1958, p. 4; Carter *et al.*, 1990, p. 9; Sherman and Nordstrom, 1995). An inexorable human attraction to the coast has resulted in dunes becoming the focus of intense pressures from development and recreational activities. Population densities within US coastal areas are five times the national average, and by the year 2025, close to 75 per cent of the US population will live in coastal areas (US Department of Commerce, 1998). In terms of US tourism, beaches are the leading destination, even more popular than national parks and historic sites – 90 per cent of all tourist spending occurs in coastal states (Houston, 1996). Beach-sand extraction, addition, relocation, and stabilisation exercises are conducted at massive scales using heavy machinery, engineered structures, wind blocks, sand fencing, and artificial, natural, or vegetative surface covers. These activities, in combination with political, legislative, and economic instruments, serve to control, restrict, and redirect the movement of sand and people in ways that differ significantly from unmanaged environments. Arguably, and somewhat ironically, it is the unmanaged environments that we know least about, and yet a fundamental understanding of the functional role of dunes as integral components of beach-dune systems seems basic and essential to sound management practices (Psuty, 1989). A recent US government report (US Department of Commerce, 1998, p. F-15) states that 'building on or in front of sand dunes, rather than behind them, are major factors contributing to increased beach erosion problems'. Thus, dunes are much more than passive elements on the coastal landscape.

The purpose of this paper is to summarise what is known about coastal dunes in the context of natural (unmanaged) beach environments and to speculate about the problems and prospects inherent to advancing this state of knowledge. We make no claims for comprehensiveness and point the reader to any of several compendia on coastal dune environments for more complete coverage (e.g. Pye, 1983; Psuty, 1988; Gimingham *et al.*, 1989; Baaker *et al.*, 1990; Davidson-Arnott, 1990; Nordstrom *et al.*, 1990; Carter *et al.*, 1992). Our strategy is to introduce what we believe to be the fundamental attributes of coastal dunes relative to other dune forms. Then, we offer a brief synthesis of how coastal dune dynamics has been conceptualised and studied, supported by some admittedly selective literatures illustrative of our thesis. At points, we move beyond the conceptual realm and introduce empirical examples of the complex reality of studying coastal dunes. Our goal is to provide a clearer vision for the future by pointing to some unifying themes and by identifying problematic areas of research where we believe progress is necessary and should be stimulated.

COASTAL DUNE ESSENTIALS

A dune is a specific type of bedform that is created when a fluid moves across a deformable surface of non-cohesive sediments. Sediments are entrained, transported, and deposited in quasi-ordered fashion to create distinctive geometric patterns and morphologies widely recognised as ripples and dunes, although several other types of bedform types are possible. Sedimentologists have adopted general criteria that identify dunes based on their size and geometry. Ashley (1990), for example, suggests that sub-aqueous dunes have spacings in the range of less than 1 m to greater than

1000 m and heights of about 0.075 m to much greater than 5 m. Within this continuum, arbitrary size classes (small, medium, large, very large) have been established, and dune identification and differentiation are further aided with the use of descriptors addressing shape (2-D, 3-D), bedform superposition (simple, compound), sediment size/sorting, slope profile, bedding geometry, and flow hydraulics. Aeolian dunes are often larger than their sub-aqueous counterparts, with extreme heights approaching 100 m and spacings up to several km, probably because flow depth is not constrained as in an aqueous system. In addition, there is a general tendency for aeolian dunes to achieve greater height-to-spacing ratios (Lancaster, 1988). Aeolian dunes typically comprise only very well to moderately well sorted sand-sized sediments (Lancaster, 1995, p. 104) with mean grain sizes in the fine to medium range (i.e. around 0.16–0.33 mm). Many classification systems for aeolian dunes have been proposed (Lancaster, 1995, p. 45) including several that are specific to coastal dunes (e.g. Short and Hesp, 1982; Pye, 1983; Goldsmith, 1989). The latter include the classic aerodynamic forms, such as barchanoid, parabolic, star and transverse dunes, as well as complex discrete forms such as pyramid dunes, coppice dunes and dune hummocks that are intimately related to the colonisation of vegetation.

Necessary, but not sufficient, conditions to create aeolian dunes (Sherman and Hotta, 1990, p.18; Pye, 1993, p. 23) include: (1) an extensive surface upon which sand can be deposited and dunes can evolve; (2) a readily available supply of sand, either from proximal sources or from re-working of underlying deposits; and (3) a competent wind field to entrain and transport sediments in a persistent direction for sustained periods of time. In the case of coastal dunes, Cooper (1958, p. 25) argues that the widespread presence of water and vegetation leads to dune forms that are more diverse and irregular than their desert counterparts, but suggests that 'too often a sharp distinction has been made between coastal and desert dunes . . . [t]he two groups have many fundamentals in common' (Cooper, 1958, p. 7). It should be noted, however, that Cooper (1958) focused a great deal of his attention on large secondary transgressive dune complexes and sand sheets that develop landward of the primary foredune ridge, and in this context he is correct. Active dunes in the middle of large dune fields, whether vegetated or not, are prone to similar process interactions regardless of coastal or desert (or, indeed, any other) environment, and several researchers have adopted this line of reasoning (e.g. Willetts, 1989). Psuty (1989, p. 291) points out, however, that most dunes inland from the beach profile should be considered as secondary dunes because they have been stranded by an accretionary phase of geomorphic development that has largely dissociated them from a dynamical interrelationship with the beach. In sympathy with this notion, it is also our contention that the only distinctive coastal dune is the foredune, because it is in many ways different from all non-coastal dunes. In particular, a foredune is a dependent form that is integrally coupled to the complex suite of nearshore processes fronting the beach-dune system (e.g. Short and Hesp, 1982; Psuty, 1989, 1992; Carter et al., 1992; •
Sherman and Bauer, 1993), and it is geomorphologically conditioned by the germination, colonisation, and succession of vegetation assemblages characteristic of coastal environments (e.g. Carter, 1988; Hesp, 1989; McCann and Byrne, 1989; Read, 1989).

During phases of onshore aeolian transport, foredunes store sediment and grow. During phases of stormy weather and wave attack, foredunes are eroded and release sediment back to the nearshore system. In this way, sediment is exchanged with the nearshore littoral system thereby ameliorating widespread beach erosion and coastal destruction. Indeed, the role of foredunes as sediment reservoirs has made them valuable for coastal protection. A balance must be struck between the coastal wind and wave climatologies over both short and long time periods so as to complement foredune growth and maintenance. Wind events must be sufficiently intense and frequent to transport substantial amounts of sediment from the foreshore to the back-beach where a locus of deposition facilitates the gradual evolution of incipient dunes into established foredune systems. Wave action during fair-weather periods must be sufficiently energetic to replenish the sediment supply stripped from the foreshore by aeolian processes. Goldsmith and Golik (1980), for example, estimate that aeolian removal of sediment from the foreshore along the coast of Israel amounts to 20–30% of the annual longshore sediment transport volume. On the other hand, wave action during storms must be sufficiently benign so as to preclude the complete erosion and overwash of incipient dunes or established foredunes. Psuty (1989, 1992) argues that foredune growth is most favoured under conditions of a negative beach budget (as with marine transgressions) because this situation leads to intermittent scarping of the dune face and remobilisation of dune sands. Exposed sand surfaces created in this way lead to persistent aeolian transfers beyond the foredune crestline. Over some finite time interval (probably decadal or longer), the net provision of sand to the foredune by aeolian processes must equal or exceed the removal of sand from the foredune by nearshore processes – the very presence of the dune deposit attests to this.

The precarious balance between wind and wave activity along coasts, overlain by vegetative controls, gives rise to dune morphologies that differ quite radically from their desert counterparts. Indeed, the classic two-dimensional dune shape, with a gentle stoss slope on the windward side and an angle-of-repose avalanche slope on the lee side, is seldom observed along coastal environments and seems particularly ill-suited as a model or characteristic form. Figure 4.1 shows several photos of linear foredunes along coasts and they clearly indicate that the classic aerodynamic form is essentially reversed (additional views may be found in McCann and Byrne, 1989, pp. 207–208, Figs 3 and 4; and in Carter et al., 1990, p. 224, Fig. 5). The windward (water-facing) slope is typically near the angle of repose, usually as a result of dune scarping by wave attack (e.g. Sherman and Nordstrom, 1985) followed by avalanching or block slumping (e.g. Carter et al., 1990; Nishi and Kraus, 1996). The leeward slope tends to be gentle and is often heavily vegetated. Only in the case of foredune blowouts and secondary parabolic dunes is the classic two-dimensional dune form approximately reproduced (Figure 4.2). This poses considerable difficulty when contemplating or modeling coastal dune dynamics according to traditional schemes. Quite simply, if the dunes are backwards, the wind can not explain everything. Analytical and numerical models that are predicated on aerodynamic principles alone (e.g. Stam, 1997), even if they include the effects of vegetation (e.g. Willetts, 1989), are therefore necessary, but incomplete.

It is this realisation that motivates a perspective of coastal dunes as something other than simple aeolian bedforms. Dunes are integral components of the beach-dune system, and it seems foolhardy to examine dunes without serious consideration

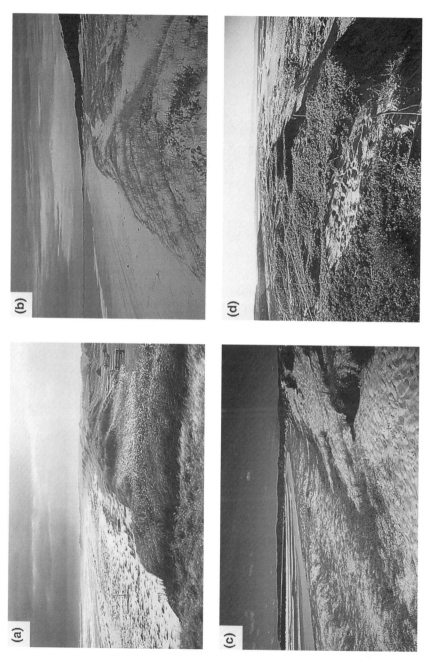

Figure 4.1 Examples of coastal foredunes from (a) Prince Edward Island, Canada, (b, c) New South Wales, Australia, and (d) California, United States. Periodic scarping of dune front by wave action and the influence of vegetation are important controls on foredune form

Figure 4.2   Coastal foredunes typically have steep seaward-facing slopes and gentle landward-facing slopes that are vegetated. Classic aerodynamic geometries, with gentle stoss slopes and steep lee slopes, are usually only apparent in blowouts

of the adjacent beach and of the overall ecological context (e.g. Carter, 1988, Ch. 8). Several conceptual models of beach-dune interaction have been proposed (e.g. Short and Hesp, 1982; Psuty, 1988, 1989, 1992; Sherman and Bauer, 1993), and these demonstrate that there is strong coupling and mutual adjustment between the sub-aqueous and sub-aerial environments. Nevertheless, the mechanical nature of this coupling, the direction and strength of the causal relationships, and the morpho-logical manifestation of the interactions, remain areas of relative ignorance. The primary reason for this state of affairs is the lack of high-quality, longitudinal data over sufficiently long periods that are consistent with the time scales of coastal dune dynamics. Short-term process studies are best suited for examining the micro-scale attributes of dune systems, but they provide relatively small windows into the dynamics of dune evolution. Moreover, they focus on quantities that are variable over the short term (e.g. wind velocity, sediment flux and topographical adjustment), and they ignore other quantities that change over intervals that are long relative to the duration of a typical field experiment (e.g. dune volume and geometry, crest height relative to mean sea level, per cent vegetation cover or plant successions, storm climatology, or tectonic/eustatic displacement). In contrast, form-based studies that focus on morphologic and sedimentologic attributes of dunes (e.g. Cooper, 1958; Pye, 1993) tend to involve larger spatial and temporal scales, and process dynamics are rarely measured. They are strongly ergodic in the sense that evolutionary processes are inferred from spatial assemblages (e.g. Cooper, 1958, Plate 14; Gares and Nordstrom, 1995).

Not surprisingly, these two types of studies (process-based *vs* form-based) correspond to, and perhaps are legitimated by, the ways we have conceptualised and theorised coastal dune dynamics. At one end of the spectrum is the *mechanistic-reductionist* approach, rooted in classic mechanics and dynamics of fluids and solids. It encourages deconstruction of integrated systems into their constituent elemental parts (i.e. reductionism), and it seeks raw and fundamental 'truths' that are revealed gradually as more and more of the outer layers are peeled away. The *holistic-constructivist* approach, on the other hand, strives to understand the essence and basic workings of the systems ensemble and it is inherently geographical. It is constructivist, not by way of simply reassembling the pieces, but rather, in an epistemological sense of searching for rational explanations of 'behaviour.' This approach seeks to discover underlying relations and meanings that are hidden to the eye, and it relies on interpretations. Both approaches are plagued by practical and methodological shortcomings that hinder our quest for comprehensive understanding of coastal dunes, and the constraints imposed by these scale-orientated issues remains a fundamental concern to all geomorphologists (e.g. Sherman, 1995). There have been very few practical attempts to bridge these scale discontinuities (e.g. Arens, 1997).

MECHANISTIC-REDUCTIONIST APPROACHES

Bagnold is probably the best-known aeolian mechanistic-reductionist – he is certainly the most widely respected. The trajectory of his career epitomises this genre of research strategy. His initial academic interests were centred on the geography and morphology of desert-dune environments in North Africa (e.g. Bagnold, 1933). Eventually he became renowned for his theoretical and empirical contributions to our understanding of the mechanics of aeolian saltation, grain-dispersive pressures, sediment size distributions, and the physics of sediment transport. The mechanistic-reductionist approach usually finds the investigator shifting interests from initial concerns with 'the system' to subsequent concern with constituent parts. The hope is that the parts can eventually be reassembled to re-create the system. This strategy is quite common in aeolian geomorphology. We are not surprised, and indeed quite pleased, to find some of our colleagues investigating the spin-rate of particles during flight, the character of elastic rebound and momentum transfer during grain collisions, the altered nature of saltation dynamics in hyperbaric and hypobaric environments, the strength of cohesion due to thin films of water, or the quasi-coherent structure of turbulent boundary-layer flows. These are all elemental pieces of the puzzle for which we have incomplete understanding.

If we adopt such a reductionist strategy for the study of coastal dunes, there are several appropriate scales at which insight into system dynamics may be revealed. For example, we might be concerned initially with the *dune field* as the entity of study. At this scale, there is particular interest in the geometrical and geographical relationships between different dune forms and in the patterns of sediment transport and vegetation assemblages relative to the morphologic patterns. At this scale, the study of coastal dune systems may not differ appreciably from approaches used in other environments. The largest constituent parts are the individual dune forms, and, for coastal systems, closely affiliated constituents might include the adjacent beach, the back-beach, and perhaps the nearshore zone. At the next smaller focus of examination – the scale of an *individual dune* – research tends to centre on processes and responses across surfaces of roughly $1\,m^2$ to perhaps $10\,000\,m^2$. The objects of study become the various morphologic dune facets, such as the stoss and lee slopes or the dune crest, as well as the inter-dune depressions and slacks. Strong emphasis is placed on sediment transport and wind flow as influenced by topographic slope, vegetation cover and, perhaps, bedform development. This is the scale at which processes are investigated using on-site instrumentation, and the aim is to understand spatial and temporal trends in aeolian transport and their relationship to topographic adjustments. At even smaller scales (*ca* $0.1-1\,m^2$), there are *point* studies. These might centre, for example, on the localised environment of a sediment trap, erosion pin, or anemometer, with the intent of quantifying several local sediment transport variables (e.g. surface moisture conditions, grain-size populations, vertical distribution of velocity and sediment flux) as well as related inputs, outputs, and internal changes (i.e. a control volume approach). The aerodynamic conditions in the immediate vicinity of an individual plant would also be appropriate to this scale.

Field investigations involving scales smaller than $0.1\,m^2$ are conceivable (e.g. point measures of turbulence using hot-film anemometry), but very few processes at this scale are of direct geomorphic application given our present approaches. For the most part, investigations at this scale or finer are relegated to the laboratory environment, and their immediate purpose is to test or enhance theory. These might include investigations of grain trajectories in saltation, initiation of particle motion, binding effects of moisture and salts, wind-field modulation in a saltation layer, or of grain shape and grain surface characteristics. Studies at this scale usually employ physics and fluid mechanics to describe simple grain behaviours, or combine those approaches with probability functions to describe population behaviour.

The sample topics presented above are neither complete nor prescriptive. No doubt, we have omitted someone's lifelong research agenda, and we rationalise such omissions by noting that these topics have been described extensively elsewhere and are the domain of other contributors to this volume. Our purpose in introducing them is merely to set the stage for examining the prospects for reassembling these reductionist elements into a model of coastal dune dynamics. Can we reasonably expect to simulate a coastal dune and mimic its dynamics based on the constituent pieces that lie before us? Let us presume, for example, that we have perfect understanding of the dynamics of an irregular grain exposed to a turbulent flow (which we do not). What barriers might there be to prevent us from predicting the initiation of motion and subsequent saltation under natural conditions? First, there is a requirement that we specify the geometry, packing, and pivot angles of all adjacent grains that might interact with the immediate flow field. Second, if the grains are not clean and dry, there is a requirement that we specify, on a grain by grain basis, the effects of sand-binding influences such as those due to moisture, salts, or freezing (e.g. McKenna Neuman and Nickling, 1989). Third, we need an expression for the balance of forces about various moment centres for every grain on the sediment surface and in the saltation layer at all stages of motion. These rather basic requirements of a mechanically deterministic approach to sediment transport are already sufficiently tenuous to confound our objective. The reductionist strategy leads us to the same conclusion reached by other experimentalists vying for an atomic description of gases or a subatomic description of matter – the concepts are sound, but the size of the population is prohibitive. It becomes imperative to change methodologies and to invoke descriptions based on probabilistic patterns of behaviour or on constitutive properties of populations (e.g. bulk density, hydraulic conductivity, eddy viscosity). The saltation models that employ splash functions to describe grain ejection speeds and angles (e.g. Anderson and Hallett, 1986) and the work on log-hyperbolic grain size distributions (e.g. Christiansen and Hartmann, 1988: Barndorff-Nielsen, 1989) are examples of the first and second type, respectively. These approaches allow us to predict or interpret most-likely scenarios for aeolian processes and sedimentary responses, but they are unable to deliver certainty. With large enough sample populations, however, it is possible to gain statistical confidence.

For the moment, let us presume that we understand perfectly such things as the initiation of motion and the saltation dynamics of naturally occurring beach sands (which we do not). Could we then model aeolian transport across a beach and predict dune initiation and growth? Additional layers of complexity are immediately apparent when we consider the dynamic nature of the beach environment and the influence of nearshore processes. These continuously alter beach slope, sediment texture and fabric, and moisture content. Such effects occur mainly at the upwind edge of the transport surface where fetch distances are short and transport disequilibria are common (e.g. Nordstrom and Jackson, 1992). They have profound implications for the nature of the sand-feed system into the dune environment (e.g. Short and Hesp, 1982; Bauer et al., 1990; Bauer, 1991). The location and characteristics of the beach zone shifts in response to nearshore processes operating across a range of spatial and temporal scales, and there is a close coupling with the wind field above (manifested mainly through changing aerodynamic roughness lengths or streamline convergence). It would be convenient, but short sighted, to overlook the delivery of obstacles, such as seaweed, logs and other debris, from the nearshore to the backbeach where they act as barriers to sand transport and alter the local wind field.

For the sake of argument, let us pretend that we can specify and model accurately such nearshore hydrodynamical and aerodynamical constraints on the coastal aeolian system (which we can not). Are there more complicating variables? The beach-groundwater table may provide moisture to surface sediments at rates that vary spatially and temporally. Similarly, evaporation rates will vary as a result of changes in moisture supply, atmospheric humidity, temperature gradients, radiative fluxes and even subtle variations in surface aspect. These differences are detectable at scales conforming to beach zonation (e.g. Nordstrom et al., 1996) and at the scale of individual bedforms. Because of the strong influence that sediment moisture has on sand entrainment rates (e.g. Namikas and Sherman, 1995; McKenna Neuman and Maljaars Scott, 1998) and on sediment release from cohesive surfaces, these factors must be considered non-trivial. Additional complexities are introduced by general meteorological conditions that include shifts in wind speed and direction (e.g. land-sea breezes, veering with fronts, seasonal climatologies, topographic trapping) and in precipitation amount, intensity, and form (rain or snow). Even if we could specify these suites of processes and responses in precise terms as continuous variables, it is sobering to note that no universally accepted transport equation currently exists that is capable of accepting this information. For these reasons, the problem of sediment flux prediction across beaches has been labelled indeterminate (Bauer *et al.*, 1996).

Nevertheless, aeolian sand transport models are at the heart of most process-orientated approaches to coastal dune modeling. The models have been developed for simple transport environments and equilibrium transport conditions (e.g. Kawamura, 1951; McEwan and Willetts, 1994). Under certain circumstances, a few of these models have been shown to replicate field observations relatively well (e.g. Sherman *et al.*, 1996). It is common to append terms to the basic models to correct for the effects of slope (e.g. Zeman and Jensen, 1988), moisture content (e.g. Belly, 1964), or vegetation (e.g. Willetts, 1989; Lancaster and Baas, 1998). These factors are important in almost all aeolian environments, but they are especially important in the coastal zone. The large sediment moisture contents (frequently exceeding 7 per cent, with even higher values in lower intertidal zones and interdune slacks), steep stoss slopes (locally exceeding the angle of repose for scarped dune fronts), and dense vegetation covers that are characteristics of many coastal aeolian environments may profoundly influence sand transport rates (e.g. Arens, 1996). Moreover, these three characteristics are fundamental attributes of coastal fore-dunes, and they are largely responsible for the distinctive form. Yet, even if we were able to incorporate these three elements into a grand transport model, (and we have not yet done this successfully) it is unlikely that we could simulate the evolution of a foredune on an initially flat surface. In modelling terms, we are uncertain what the correct perturbation should be. We also lack another key piece of the reconstructive puzzle – we do not know what an equilibrium coastal foredune is supposed to look like. Although we can easily recognise forms that we might agree to designate to the class 'foredune,' there would likely be disagreement if it came to specifying characteristic geometries and locations (vis-à-vis other geomorphological environments). This remains perhaps the next great challenge for that part of our community seeking to simulate dune evolution using transport rate relationships (e.g. Namikas and Sherman, 1998). For such an endeavour, we need to seek knowledge generated outside the sphere of the mechanistic-reductionist approach.

HOLISTIC-CONSTRUCTIVIST APPROACH

The epitome of the holistic-constructivist approach in geomorphology is 'The Geographical Cycle' of William Morris Davis (Davis, 1899). This approach presumes that landforms and landscapes behave in certain predictable ways with governing principles derived through inferential reasoning and via analogy. Such models are useful as conceptual constructs by which to understand geomorphic tendencies, but their ability to admit errant behaviour or intervening opportunities make them impractical for predictive purposes. A great deal of effort in geomorphology continues to be expended on finding 'type' examples of the classic models and on explaining why they are so difficult to find. The same is true for coastal dunes, as we shall see. There are several aspects of coastal dunes that are particularly difficult to resolve, and these include how coastal dunes are initiated, how they grow, how they migrate, how they stabilise and why they reactivate. Central to these concerns is the notion of an equilibrium dune field. Does it exist? What does it look like? What are the fundamental process-form interactions? What are the characteristic deviations? And, what are the time scales over which the equilibria are maintained? Given that dunes are integral components of a larger beach-dune system, these questions apply equally to a single dune, to the entire dune field (including the foredunes, blowouts, dune slacks, and secondary dunes), the ecological environment of the dune field (vegetation, macro-fauna, micro-fauna, nutrients, moisture and micro-meteorological conditions), and to the beach and nearshore adjacent to the dunes. Without answers to these questions, it is difficult to generate a comprehensive understanding of coastal dune dynamics because it will never be certain whether the dunes in question are equilibrium or disequilibrium forms. Are they active but stable features that respond to contemporary conditions, or are they remnant features inherited from processes active during past low sea-level stands with only cosmetic adjustments imparted recently?

Coastal Dune Initiation

The mystery and controversy surrounding the origin of aeolian ripples (e.g. Bagnold, 1941; Anderson, 1987; Werner, 1988) seems not to have extended to the study of coastal dunes. It is generally accepted that their initiation can be traced to the influence of vegetation (e.g. Hesp, 1984, 1989; Carter and Wilson, 1990) because plant covers will enhance surface roughness, retard the local flow field, and thereby prevent surface erosion or induce sediment deposition (e.g. Wolfe and Nickling, 1993). Hesp (1984, 1989) describes several types of incipient foredune that he relates convincingly to the type of vegetation cover (germination, growth pattern, density/ height and succession). For example, an extensive zone of perennial grass is likely to lead to aeolian deposition in laterally continuous (alongshore) ridges. Seedlings often germinate within the high-tide strandline (wrack line) deposited at the upper limit of swash, and as aeolian accretion becomes more pronounced, the plants colonise and grow to keep pace with deposition. Alternatively, deposits of aeolian sand may accumulate in conjunction with isolated plants or plant clumps that germinate from individual seeds or storm-deposited vegetative material. These discrete features are called pyramidal dunes or shadow dunes depending on whether deposition takes place predominantly within the vegetation mat or immediately downwind of the obstruction (Hesp, 1981, 1989). More generally, these incipient forms are called embryo dunes. Note that shadow dunes may also be spawned in the wake of stray pieces of beached flotsam, kelp, woody debris, or anthropogenic garbage.

It is widely appreciated that horseshoe vortices may develop in response to air flowing around obstacles (e.g. Greeley and Iverson, 1985). Enhanced flow velocities and potential scouring action associated with the lateral arms of these vortices occur to either side of an obstacle and may extend a considerable distance downwind. In the lee, a wake zone develops that is usually associated with reduced flow velocities and sediment deposition (Figure 4.3(a)). Downwind extension of the scour arms may be accompanied by linear depositional streaks (Figure 4.3(b)). These sediment deposits are anchored in place by the debris or vegetation and, thus, are not easily removed. Indeed, reduced wind velocities often lead to depositional sand ramps at the front of the object (Figures 4.3(a) and (b)), and these elements perturb the wind field in a way that reinforces the positive feedback between flow contraction-expansion and sediment erosion-deposition. Ultimately, a substantive protodune may be created for which flow separation from the dune crest and eddy recirculation in the lee of the dune dominate the flow dynamics, as in any of several models of dune growth (e.g. Kocurek *et al.*, 1992; Stam, 1997).

Vegetation and scattered debris are likely the most common nucleation sites from which coastal dune fields are spawned. But is this the only possible mechanism? Goldsmith (1989, p. 6) differentiates between 'artificial' dunes (those that are constructed by human actions) and 'artificially-inseminated' dunes (those that form around some 'seed' emplaced either accidentally or on purpose), and notes that neither of these is 'completely natural.' Implicitly, he is suggesting that there exists such a 'natural' dune and he clearly associates these with dune-forming processes in desert-like circumstances, although no details are provided. The spontaneous evolution of ripples on carefully flattened and smoothed sediment surfaces in wind tunnels suggests that the presence of debris or vegetation is, in fact, not an essential requirement for bedform initiation. Ripples are believed to form in association with inherent bed defects, either as a result of micro-scale topographical differences or due to non-uniform grain size distributions and clustering effects that lead to differences in entrainment potential (e.g. Bagnold, 1941; Nickling, 1988). Once such a small bedform exists, a downwind-propagating pattern is reproduced (e.g. Wilson, 1972; Anderson, 1990). The emergence of discrete ripple patches may impose sufficient alteration of surface roughness to influence the local flow dynamics in favour of larger scale patterns of enhanced deposition, which may ultimately lead to protodune formation. As Bagnold (1941, p. 437) suggested, 'The growth of a dune need not depend on the deformation of the wind velocity distribution by the shape of the dune as an existing obstacle in the wind's path, but merely on a change of surface texture'.

A complementary component of such a conceptual model of coastal dune initiation may involve aeolian decoupling of sediment populations, a process through which various size fractions in a source mixture of sediments are separated and redeposited in different sedimentological environments (e.g. Carter, 1976; Bauer, 1991). Such decoupling is readily observed in aeolian ripples (Figure 4.4), and the classic example of sediment decoupling along coasts comprises a 'lag' deposit of coarse sea-shell fragments (or fine-grained heavy minerals) and a 'winnowed' fraction of fine-grained or light materials deposited somewhere downwind (e.g. Davidson-Arnott *et al.*, 1998; van der Wal, 1998). Consider a relatively flat overwash barrier and an onshore wind that moves from the ocean onto the beach. The wind field experiences pronounced changes in surface roughness conditions (e.g. Rasmussen, 1989), and in consequence, a sequence of boundary-layer adjustments takes place in the downwind direction. All other things being equal, these include

Figure 4.3   Aeolian deposition and shadow dune initiation often occurs in association with (a) clumps of vegetation (wind from left to right) or (b) woody debris scattered on the back-beach or overwash flat (wind from right to left). Small, depositional sand ramps can be found on the windward 'fronts' of such obstacles. The lateral sides and leeward 'backs' of obstacles may experience deposition or scour depending on the nature of vortices in the wake

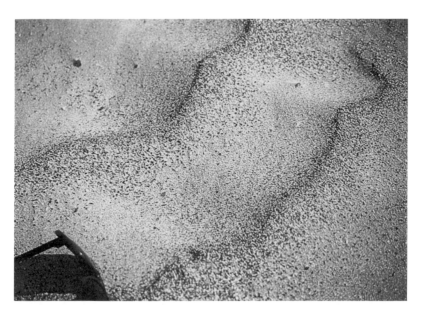

Figure 4.4 Aeolian sediment decoupling is strikingly apparent in aeolian ripples that have accumulations of coarse sediments on the crest and fine sediments in the intervening troughs. Wind is from lower right to upper left of photo

vertical growth of the internal boundary layer, reduction in surface shear stress, and reduction in flow competence (Bauer, 1991, pp. 298–299). The medium- and fine-grained sediments entrained from the foreshore will move downwind as saltation and suspension load, and will continue to be transported as long as the winds are sustained. The coarsest sediments, however, will remain immobile or move along the surface as traction load. At some point downwind, these coarse particles may stall because of the reduced competence of the flow field. This represents a locus of deposition for the coarse sediments that can only be removed by stronger winds, which in turn have a locus of deposition somewhere farther inland. This long-term interaction of the average wind climatology with the foreshore sediment mixture may create a zone of coarse-grained aeolian ripples. Enhanced surface roughness (due to coarse-grain accumulation and ripple evolution) may lead to near-surface flow reduction, progressive deposition, and incipient foredune growth. Ultimately, flow separation at the ridge crest during periods of vigorous winds may occur (as described by Cooper, 1958, p. 38), and an asymmetric dune profile may evolve.

Given the energetic nature of coastal beach environments, it is likely that the effects of such subtle decoupling processes will be overwhelmed by more pronounced processes such as berm construction and foreshore erosion by wave action. Davies (1957) advanced the thesis that cut-and-fill processes on the berm are the basis for foredune initiation, although some controversy surrounding this mechanism persists (see summary in Hesp, 1984). As with many other controversial ideas about natural processes, the central points of contention are usually different, with one proponent arguing for the conceptual feasibility of a process whereas the detractors argue against it on the basis that other processes are much more common and effective.

Coastal Dune Evolution

As obvious as it may sound, in order to observe coastal dune evolution, one must find contemporary environments where dunes are currently evolving, and where the relative state of that evolution can be recognised. This is not as easy as it seems. Most coastal dune fields comprise immature, mature and relict features, and increasingly they are being managed or altered by humans. To capture the true dynamics of coastal dune evolution, it is necessary to find sedimentary environments where dunes are growing on newly created or freshly exposed sediment surfaces. Groomed beaches, such as those along the coast of southern California, might provide such surfaces, but these heavily frequented beaches are rarely left untended long enough for dunes to evolve. Dunes also evolve opportunistically on new sand surfaces associated with emplacement of large jetties or groins that impede littoral currents and cause updrift sedimentary wedges to form (Figure 4.5(a)). Nordstrom (1988) discusses an interesting case study that assessed the effects of new policies regarding beach grading and other management practices on the integrity of coastal dunes in two coastal communities in Oregon where jetties created such sedimentary wedges. Sanjaume and Pardo (1991) monitored dune evolution on a portion of a barrier island that was flattened for development purposes in 1970 and eventually reclaimed and fenced off in 1979. Environments where protodunes may be created and destroyed periodically by natural processes include high-energy (morphodynamically intermediate) beaches, barrier islands (Figure 4.5(b)), and bay-mouth or river-mouth barrier bars.

The standard conceptual model of coastal foredune evolution involves the germination and colonisation of pioneer plants in the litter layer deposited at the upper limit of spring-tide swash. Hesp (1984) provides an informative overview of these ideas and presents extensive field data to validate his notions about incipient foredunes and beach ridges (which he claims are really relict foredunes) in some environments. His observations indicate that incipient foredunes tend to evolve high on the beach in the backshore rather than near the berm crest. Pioneer plant seedlings rarely colonise the berm crest, presumably because the berm is reworked frequently by waves or because the micro-environment (moisture, nutrients, grain size) is not ideal for plant growth. As the plant seedlings colonise the backshore, they induce sediment deposition and incipient foredune growth. Carter and Wilson (1990) describe the development of coastal foredunes at Magilligan Point, Northern Ireland, and they provide a sequence of three photos (Carter and Wilson, 1990, pp. 138–139, Fig. 5) that are quite persuasive in this regard. Hesp (1984) argues that foredune ridges have the propensity to grow seaward, but often such expansion is limited by scarping events. Thus, it is uncommon to find the continuous seaward migration of a single foredune system keeping pace with a prograding shoreline. Rather, there is a tendency for foredune ridges to be initiated in new seaward locations because of seedling germination and plant colonisation at new strandline

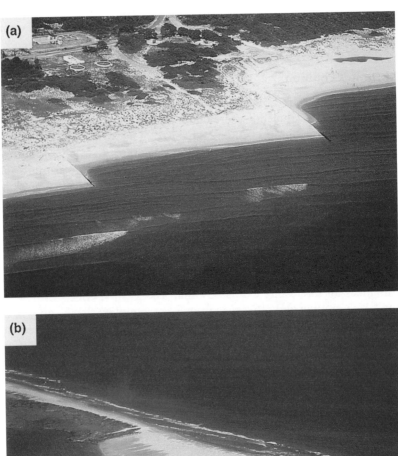

Figure 4.5 Freshly exposed surfaces such as those associated with (a) groin embayments and (b) overwash lobes on barrier islands are ideal places to observe coastal dune evolution. Sediments moving alongshore are trapped updrift of groins producing this characteristic wedge-shaped geometry. Incipient hummocky dunes are visible landward of the groin embayments. Storm related overwash erases existing dunes and eliminates vegetation, thereby exposing fresh sediment to aeolian processes. Such overwash processes cause depositional lobes to extend into the back-barrier lagoon, and these are essential to the landward migration of barrier islands

positions (Hesp, 1984, Fig. 11). Hesp (1989, Figure 2) presents an 11-year sequence of cross-sectional profiles that shows such 'jumping' behaviour in foredune initiation, and this process is suggested to be the origin of beach ridges.

Carter *et al.* (1992) proposed an alternative conceptual model of coastal foredune evolution, which occurs in the aftermath of stream-outlet closure. The model is based on sedimentary evidence and morphological observations made over several years in Ireland and California. Although it envisions a characteristic sequence of stages, it admits several alternative paths such as those described by Hesp (1984). During wet periods, sustained stream discharge delivers substantial quantities of sediment to the nearshore system where a prograding deltaic feature may be present. These pronounced river flows prevent the onshore movement of sediment by waves in the vicinity of the stream outlet. During dry periods, stream discharge is reduced and wave activity dominates the interaction. A river-mouth barrier bar is built, and it cuts off sediment and water discharge from the stream to the ocean (Figure 4.6(a)). Shoaling waves will continue to accrete sediments to the seaward side of the barrier, and swash processes on the foreshore of the barrier bar will eventually construct a pronounced ridge crest. The barrier ridge crest is a locus of deposition for both sediments and strandline debris. The passage of occasional storms will create steep waves that tend to erode the seaward margin of the barrier making it more susceptible to breaching. During high tide stages, the largest waves have the propensity to overwash the barrier, and this can lead to landward extension of the barrier in the form of depositional lobes and thin sheets of sand. These are the fresh sediment surfaces upon which incipient dunes will first evolve.

As part of these overwash processes, strandline debris is transported across the barrier ridge crest and deposited randomly on the back-barrier flat. When aeolian transport is initiated, these obstacles form the nuclei of shadow dunes. At this point, the system is in a delicate balance. If overwash processes predominate, the incipient dunes cannot establish a solid base. However, if overwash processes wane, either because the stormy season has passed or because a substantive swash berm has aggraded, aeolian processes can dominate and a shadow dunefield may evolve on the back-barrier flat (Figure 4.6(b)). Pyramidal dunes and hummocky dunes are often stabilised and reinforced through the growth of vegetation mats. If this happens, the dune hummocks become very difficult to remove. Hesp (1989) suggests that a succession of plants from annuals to perennials is essential to the overall long-term viability of these incipient dunes. Subsequent overwash processes will erode the fronts and lateral margins of dune hummocks, but will rarely remove them entirely (Figure 4.7). Indeed, the presence of the hummocks serves to channelise the overwash flows thereby accentuating local relief. This serves several purposes. Back-barrier sediments are reworked and fresh sediment surfaces are exposed, thereby making them more susceptible to aeolian entrainment and to invasion by vegetation. Nearshore sediments and additional debris are transported to the back-barrier flat and added to the landward margins of the barrier. This extends the surface area over which dunes can grow and provide new nucleation sites for dunes. Most importantly, the scarped frontal margins of the dune hummocks become zones of aeolian deposition, in the same fashion as the aeolian sand ramps that form in conjunction with other wave-scarped features or at the base of cliffs and mountains.

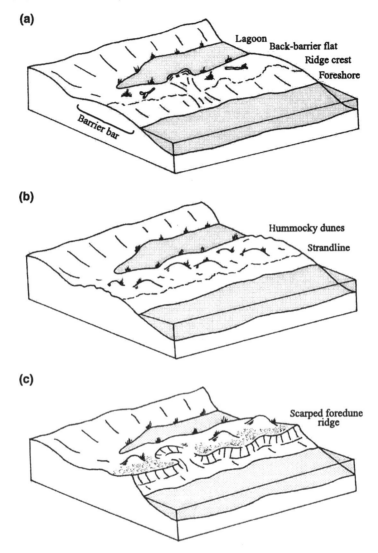

Figure 4.6 Conceptual model of foredune evolution proposed by Carter *et al.* (1992). (a) Shortly after river-mouth closure, the wave-formed barrier bar is easily overwashed by storm waves that carry sediment and assorted debris into the lagoon and back-barrier flat. The ridge crest is initially ill defined, but eventually becomes more pronounced and it is often marked by a high-tide strandline. (b) Shadow dunes begin to form on the back-barrier flat in association with stranded debris. The barrier ridge crest has accreted and moved seaward thereby reducing the frequency of overwash and establishing the dominance of aeolian processes on the back-barrier flat. Gradual colonisation by vegetation leads to the development of hummocky dunes that begin to coalesce, and this strengthens the integrity of the evolving dune field. (c) The mature stage of foredune evolution is a manifestation of the quasi-equilibrium interrelationship between aeolian processes, which tend to build dunes upward and seaward, and nearshore processes, which tend to scarp the fronts of dunes or breach the foredune through blowout channels

Figure 4.7 Overwash processes deliver fresh, nearshore sediments to the back-barrier flat, as well as rework the existing hummocky surface by eroding channels and scarping the fronts and lateral margins of shadow dunes. Subsequent aeolian activity will deposit sand ramps at the fronts of these dunes, which can be invaded by vegetation, and these processes, in combination, lead to the seaward growth of these dune forms

As a result, the dunes are actually able to grow in the seaward direction, and in this aspect of their dynamics, coastal dunes are very dissimilar to their dry-land cousins.

Carter *et al.* (1992) suggest that the dune hummocks on the back-barrier flat will continue to grow and progressively coalesce. At the same time the locus of deposition shifts seaward toward the ridge crest, and a true foredune eventually evolves (Figure 4.6(c)). Exactly how and why this occurs is not certain, but it seems worthy of long-term study. The ever-strengthening role of vegetation as it takes hold and expands its domain is one key factor. Goldsmith (1989, p. 8) observes that dunes tend to expand and advance in the direction of vegetation growth. Hesp (1989) describes such species-specific interactions, and suggests that the seaward advance of established foredunes on prograding coasts is critically dependent on vegetation growth. Rhizome-stolon colonisation and shoot growth into the unoccupied sand ramps fronting the foredune ridge will stabilise these stoss slopes and lead to enhanced aeolian deposition. In a similar way, the hummocky dunes occupying the back-barrier flat may expand, consolidate, and advance seaward, presuming that overwash processes are ineffective in eliminating aeolian accretion. The subtle interaction of all these system components and the gradual evolution of a true foredune will eventually force a transition to a new state or regime – one that might be interpreted as 'mature'. This state can be recognised by several features. The general effectiveness of overwash processes that penetrate the back-barrier is reduced in favour of dune-scarping processes by waves that attack only the front of the foredune (Figure 4.6(c)). The growing foredune also yields complex aerodynamic flow patterns. In particular, flow separation at the foredune crest leads to wake zones on the lee of the crest, and this encourages grain-fall deposition (e.g. Bauer, 1991; Arens, 1996). Often, a dead zone appears at the front of scarped foredunes, and sand ramps have a tendency to accumulate there. The medium and finest size fractions, which move as saltation and suspension load, will be transported past the foredune crest and deposited on the back-barrier surface. Such widespread deposition encourages a more diverse dune ecology, and this fosters even greater deposition and dune growth.

Despite the appealing nature of this conceptual model, the entire sequence of evolutionary stages, from overwash flat to shadow dunefield to foredune establishment and migration, has never been observed or documented to our knowledge. Examples of each of the stages and their associated characteristics have been reported, but the time-sequence of changes has only been inferred from morphologic and sedimentary evidence. The coastal equivalent of revisiting Bagnold's barchan dune after 57 years (Haynes, 1989) might prove quite interesting in this regard. Our attempts at documenting dune evolution have also been stymied by intervening events that have essentially re-set the system or altered its course. Figure 4.8 shows four photos of the overwash flat just north of the Salinas River that we have been visiting for the last decade. The first photo (Figure 4.8(a)), taken in 1989, shows a freshly overwashed barrier surface replete with coarse-grained aeolian ripples. A few pieces of debris are lodged in the sand, but no incipient dunes are evident. The next photo (Figure 4.8(b)) is taken only one year later, and several clumps of vegetation and embryonic dune forms are visible. Given these rather rapid developments, one might anticipate that a proper foredune was imminent, but even by 1995, the situation had changed very little (Figure 4.8(c)). The evolution of the system seems to have been 'arrested' by frequently recurring overwash. Note, however, that a fairly large dune hummock (centre of the photo) has grown, and it

---

Figure 4.8 Temporal sequence of photos showing the complex nature of dune-forming processes on an overwash barrier north of the Salinas River outlet into Monterey Bay, California. All photos are taken in late winter and in roughly the same area – the solid arrows point to a large, immobile log that serves as a reference. (a) In 1989, the barrier was freshly overwashed and all topographical elements were removed. The sand surface is smooth except for a few pieces of woody debris and a covering of coarse-grained aeolian ripples. (b) By 1990, small dune hummocks appeared and these were stabilised by clumps of vegetation. (c) This 1995 photo shows that evolution of a proper shadow dunefield has been arrested by frequent overwash activity during the previous 5-year period. The extreme relief between the back-beach surface in the foreground relative to the foredunes in the background suggests that a great deal of sediment has been stripped off the barrier surface by overwash activity. In the centre-left of the photo, a fairly substantial dune hummock has resisted complete removal. (d) Major storms during the winter of 1997 resulted in widespread deposition of woody debris on the beaches of Monterey Bay, and the evolution of the foredune system has been perturbed. Substantial aeolian deposition is already apparent in the lee of large logs and vegetation is vigorously invading these new deposits

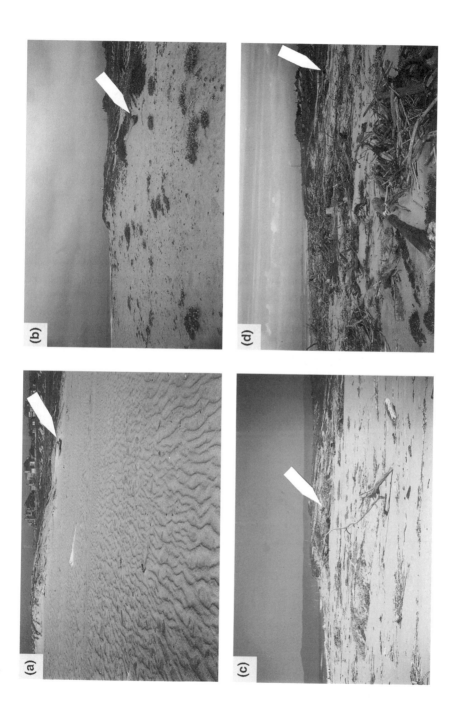

is of a size that will be difficult to remove. McCann and Byrne (1989, p. 209, Fig. 5) describe a barrier system on Cavendish Spit, Prince Edward Island that also has dune hummocks and is in a similar state of arrest. Finally, in 1997, a major storm attacked the central California coast in conjunction with a wetter-than-normal hydrologic year. The Salinas River breached its barrier and huge amounts of scattered woody debris were stranded on the beach as a result (Figure 4.8(d)). It is clear that the aerodynamic properties of the overwash surface have been altered radically and significant sediment accumulation has already taken place amongst the debris. Although we have not had the opportunity to re-visit the site, it is likely that even more sedimentation by aeolian processes will have occurred. One is left to wonder whether sequences of gradual evolution punctuated by critical, state-altering events are the norm in coastal dune dynamics.

Coastal Dune Equilibrium

Problems of Definition and Scale

Other than pure academic curiosity, perhaps the most compelling reason to understand dune equilibrium is the implications for coastal management purposes. If we do not know what an equilibrium dune is and what conditions lead to disequilibrium, then any attempts at management intervention are likely to be ineffective at best, and damaging at worst. In either case, they are likely to be expensive. Psuty (1989, p. 289) made this point when he stated,

> Part of the problem certainly lies with the inadequate knowledge of the processes of coastal dune development and what constitutes a coastal dune and what is dynamic equilibrium in coastal dune development. There is a fundamental question about whether there is a single coastal dune model that can be applied in a variety of situations and locations that will serve as a basis for evaluation. If so, what is it? If not, what is lacking in the creation of a dune development model?

We contend that what is lacking is a robust conceptual framework into which we can place limited pieces of knowledge about dune dynamics (notwithstanding the efforts of Short and Hesp, 1982; Psuty, 1988, 1989; Sherman and Bauer, 1993; among others). We now engage a rather general, conceptual discussion that highlights some of the uncertainty that pervades this aspect of our science.

Problematic issues surrounding the notion of equilibrium aside (Thorn and Welford, 1994), it is of interest to contemplate which fundamental attributes of dunes we expect to display equilibrium. Most dictionary-derived definitions of 'dune' offer little insight because they usually suggest that aeolian dunes are nothing more than hills, mounds, or ridges of sand piled up by wind, which is in essence the definition provided by Bagnold (1941, p. 188). As we have seen, these semi-discrete piles of sand can take many forms and can be classified in many ways. In contrast, Cooper (1958, p. 25) argues that 'pattern is a more useful concept than the individual "dune"', and focuses his attention on the description of transverse-ridge and oblique-ridge patterns. How might these alternative foci on dune form *vs* dune pattern lead to different conclusions about dune equilibrium? Cooper (1958, p. 27) argues (implicitly) that transverse-ridge patterns are equilibrium entities by pointing to their widespread occurrence in coastal environments and describing their relationship to prevailing wind conditions. At the same time, he recognises that alternative viewpoints seem to suggest that the transverse ridge is an unstable and transitional form, and he quotes Bagnold (1941, p. 205) as saying that 'dunes, though they may be isolated or longitudinal, are never transverse to the wind'. He concedes Bagnold's explanation that a local increase in wind velocity through localised low spots in the crest may lead to enhanced erosion and to the development of large gaps or blowouts. However, he will not concede that the isolated barchans developed from the consequent break up of a transverse ridge are distinct features, but rather insists that they are fundamentally transverse ridge units. Of course, depending on the scale of examination, both Bagnold and Cooper can be right, and Lancaster (1995, Plates 2 and 3) makes the point that satellite and air photos are able to reveal general patterns that are obscure from the ground. Indeed, it is somewhat ironic that Cooper (1958, p. 147, Plate 7) presents two air photos that are representative of the transverse-ridge patterns and a third, ground-based oblique-view photo from the same area that shows individual dune forms but no evidence of the larger scale transverse-ridge pattern.

Whether we focus attention on an individual dune or the larger scale pattern created by an ensemble of dunes is not a simple matter of personal preference. The view we receive and the insights we generate are distinctly different, and yet, all are part of the broader dynamical system. It is well known that a single grain of sand travelling across a surface occupied by bedforms will pass through brief stages of transport followed by extensive periods of dormancy, even when the overall transport field is continuously active. The grain is deposited in the bedform trough and buried by other sediments avalanching on top. Not until the entire bedform has passed, will the grain again be unearthed and become susceptible to entrainment and transport. Evidently, this is a characteristic sequence of movements, but it challenges our notions of equilibrium, especially as they pertain to the grain *vs* the bedform. In a similar vein, it is well known that individual bedforms migrate and change shape during the transport process, and this makes them difficult to track over extensive distances. Consider, for example, a composite set of bedforms with ripples climbing up the stoss slope of a dune. New ripples are spawned near the trough and they advance up the stoss slope. Eventually, the ripples reach the dune crest where they effectively disappear as the grains avalanche down the slip face. An individual grain of sand moving across the stoss slope of the dune will experience short periods of activity and burial in association with each of the ripples occupying the back of the dune, but eventually the grain will avalanche down the lee slope of the dune with the potential of being buried for an extensive period. The sequence of events characterising the grain differs from the transformations characterising the ripples and despite their discontinuous and aperiodic nature, they are all fundamental attributes of equilibrium dune migration and shape. Can the same notion of equilibrium be applied equally to the grains, ripples, and dunes at the same time? If so, how general must it be, and if not, how useful is the concept?

Temporal Views of Dune Equilibrium

The identification of an equilibrium state or characteristic sequence of events for coastal foredunes is problematic. We are tempted to invoke a single equilibrium form and ascribe to it the features evident in examples of foredune systems that have been stabilised by vegetation and that are experiencing relatively little morphological change. However, in most foredune systems, the individual dune forms are mobile and continually evolving. They display dynamic equilibrium because their evolution keeps pace with imposed conditions, such as those driven by coastal progradation/ transgression or relative sea-level changes due to eustatic or isostatic effects (e.g. Sherman and Bauer, 1993).

Foredune forms can differ radically. At one end of the spectrum are the linearly continuous, seaward-advancing forms, such as those described by Hesp (1989) and Carter and Wilson (1990). As noted earlier, these are generally found on prograding coasts and they appear to evolve in a sequence of 'jumps' that leave behind stranded beach ridges. Each 'jump' has a characteristic sequence of stages, beginning with strandline deposition of litter and seeds, germination and colonisation of pioneer seedlings, deposition of aeolian sediment, subsequent growth of a foredune ridge, and eventual progradation of the shoreline. At the other end of the spectrum are the landward-advancing forms such as those found in conjunction with marine trans-gressions and migrating barrier islands. These are strongly influenced by wave activity and typically have scarped fronts. The crestlines of such foredune ridges tend to be crenulate and dissected, and there is some evidence for alongshore rhythmic tendencies that can be linked to nearshore phenomena such as edge waves (Allen and Psuty, 1987; Psuty, 1990). At the local scale, foredune ridge geometry is intimately linked to blowout evolution, and in order to understand how overall equilibrium is sustained in these foredune systems, we need insight into blowout dynamics.

Blowouts are erosional hollows, depressions, troughs, or swales that develop in a dune complex as a result of the deflation of unconsolidated sediments by aeolian processes (Gares and Nordstrom, 1995). There has been a long tradition of research on dune blowouts (see summaries in Carter *et al.*, 1990; or Fraser *et al.*, 1998), and several recent studies have yielded insight into the complex interactions between topography, wind patterns, and sediment transport in mature blowouts (e.g. Gares, 1992; Hesp and Hyde, 1996; Fraser *et al.*, 1998). They occur readily in vegetated dunes, and their primary morphologic features include: (1) an erosional throat; (2) steep lateral walls adjacent to the throat; and (3) a large depositional fan at the landward end of the blowout consisting of a gently-to-steeply inclined sand ramp leading to a distinct semi-circular rim. Several extensional lobes may emanate from the fan (Fraser *et al.*, 1998), and these often represent the incipient stages of parabolic dune formation. It is common to associate the presence of blowouts as diagnostic of advanced states of destabilisation and reactivation, usually in response to external influences that reduce or eliminate vegetation covers. Such influences would include poor land management practices, recreational activities, animal grazing, significant shifts in meteorological conditions, or natural cycles of nutrient exchange and depletion (e.g. Carter *et al.*, 1990, p. 231). This is implicit to

Goldsmith's (1989, p. 5) assertion that, 'Unlike desert dunes which form and grow by advancing horizontally, coastal vegetated dunes generally grow upward in place'. If this were strictly true, vegetated coastal dunes would never achieve equilibrium heights nor would they migrate. To the contrary, mobile vegetated coastal dunes are quite common, and blowout initiation and evolution may be a process inherent to coastal dune dynamics.

An intriguing quasi-equilibrium model of cyclic blowout evolution in coastal foredunes was proposed by Gares and Nordstrom (1995). Their conceptual ideas are provisional and still require empirical scrutiny because the model is predicated on short-term observations (*ca* 10 years) of topographical adjustments in several neighboring blowouts, each apparently in a different stage of evolution. Thus, the authors have substituted spatial conditions, which are thought to represent distinct stages in the blowout cycle, for a temporal sequence that is expected to last 20 years or more. There is also some concern for the broader applicability of the model because their observations of blowouts were taken on the leeward (east) coast of North America where the most frequently recurring winds are offshore (there has been increasing interest recently in the overall importance of offshore winds to dune dynamics, e.g. Wal and McManus, 1993; Gares *et al.*, 1996; Nordstrom *et al.*, 1996). In any case, Gares and Nordstrom (1995) propose that blowouts proceed through an evolutionary sequence that is cyclical in the sense that blowouts have the propensity to 'heal'. The stages in this model include: (1) initial notch development in places where vegetation cover is thin or minimal; (2) notch widening and deepening and gradual formation of a downwind depositional lobe; (3) blowout deflation and creation of a blowout floor at the level of the back beach; and (4) blowout healing via deposition in the form of a shadow dune and eventual throat closure.

The erosional stages of blowout evolution are facilitated by convergence and acceleration of onshore winds through the notch, which leads to an erosive 'jet' that scours sediment from the base and walls of the incipient blowout. This sediment is transported to landward and deposited as an expanding depositional lobe. There is often distinct asymmetry in the lateral walls and depositional lobe, depending on the character of the wind climatology. One flank tends to recede more rapidly than the other – vertical scarps at the top of this flank and avalanche ramps at its base are common. The other flank is often dominated by accretion or is in the process of revegetation (Carter *et al.*, 1990, p. 239). Ritchie (1972) notes that there are lower limits to the erosion of coastal blowout floors because of the development of lag deposits and the presence of fluctuating groundwater tables. Whether this is a critical precondition for the initiation of healing is unknown, but this state seems to be diagnostic of a potential process reversal from erosion to deposition. The environment in the blowout floor may simply be more conducive to colonisation of vegetation (e.g. Carter *et al.*, 1990, p. 236. Fig. 12A) or, alternatively, the attainment of a 'base level' blowout floor may have consequences for enhanced erosion of the lateral flanks. At some point, and usually in association with a seasonal shift in the wind regime, deposition occurs in the vicinity of the blowout floor creating a shadow dune directly in the throat or immediately in front of it (Gares and Nordstrom, 1995, p. 17; Fraser *et al.*, 1998, p. 457, Fig. 8). Sand deposition leads to new micro-environmental conditions that may encourage positive feedback between vegetation growth and enhanced sediment deposition, and in this way, healing is achieved and the cycle is closed. Gares and Nordstrom (1995) note that progressive evolution can potentially be arrested at any stage due to intervening processes such as wave overwash, vegetation die-off, or human intervention.

It is evident that such a cyclic model envisions foredune blowouts as integral components of a coastal dune system in a state of dynamic equilibrium. Stable and unstable morphologies co-exist in the same landscape, and one might expect to find incipient notches, dish-shaped basins, as well as deep, linear troughs that dissect the foredune ridge. The latter have been most intensely studied, partly because they are the most dramatic features of transgressive dunescapes, but more importantly, because they are the conduits through which sand from the beach is fed to the beach hinterland (e.g. Ritchie, 1972; Carter *et al.*, 1990). It is intriguing to consider that the only difference between a coastal foredune system displaying cyclic response and comprising relatively stable, vegetated, secondary dunes, and a 'mobile' sand-sheet system with a strongly dissected foredune and active parabolic dunes may be something as subtle as whether the blowouts, at some critical stage, were able to breach the host foredune. In this context, it should be noted that Carter *et al.* (1992, p. 66, Fig. 8) postulated that duneline breaching through blowout throats on an overwash barrier could lead to the establishment of a new foredune line inland of the old dissected foredune. This new foredune will be fed by sediments eroded from the beach and cannibalised from the initial foredune. In time, the coast will recede to the new duneline. New blowouts and washover activity will eventually breach it as well, and the sequence will be repeated in conjunction with barrier migration (Carter *et al.*, 1991). In many regards, this quasi-equilibrium model of foredune evolution is also cyclic, except that it postulates a sequence of *relocation and replacement* rather than *in situ reworking and regrowth* as envisioned by Gares and Nordstrom (1995). A simple aerodynamic model, on the other hand, is likely to produce only simple form *migration*, as with barchans marching across desert surfaces, and this seems wholly inadequate for understanding coastal dune dynamics.

Spatial Views of Dune Equilibrium

One of the most compelling models of coastal dune equilibrium is that of Short and Hesp (1982), which relates coastal dune geometry to nearshore morphodynamic state in an associative, holistic manner. There are implicit temporal dimensions, but the focus is dominantly spatial. Table 4.1 provides an overview of this model, and it is apparent that dune morphology tends to be related to the modal, nearshore morphodynamic state. Much has been written on this association (e.g. Psuty, 1988; Sherman and Bauer, 1993), but there are several points worth reiterating. Distinctive and durable coastal foredunes (of the type described herein and shown in Figure 4.2) tend to be found on reflective beaches. The occurrence of wave scarping processes is frequent, and the coupling between foredune and reflective nearshore is bilateral, intimate and not dominated by unidirectional onshore transport of sediment (as in dissipative systems). Despite their small size, foredunes on reflective beaches are least susceptible to destruction. They appear to be more 'in tune' with the nearshore processes fronting them than the foredunes and sand sheets found on

Table 4.1 Summary of the Short and Hesp (1982) conceptual model relating coastal dune dynamics to modal nearshore morphodynamic state

	Modal state				
	Dissipative	Intermediate	Intermediate	Intermediate	Reflective
Sand transport rates	High	High to moderate	Moderate	Moderate to low	Low
Dominant dunes	Large-scale transgressive dune sheets	Large parabolics to dune sheets	Large parabolics and large blowouts	Crenulate, dissected foredune with pronounced blowouts	Linear foredune with small blowouts
Foredune size	Large	Large to moderate	Moderate	Moderate to small	Small
Dune scarp	Continuous	In rip embayments (>1 km spacings)	In rip embayments (0.5–1 km spacings)	In rip embayments (< 0.5 km spacings)	Continuous
Frequency of wave attack	Low	Moderate	Moderate	Moderate	High
Beach mobility index	Low	Low to moderate	High	Low to moderate	Low
Probability of foredune destruction per century	Moderate	Moderate to high	High	Moderate to low	Low

dissipative beaches for which the equilibrium is longer term. Interestingly, the foredunes of the intermediate modal state beaches are most susceptible to damage, primarily because these beaches are characteristically mobile – that is, sequences of beach erosion and accretion are frequent and the envelope of cross-sectional profiles is wide (Short and Hesp, 1982, p. 268, Fig. 4). These observations led Short and Hesp (1982, p. 280) to conclude that there is 'greater importance of the level of wave energy in determining beach and foredune instability . . . [while] initiation of major instability and dune transgressions by wind is minimal'. This has profound implications for our ability to model foredune evolution and equilibrium because it implies that we need to have a sound understanding of nearshore processes as a necessary complement to the aeolian processes we traditionally focus on.

The Short and Hesp (1982) model was developed for the microtidal, low- to high-energy beaches of southeast Australia, where large sediment contributions from the continental shelf during the Holocene transgression were followed by recent shoreline stability. This model has not been widely tested in other regions of the globe. However, in studying dune systems in the Great Lakes, Davidson-Arnott and Fisher (1992) and Davidson-Arnott and Law (1996) have shown how overwash processes and foredune evolution can be linked to local beach morphodynamics and changing lake levels. In particular, they note the importance of beach width as a dominant seasonal control on sediment delivery to the dunes, and conclude that (Davidson-Arnott and Law, 1996, p. 654),

> Variations in sediment deposition from year to year, and between sites, were controlled primarily by variations in beach width, related to changes in lake levels and to local beach morphodynamics, rather than by variations in potential sediment transport based on wind velocity.

Sherman and Bauer (1993) and Sherman and Lyons (1994) demonstrated how the Short and Hesp (1982) model was consistent with a mechanistic-reductionist approach by using standard transport equations (driven by characteristic values for the general topographic, sedimentologic, and meteorologic variables found on dissipative and reflective beaches) to predict sediment flux rates. One can envision a more generally applicable version of the Short and Hesp (1982) model by creating a three-dimensional phase space or existence diagram with the following axes: (1) dissipative through reflective morphodynamic nearshore state; (2) progradational (emergent/regressive) through retrogradational (submergent/transgressive) shoreline tendency; and (3) short-term through long-term process dynamics. A great deal of the research we have discussed above can be placed into such a phase space and thereby situated in its appropriate context. Whether this might prove useful remains to be seen.

SUMMARY AND CONCLUSIONS

In the course of synthesising a selected literature on the topic of coastal dune dynamics, we have attempted to highlight some fundamental differences between coastal dunes and their dry-land counterparts. We have argued that the only distinctive coastal dune is the foredune because, unlike any other aeolian dune, its form is closely coupled to wave processes and sediment supply in the nearshore. This reality, with added diversity introduced by way of moisture, vegetation and slope effects, confounds our ability to conceptualise and model coastal dune dynamics. Studies of coastal dunes have proceeded along two divergent paths – *mechanistic-reductionist* and *holistic-constructivist* – each with its own objectives, methods, and classes of knowledge. Although each strategy provides essential insight into coastal dune dynamics, it is not immediately evident how one might combine this knowledge. We are left with the following, somewhat unfulfilling conclusions:

1. Foredunes are integral components of coastal systems and because of this, they display complex forms and behaviours.

2. Reductionist approaches to the study of coastal dunes lead toward detailed understanding of such processes as the mechanics of sediment transport in saltation curtains, the entrainment of sediments from partially cohesive surfaces, the development of internal boundary layers across variable roughness elements or sloped surfaces, the modulating effects of vegetation covers on wind and sand transport, the cycling of nutrients in coastal environments, the germination, colonisation, and succession of vegetative species, the role of dune micro and macro fauna, the dynamics of beach groundwater, and the influence of moisture, heat, and radiation fluxes on sediment transport, among many others. In many cases, these studies converge on knowledge similar to that generated by those working in other aeolian systems, including arid and semi-arid environments and laboratory wind tunnels. This elemental knowledge necessarily represents the foundation of any physically-based predictive model of dune evolution. Nevertheless, because of the large number of combinations of variables and in light of intervening influences and opportunities, it remains uncertain whether such models can be used for anything other than providing constraints and guidance to our thinking. At the moment, we can not model the evolution of a coastal foredune from first principles.

3. Constructivist approaches to the study of coastal dunes span a range of spatial and temporal scales, with the dominant objective being explication of the existence, association, and behaviour of dunes and dune assemblages. Several models of the initiation and evolution of coastal dune systems have been proposed, and each is graced with unique insights and varying degrees of applicability. The overall viability of these models remains moot in many cases because the ideas are inferential and based on studies that have examined dune fields from the perspective of a short-term observer intruding on the landscape during a specific stage in its evolution. However skilled the observer, inferences can be incorrect, experiences will differ, circumstances may change, and equifinality abounds.

In our view, there are two, most pressing needs for advancing the state of the science. First is the development of a robust conceptual framework or grand, unifying theory that can serve as the template upon which we may inscribe our piecemeal contributions. Second is the collection of long-term, longitudinal data bases on coastal dune evolution that are sufficiently data-dense to be useful in driving and validating physically-based, predictive models. The last half of the twentieth century has seen several fundamental improvements in our understanding of coastal dune dynamics, and even more substantial advances in our understanding of what else we need to know. The challenges for the next few decades are daunting. *Quo Vadis, Aeolus?*

ACKNOWLEDGEMENTS

We thank the members of the AEOLUS Project for many fruitful interactions, all of which have contributed to shaping the thoughts expressed in this paper. We also gratefully acknowledge continued financial support provided by the California Department of Boating and Waterways and by the National Science Foundation (SBE-9511529).

REFERENCES

Allen, J.R. and Psuty, N.P., 1987. Morphodynamics of a single-barred beach with a rip channel, Fire Island, NY. *Proceedings Coastal Sediments '87*. ASCE, New York, NY. 1964–1975.

Anderson, R.S., 1987. A theoretical model for aeolian impact ripples. *Sedimentology*, **34**, 943–956.

Anderson, R.S., 1990. Eolian ripples as examples of self-organization in geomorphological systems. *Earth-Science Reviews*, **29**, 77–96.

Anderson, R.S. and Hallet, B., 1986. Sediment transport by wind: toward a general model. *Geological Society of America Bulletin*, **97**, 523–535.

Arens, S.M., 1996. Patterns of sand transport on vegetated foredunes. *Geomorphology*, **17**, 339–350.

Arens, S.M., 1997. Transport rates and volume changes in a coastal foredune on a Dutch Wadden island. *Journal of Coastal Conservation*, **3**, 49–56.

Ashley, G.M., 1990. Classification of large-scale subaqueous bedforms: a new look at an old problem. *Journal of Sedimentary Petrology*, **60**, 160–172.

Baaker, Th.W., Jungeruis, P.D. and Klijn, J.A. (Eds), 1990. Dunes of the European Coasts. *Catena, Suppl.* 18.

Bagnold, R.A., 1933. A further journey in the Libyan Desert. *Geographical Journal*, **82**, 103–129, 211–235.

Bagnold, R.A., 1941. *The Physics of Blown Sand and Desert Dunes*. Chapman and Hall, London.

Barndorff-Nielson, O.E., 1989. Sorting, texture, and structure. *Proceedings of the Royal Society of Edinburgh*, **96B**, 167–179.

Bauer, B.O., 1991. Aeolian decoupling of beach sediments. *Annals of the Association of American Geographers*, **8**, 290–303.

Bauer, B.O., Davidson-Arnott, R.G.D., Nordstrom, K.F., Ollerhead, J. and Jackson, N.L., 1996. Indeterminacy in aeolian sediment transport across beaches. *Journal of Coastal Research*, **12**, 641–653.

Bauer, B.O., Sherman, D.J., Nordstrom, K.F. and Gares, P.A., 1990. Aeolian transport measurement and prediction across a beach and dune at Castroville, California. *Coastal Dunes: Form and Process*, K.F. Nordstrom, N.P. Psuty and R.W.G. Carter (Eds). John Wiley, Chichester, pp. 39–56.

Belly, P.-Y., 1964. Sand Movement by Wind. U.S. Army Corps of Engineers, *CERC Technical Memorandum 1*. Washington, DC, 38 pp.

Carter, R.W.G., 1976. Formation, maintenance and geomorphological significance of an eolian shell pavement. *Journal of Sedimentary Petrology*, **47**, 331–338.

Carter, R.W.G., 1988. *Coastal Environments: An Introduction to the Physical, Ecological, and Cultural Systems of Coastlines*. Academic Press, New York, NY.

Carter, R.W.G. and Wilson, P., 1990. The geomorphological, ecological, and pedological development of coastal foredunes at Magilligan Point, Northern Ireland. In *Coastal Dunes: Form and Process*, K.F. Nordstrom, N.P. Psuty and R.W.G. Carter (Eds). John Wiley, Chichester, pp. 129–157.

Carter, R.W.G., Hesp, P.A. and Nordstrom, K.F., 1990. Erosional landforms in coastal dunes. In *Coastal Dunes: Form and Process*, K.F. Nordstrom, N.P. Psuty and R.W.G. Carter (Eds). John Wiley, Chichester, pp. 217–250.

Carter, R.W.G., Allen, J.R.L., Carr, A.P., Nicholls, R.J. and Orford, J.D., 1991. *Coastal Sedimentary Environments of Southern England, South Wales and Southeast Ireland*. BSRG, Cambridge.

Carter, R.W.G., Bauer, B.O., Sherman, D.J., Davidson-Arnott, R.G.D., Gares, P.A., Nordstrom, K.F. and Orford, J.D., 1992. Dune development in the aftermath of stream outlet closure: examples from Ireland and California. In *Coastal Dunes: Geomorphology, Ecology, and Management for Conservation*, R.W.G. Carter, T.G.F. Curtis and M.S. Sheehy-Skeffington (Eds). Balkema, Rotterdam, pp. 57–69.

Carter, R.W.G., Curtis, T.G.F. and Sheehy-Skeffington, M.J. (Eds), 1992. *Coastal Dunes: Geomorphology, Ecology, and Management for Conservation*. Balkema, Rotterdam.

Christiansen, C. and Hartmann, D., 1988. Settling-velocity distributions and sorting on a longitudinal dune: a case study. *Earth Surface Processes and Landforms*, **13**, 649–656.

Cooper, W.S., 1958. Coastal Sand Dunes of Oregon and Washington. *Geological Society of America Memoir 72*.

Davidson-Arnott, R.G.D. (Ed.), 1990. *Proceedings of the Symposium on Coastal Sand Dunes*, Ottawa, Canada, National Research Council.

Davidson-Arnott, R.G.D. and Fisher, J.D., 1992. Spatial and temporal controls on overwash occurrence on a Great Lakes barrier spit. *Canadian Journal of Earth Sciences*, **29**, 102–117.

Davidson-Arnott, R.G.D. and Law, M.D., 1996. Measurement and prediction of long-term sediment supply to coastal foredunes. *Journal of Coastal Research*, **12**, 654–663.

Davidson-Arnott, R.G.D., White, D.C. and Ollerhead, J., 1998. The effects of artificial pebble concentrations on eolian sand transport on a beach. *Canadian Journal of Earth Sciences*, **34**, 1499–1508.

Davies, J.L., 1957. The importance of cut and fill in the development of sand beach ridges. *Australian Journal of Science*, **20**, 105–111.

Davis, W.M., 1899. The geographical cycle. *Geographical Journal*, **14**, 481–504.

Fraser, G.S., Bennett, S.W., Olyphant, G.A. *et al.*, 1998. Windflow circulation patterns in a coastal dune blowout, South Coast of Lake Michigan. *Journal of Coastal Research*, **14**, 451–460.

Gares, P.A., 1992. Topographic changes associated with coastal dune blowouts at Island Beach State Park, New Jersey. *Earth Surface Processes and Landforms*, **17**, 589–604.

Gares, P.A. and Nordstrom, K.F., 1995. A cyclic model of foredune blowout evolution for a leeward coast: Island Beach, New Jersey. *Annals of the Association of American Geographers*, **85**, 1–20.

Gares, P.A., Davidson-Arnott, R.G.D., Bauer, B.O. *et al.*, 1996. Alongshore variations in aeolian sediment transport: Carrick Finn Strand, Ireland. *Journal of Coastal Research*, **12**, 673–682.

Gimingham, C.H., Ritchie, W., Willetts, B.B. and Willis, A.J. (Eds), 1989. Coastal Sand Dunes, *Proceedings of the Royal Society of Edinburgh*, **96B**, 1–314.

Goldsmith, V., 1989. Coastal sand dunes as geomorphological systems. *Proceedings of the Royal Society of Edinburgh*, **96B**, 3–15.

Goldsmith, V. and Golik, A., 1980. Sediment transport model of the southeastern Mediterranean coast. *Marine Geology*, **37**, 147–175.

Haynes, Jr, C.V., 1989. Bagnold's Barchan: A 57-year record of dune movement in the Eastern Sahara and implications for dune origin and paleoclimate since Neolithic times. *Quaternary Research*, **32**, 153–167.

Hesp, P.A., 1981. The formation of shadow dunes. *Journal of Sedimentary Petrology*, **51**, 101–111.

Hesp, P.A., 1984. Foredune formation in southeast Australia. pp. 69-97. In *Coastal Geomorphology in Australia*, B.A. Thom (Ed.). Academic Press, Sydney.

Hesp, P.A., 1989. A review of biological and geomorphological processes involved in the initiation and development of incipient foredunes. *Proceedings of the Royal Society of Edinburgh*, **96B**, 181–201.

Hesp, P.A. and Hyde, R., 1996. Flow dynamics and geomorphology of a trough blowout. *Sedimentology*, **43**, 505–525.

Houston, J.R., 1996. International Tourism and US Beaches, *Shore and Beach*, **64**, 3–4.

Kawamura, R., 1951. *Study of Sand Movement by Wind.* Translated (1965) as University of California Hydraulics Engineering Laboratory Report HEL 2-8, Berkeley, California.

Kocurek, G., Townsley, M., Yeh, E. *et al.*, 1992. Dune and dunefield development on Padre Island, Texas, with implications for interdune deposition and water-table-controlled accumulation. *Journal of Sedimentary Petrology*, **62**, 622–635.

Lancaster, N., 1988. Controls of eolian dune size and spacing. *Geology*, **16**, 972–975.

Lancaster, N., 1995. *Geomorphology of Desert Dunes.* Routledge, New York.

Lancaster, N. and Baas, A., 1998. Influence of vegetation cover on sand transport by wind: field studies at Owens Lake, California. *Earth Surface Processes and Landforms*, **23**, 69–82.

McCann, S.B. and Byrne, M.-L., 1989. Stratification models for vegetated coastal dunes in Atlantic Canada. *Proceedings of the Royal Society of Edinburgh*, **96B**, 203–215.

McEwan, I.K. and Willetts, B.B., 1994. On the prediction of bed-load sand transport rate in air. *Sedimentology*, **41**, 1241–1251.

McKenna Neuman, C. and Nickling, W.G., 1989. A theoretical and wind tunnel investigation of the effect of capillary water on the entrainment of sediment by wind. *Canadian Journal of Soil Science*, **69**, 79–96.

McKenna Neuman, C. and Maljars Scott, M., 1998. A wind tunnel study of the influence of pore water on aeolian sediment transport. *Journal of Arid Environments*, **39**, 403–419.

Namikas, S. and Sherman, D.J., 1995. A review of the effects of surface moisture content on aeolian sand transport. In *Desert Aeolian Processes*, V. Tchakerian (Ed.). Chapman and Hall, London, pp. 269–292.

Namikas, S. and Sherman, D.J., 1998. AEOLUS II: an interactive program for the simulation of aeolian sedimentation. *Geomorphology*, **22**, 135–149.

Nickling, W.G., 1988. The initiation of particle movement by wind. *Sedimentology*, **35**, 499–511.

Nickling, W.G. and Davidson-Arnott, R.G.D., 1990. Aeolian sediment transport on beaches and coastal sand dunes. *Proceedings, Canadian Symposium on Coastal Sand Dunes 1990.* Guelph, Ontario, pp. 1–35.

Nishi, R. and Kraus, N.C., 1996. Mechanism and calculation of sand dune erosion by storms. *Proceedings of the 25th International Conference on Coastal Engineering 1996.* ASCE, New York, pp. 3034–3047.

Nordstrom, K.F., 1988. Dune grading along the Oregon coast, USA: a changing environmental policy. *Applied Geography*, **8**, 101–116.

Nordstrom, K.F. and Jackson, N.L., 1992. Effect of source width and tidal elevation changes on eolian transport on an estuarine beach. *Sedimentology*, **40**, 769–778.

Nordstrom, K.F., Psuty, N.P. and Carter, R.W.G. (Eds), 1990. *Castal Dunes: Form and Process.* John Wiley, Chichester.

Nordstrom, K.F., Bauer, B.O., Davidson-Arnott, R.G.D. *et al.*, 1996. Offshore aeolian transport across a beach: Carrick Finn Strand, Ireland. *Journal of Coastal Research*, **12**, 664–672.

Psuty, N.P. (Ed.), 1988. Dune/Beach Interaction. Special Issue No. 3. *Journal of Coastal Research*, Charlottesville, VA.

Psuty, N.P., 1989. Management of coastal dunes along the Atlantic coast of the U.S.A. *Proceedings of the Royal Society of Edinburgh*, **96B**, 289–301.

Psuty, N.P., 1990. Foredune mobility and stability, Fire Island, New York. In *Coastal Dunes: Form and Process*, K.F. Nordstrom, N.P. Psuty and R.W.G. Carter (Eds). John Wiley, Chichester, pp. 159–176.

Psuty, N.P., 1992. Spatial variation in coastal foredune development. In *Coastal Dunes: Geomorphology, Ecology, and Management for Conservation*, R.W.G. Carter, T.G.F. Curtis and M.J. Sheehy-Skeffington (Eds). Balkema: Rotterdam, pp. 3–13.

Pye, K., 1983. Coastal dunes. *Progress in Physical Geography*, **7**, 531–557.

Rasmussen, K.R., 1989. Some aspects of flow over coastal dunes. *Proceedings of the Royal Society of Edinburgh*, **96B**, 129–147.

Read, D.J., 1989. Mycorrhizas and nutrient cycling in sand dune ecosystems. *Proceedings of the Royal Society of Edinburgh*, **96B**, 89–110.

Ritchie, W. (1972). The evolution of coastal sand dunes. *Scottish Geographical Magazine*, **88**, 19–35.

Sanjaume, E. and Pardo, J., 1991. Dune regeneration on a previously destroyed dune field, Devesa del Saler, Valencia, Spain. *Zeitschrift für Geomorphologie N.F.*, Supplement-Band 81, 125–134.

Sherman, D.J., 1995. Problems of scale in the modeling and interpretation of coastal dunes. *Marine Geology*, **124**, 339–349.

Sherman, D.J. and Bauer, B.O., 1993. Dynamics of beach-dune interaction. *Progress in Physical Geography*, **17**, 413–447.

Sherman, D.J. and Hotta, S., 1990. Aeolian sediment transport: theory and measurement. In *Coastal Dunes: Form and Process*, K.F. Nordstrom, N.P. Psuty and R.W.G. Carter (Eds). John Wiley, Chichester, pp. 17–37.

Sherman, D.J. and Lyons, W., 1994. Beach-state controls on aeolian sand delivery to coastal dunes. *Physical Geography*, **15**, 381–395.

Sherman, D.J. and Nordstrom, K.F., 1985. Beach scarps. *Zeitschrift für Geomorphologie N.F.*, **29**, 139–152.

Sherman, D.J. and Nordstrom, K.F., 1995. Wind-blown sand hazard in the coastal zone: a review. *Journal of Coastal Research*, **SI12**, 263–275.

Sherman, D.J., Bauer, B.O., Gares, P.A. and Jackson, D.W.T., 1996. Wind blown sand at Castroville, California. *Proceedings, 25th International Conference on Coastal Engineering*, CERC/ASCE, Sept. 2–6, Orlando, Florida, pp. 4214–4226.

Short, A.D. and Hesp, P.A., 1982. Wave, beach and dune interactions in southeastern Australia. *Marine Geology*, **48**, 259–284.

Stam, J.M.T., 1997. On the modelling of two-dimensional aeolian dunes. *Sedimentology*, **4**, 127–141.

Thorn, C.E. and Welford, M.R., 1994. The equilibrium concept in geomorphology. *Annals of the Association of American Geographers*, **84**, 666–696.

US Department of Commerce, 1998. Year of the Ocean Discussion Papers, March 1998. Prepared by the U.S. Federal Agencies with ocean-related programs. Available from: http://www.yoto98.noaa.gov/ or Office of the Chief Scientist, NOAA., Washington, DC.

van der Wal, D., 1998. Effects of fetch and surface texture on aeolian sand transport on two nourished beaches. *Journal of Arid Environments*, **39**, 533–547.

Wal, A. and McManus, J., 1993. Wind regime and sand transport on a coastal beach-dune complex, Tentsmuir, eastern Scotland. In *The Dynamics and Environmental Context of Aeolian Sedimentary Systems*, K. Pye (Ed.). Geological Society Special Publication No. 72, pp. 159–171.

Werner, B.T., 1988. A steady-state model of wind-blown sand transport. *Journal of Geology*, **98**, 1–17.

Willetts, B.B., 1989. Physics of sand movement in vegetated systems. *Proceedings of the Royal Society of Edinburgh*, **96B**, 37–49.

Wilson, I.G., 1972. Aeolian bedforms – their development and origins. *Sedimentology*, **19**, 173–210.

Wolfe, S.A. and Nickling, W.G., 1993. The protective role of sparse vegetation in wind erosion. *Progress in Physical Geography*, **17**, 50–68.

Zeman, O. and Jensen, N.O., 1987. Modification of turbulence characteristics in flow over hills. *Quarterly Journal of the Royal Meteorological Society*, **113**, 55–80.

5 Coastal and Continental Dune Management into the Twenty-first Century

DAVID S. G. THOMAS

Department of Geography, University of Sheffield, UK

The need to manage dunes in both coastal and continental contexts is significant as human pressures attempt to extract a utility from dune systems. Management is a nebulous term but is taken to mean direct human actions taken to control, maintain or enhance the utility of dunes in the face of a range of external forcing factors. This chapter concentres on key issues in dune usage and management, and examines ways in which science, primarily focusing on geomorphology, can enhance the knowledge-base of management activities.

Coastal dune systems cover far smaller areas than continental dunefields, but human use is intense and diverse. The complexity of morpho-dynamic processes and the high dynamism of beach–coast sediment systems present marked challenges for management, especially when human use involves attempts to stabilise dune systems, whether for construction and development or for marine protection purposes. Management issues relate to vegetation change and sand movement, dune elimination and reconstruction, dune water tables and marine breaching.

Continental dunes include systems both in presently hyper-arid, active sand seas and those that in arid or wetter situations that have either limited or no current natural activity and result from development during drier periods of the Quaternary. In the case of the latter, pressures of land use may lead to dune reactivation, while in the former, issues relate to reducing the hazards of natural dune activity to constructions and communication routes. Understanding the levels and patterns of activity of different dune types can enhance effective management and allow predictions of the potential impacts of different human uses.

A number of research advances that link to key issues in the management agenda for coastal and continental dune systems are explored. These include scaling and data issues, the relationship between coastal dune and beach sediment systems, temporal variability in dune system behaviour and the prediction of future climate change impacts on dune system behaviour, and conflicts between short-term and long-term management goals.

INTRODUCTION

Greeley and Iversen (1985) note that sand dunes are great fun, but that from a practical perspective they can be beneficial or a bane to humankind. In some situations dunes have for centuries been used in passive ways by societies, for

Aeolian Environments, Sediments and Landforms. Edited by A.S. Goudie, I. Livingstone and S. Stokes.
© 1999 John Wiley & Sons, Ltd.

example as natural defences against marine incursions, or as a source of moisture for communities/agriculture in dry environments (Hack, 1941). In the twentieth century enhanced human pressures have affected dune systems, including an expansion and diversification of uses. Now there is the likelihood of significant environmental changes in the twenty-first century. Together, these provide contexts for a growth in management needs and expectations, that are best informed by reliable scientific inputs.

The need to manage dunes in both coastal and inland environments tends to arise from one of two root causes: either attempts to change or enhance the utility of dunes, or attempts to rectify the effects of negative aspects of previous or current human usage. As such, dune management has dimensions that relate to the effective application of state-of-the-art knowledge of aeolian processes and in some cases hydrological knowledge, and others that relate to human behaviour: take away the human action and in some cases the need to manage dissipates. In the latter case issues of dune management may not differ from general issues related to the human usage of a range of physical environments. As such they are not discussed here, but it must be noted that, in coastal environments, dunes are a popular location for a range of human activities and demands for usage are unlikely to decline, while in inland locations population growth in marginal dryland areas today exerts a range of growing pressures on dune landscapes.

Dune management is therefore unlikely to diminish in importance, and the purpose of this chapter is to consider key issues in dune usage and management, before examining ways in which scientific advances can enhance the knowledge-base of management activities. The focus is primarily geomorphological, and coverage is of both coastal and continental dune systems. Management is in itself a nebulous term, but for the purpose of this chapter it is taken to mean direct human actions to control, maintain or enhance the utility of dunes in the face of a range of external forcing factors. In both coastal and continental dune environments management has two general goals: to overcome negative aspects of human impacts on natural systems, or the modification of natural systems to achieve human goals. Whether dealing with coastal or continental dunes, management practices attempt to modify one or more of four factors that control dune behaviour: the supply of sediment to or within the dune system, the characteristics of the wind in the boundary layer, the nature of vegetation (whether its presence, extent, type or distribution) and human actions.

DUNES IN A DYNAMIC WORLD: THE CONTEXT FOR MANAGEMENT NEEDS

Globally, sand dunes are relatively insignificant landscape components, covering only *ca* 5–6 per cent of the earth's land area (Thomas, 1997a; Pye and Tsoar, 1990). However, approximately 20 per cent of the world's drylands are covered by aeolian sand deposits (sand seas). This includes both dunefields responsive to aeolian processes today and extensive dune systems that are vegetated to varying degrees (Thomas, 1997b). Such 'relict' dunefields are also found beyond today's drylands in more humid areas, for example in western Zambia where rainfall today is in excess of 1000 mm per annum. Most of the latter are largely inactive under present environmental conditions and are the outcome of periods during the Quaternary when environmental conditions were more conducive to the operation of aeolian processes (e.g. Stokes *et al.*, 1997). Very few coastal dune systems are of sufficient size to be classified as sand seas (Pye and Tsoar, 1990). Coastal dunes are, however, found in most latitudinal belts, with the humid-tropical zone being the only climatic zone where they are almost totally absent (Goldsmith, 1978), probably because of intense chemical weathering inhibiting sand supply and/or extensive vegetation in the near shore zone (Jennings, 1964; Davies, 1973). In some areas the distinction between coastal and continental dunes is blurred, as sands blown inland from marine contexts ultimately provides an important source of sediment for sand seas, for example along the coasts of Namibia, Oman and other parts of the Arabian Gulf.

The largest continental sand seas cover areas in excess of 300 000 km^2, for example the Kalahari in southern Africa, the Rub al Khali of Arabia and the Great Sandy-Gibson Desert of Australia, with over 50 sand seas with an area in excess of 12 000 km^2 (Wilson, 1973). In contrast, the largest coastal dune complexes are on the order of only hundreds of km^2, for example Coos Bay, Oregon (300 km^2; Goldsmith, 1978); Alexandria, near Port Elizabeth, South Africa (Illenberger and Rust, 1988) and Cape Bedford/Cape Lattery, Queensland (600 km^2; Pye, 1982). However, in many management/usage contexts it is the length of the littoral front of coastal dune systems that is more significant. In the UK, only 7.4 per cent of the coast is fronted by dune systems (Doody, 1989); for western Europe as a whole this figure rises to about 20 per cent (Hansom, 1988), including 80 per cent of the coastline of the low lying territories of the Netherlands and 100 per cent of the Belgian coast, and for Australia is 24 per cent (Zann, 1995).

Coastal Dune Environments

Coastal dune landscapes are extremely complex in terms of both morpho-dynamical processes and the resultant landform diversity (Carter *et al.*, 1990b). When this is set against the variability and in some instances longevity of human usage of coastal dune systems (Nordstrom, 1994), a bewildering range of coastal dune management issues are apparent. The context for human-induced changes and management requirements within coastal dune systems has to be placed within the framework of the natural system itself. Worldwide, the evolution of coastal dunes is closely linked to post-glacial sediment supply and Holocene sea level change. Thus a general global model suggests (i) onlapping dune sequences successively built and destroyed as a result of post-glacial sea level rise; followed by (ii) a post-transgressive phase of dune stabilisation (Orme, 1990). Phase (iii), the subsequent destabilisation of coastal dunes, incorporates the impacts of human attention and the uncertainties of future global change impacts. It is within the context of phase (iii) that management needs arise. For individual dune systems, coastal sediment budgets, aeolian sand transport and destructive marine events hold the key to understanding the relative importance of controlling processes. For example, Ritche and Penland (1990) note that on the

Louisiana coast, dunes have only a decadal scale life span due to the frequent impact of hurricanes. In this situation, human impacts are relatively insignificant or likely only to be short-lived. In some contexts a *ca* 50-year major disturbance cycle (Gares and Nordstrom, 1991) may affect coastal dune systems and provides a framework in which human impacts should be viewed.

Nordstrom (1994) notes that despite rapidly escalating costs, there is no indication that development and utilisation of coastal dunes will cease; indeed, much of the evidence suggests that human manipulation of dunes will grow, especially through the combined pressures of increasing recreational time in the western world and the impacts of anthropogenic global-warming induced sea level rise (Van der Meulen, 1990). Seventy per cent of the coastal dunes of Belgium are built on, and much of San Francisco (Sherman, 1997), Tel Aviv and Durban (Livingstone and Warren, 1996) are sited on former coastal dunes. The conversion of coastal dune systems to built environments has been exceptionally rapid in some cases, with areas going from wilderness to fully developed in less than two decades (e.g. Reynolds, 1987). Although development may wax and wane in response to changing social and economic demands, once in place a construction remains to alter the operation of geomorphological and ecological processes (Nordstrom, 1994) and requires management to afford it protection. The latter tends to imply inducing stability into the environmental system, despite beaches and coastal dunes being amongst the most dynamic of all geomorphological environments (Sherman and Bauer, 1993).

Where construction is not the main form of human use, management needs may arise through the impacts of recreational activities. Uses are diverse, ranging from the appreciation of the ecological and wilderness value of the dune system, high intensity but seasonal passage through dune systems for access to beaches, to off-road vehicle use that can severely and rapidly destroy or modify the environment (Polis, 1989). A further highly destructive form of use is coastal dune mining, either simply to extract sand for the construction industry (e.g. on the Sefton coast, north west England; Pye, 1990) or to access high concentrations of heavy minerals for example at Richards Bay, South Africa (Camp, 1989) and in various locations on the north east and south east coasts of Australia (Clark, 1975).

Continental Dune Systems

It has sometimes been suggested that coastal dunes are of greater potential value to society than desert dunes (e.g. Sherman, 1995). However, human use within areas of continental dunes is marked and growing. In continental dune systems that are naturally active today (i.e. in hyper-arid areas) management tends to relate to controlling natural sand movement to permit or improve a specific human action, frequently in the form of mineral extraction, for example oil in the states of the Arabian Peninsula and diamonds in Namibia (Figure 5.1), or for transport pathways that cross areas of dynamic aeolian sand. In these locations, Watson (1990) notes that the supply of sand must be considered inexorable: consequently solutions are essentially short term. Management issues therefore relate to determining the most appropriate method of sand activity control, of which many are available (Watson, 1985). However, deciding which is most effective in a particular situation

Figure 5.1   In sand seas in hyper-arid settings, the encroachment of mobile sand on constructions may be a significant problem requiring constant management actions. At Kolmanskop in Namibia, buildings abandoned in the early 20th century, when diamond mining ceased, were soon engulfed by mobile dunes

has frequently been a matter of trial and error. To a large extent success depends on understanding the dynamics of a system, the rate of sand flux, and the seasonal behaviour of the wind regime at the location of concern. For many dry areas the necessary data do not exist prior to the onset of human use; were this the case, the design and siting of structures and protective measures would be better informed.

Leaving aside the general issue of wind erosion from agricultural land, which can occur in many environments, the second set of issues relates to human actions leading to the reactiviation of relict aeolian sands. In semi-arid and dry-subhumid areas, for example in many parts of the African Sahel (Talbot, 1984), in the Kalahari of Botswana (Thomas and Shaw, 1991), Rajasthan in India (Wasson *et al.*, 1983), the High Plains of the USA (Muhs and Maat, 1993) and the Eyre Peninsula of Australia (Heathcote, 1996), such areas are vulnerable to human-induced aeolian activity on several counts. In the developing world in particular, populations pressures are marked. The Thar Desert of India, a large part of which consists of semi-stable dune sediments, has a remarkably high population density and is subject to a range of agricultural activities that can trigger sand movement (Harsh *et al.*, 1995). The Sahel belt, on the southern margin of the Sahara Desert, has seen an extension and intensification of human activities over the last three decades, and a tendency towards the replacement of extensive transhumance pastoralism with more sedentary practices that are more susceptible to causing destabilisation (e.g. Koechlin, 1997; Raynaut, 1997). The arrival of Europeans during the nineteenth century and the development of agriculture in parts of the American interior and Australia underlain by ancient dune deposits brought similar issues and

susceptibilities to these environments (e.g. Heathcote, 1996). An additional destabil-
ising influence that is growing in occurrence in some areas is the use of off-road
recreational vehicles (e.g. Lyners *et al.*, 1980).

Semi-arid areas with relict dune systems are commonly susceptible to significant
inter-annual climatic variability and drought incidence. The natural stresses these
bring to vegetation systems may be enhanced by human impacts, resulting in an
increased incidence of aeolian activity. While the case has often been misrepresented
and over exaggerated (Thomas and Middleton, 1994), this represents one small
aspect of the general issue of desertification. The predicted global warming impacts
on presently semi-arid environments (Williams and Balling, 1995) may have great
significance for future levels of aeolian activity. As the Pleistocene development of
these ancient erg areas becomes better understood and dated, it becomes increasingly apparent that such systems are capable of rapid geomorphological responses
to relatively modest forcing stimuli (Goudie, 1994). This has great significance for
future management needs.

COASTAL DUNES: MANAGEMENT ISSUES

In coastal environments the specific purpose of any dune management activity may
be blurred because many coastal dune systems are subject to multiple human impacts
and expectations. For example, Pye (1990) notes that the 4-km wide dune belt of
the Formby/Sefton coast in Lancashire has been subject to multifarious human
impacts: levelling for agriculture, sand extraction, waste disposal, recreational uses
and plantations. Direct attempts at management to mitigate human actions on
coastal dunes date in Europe from at least the middle ages (Pye and Tsoar, 1990).
Coastal dunes can be classified in many ways, but from a management perspective
the simple division of Psuty (1989) into primary and secondary dunes is perhaps
most useful. Primary dunes are those that are the direct result of coastal processes
and therefore largely consist of the foredune ridge and embryo dunes. Secondary
dunes are inland of the foredune ridge and are coastal dunes in the sense of the origin
of their sediments and there initial development. Secondary dunes may have undergone marked re-modelling and the operation of pedogenic processes. In temperate
environments secondary dunes are almost certainly vegetated to a significant degree,
while the degree of vegetation on primary dunes may vary markedly over time in
response to the variability of marine processes and the rate of aeolian sand supply.

Vegetation change, dune erosion and sand movement

Some of the most widespread impacts of a range of human uses of coastal dunes are
changes within dune vegetation systems. In Europe, the destruction or clearance of
coastal dune vegetation dates back many centuries (e.g. Maarel, 1979). The introduction of rabbits to Britain after the Norman invasion of 1066 AD, and the creation
of rabbit warrens in coastal dunefields, contributed to increased grazing of dune
vegetation and sand movement. Devegetation of coastal dunes, to provide timber
for construction, to create agricultural land, or as a result of grazing pressures or

accidental fires, appears to have been a feature of the arrival of Europeans in southern Africa, eastern Australia and parts of North America (Orme, 1990). In New Zealand the arrival of Polynesians may have had a similar effect (Hicks, 1983). Today, vegetation destruction occurs widely as a response to recreational pressures (Eastwood and Carter, 1981).

One impact of vegetation removal on secondary dunes is an enhanced aeolian sand transport hazard, particularly where coastal dunes are located in wetter environments such as north west Europe and north west USA, and under natural conditions possess a stabilising vegetation cover. However, apocryphal stories of the burial of villages inland of European coastal dunes need to be treated with caution (Livingstone and Warren, 1996), and generalisations about the effects of vegetation removal avoided. Today, many concerns about vegetation destruction in coastal dunes relate to ecological rather than geomorphological impacts, in which context planned afforestation can be as detrimental as the creation of bare ground.

From an aeolian dynamics perspective the pattern and patchiness of vegetation destruction impacts needs to be understood if erosion risk hazards are to be properly charted and management strategies developed. The patchy development of blow-outs and parabolic dunes in coastal dune systems may be enhanced by human pressures, though in many situations they are natural features resulting from wave-induced scarping of foredunes and the impact of storm events (Junerius and van der Meulen, 1989; Carter et al., 1990a; Sherman and Bauer, 1993). The creation of footpaths across secondary coastal dunes to gain access to beaches may permit the penetration of strong on-shore winds into the heart of dune systems, enhancing the potential for wind erosion, whereas shore-parallel paths may have little impact. Vegetation destruction on dune high-points may also be very significant due to speed-up induced higher wind velocities affecting these locations. Even partial vegetation destruction, or the selective grazing of particular plants, can be detrimental by leading to enhanced turbulence. However, in all these examples, the details of vegetation-surface conditions are critical in determining the response of aeolian processes to surface changes (Wolfe and Nickling, 1993; Wiggs et al., 1994), both in terms of the impacts of plant destruction and those of any remedial planting activities (Van der Meulen and Jungerius, 1989).

Vegetation destruction can enhance water as well as wind erosion. The destruction of foredune vegetation can result in a greater susceptibility to marine erosion. Where the sediments of secondary dunes sands have a high organic content or a significant algal crust, overland flow can increase significantly when vegetation is removed or destroyed (Jungerius and de Jong, 1989).

Dune Elimination and Reconstruction

In most respects coastal dunes are not logical places to construct buildings: soft sediments provide poor foundations, they are highly dynamic systems subject to rapid, periodic sediment fluxes, and by definition they occupy vulnerable coastal locations. Yet coastal dunes in many locations are popular places to reside and build leisure resorts, as witnessed by the development of homes, holiday parks etc. along the east coast of the USA, in Belgium and in dune zones of the UK coast.

Such developments inevitably involve management in two stages, to facilitate construction in the first instance and subsequently and repetitively to maintain these constructions against the inevitable forces of nature.

The sequence of management actions commonly involves an initial destruction or levelling of dunes, to provide stable terrain suitable for construction, followed by their reconstruction at reduced size and extent to afford some coastal protection (Sorensen et al., 1984; Nordstrom, 1988). However, 'the size and location of these (new) dunes are different because they are dictated by human preference rather than the interplay among vegetation growth, sand supply and wave erosion' (Nordstrom, 1994, p. 501). This process of attempted stabilisation has been termed 'New Jerseyation' (e.g. Pilkey et al., 1983), and the dunes that are reconstructed as part of the managed environment are commonly smaller and more linear than the natural dunes they replaced (Nordstrom, 1994). This can make them more vulnerable to storm surges, enhancing the need for constant sediment replenishment, and having marked implications for issues related to sea level rise (see below). In areas where construction extends close to the littoral zone, artificial foredunes are frequently constructed with flat tops (to improve the view). One reported outcome of this at Salina Beach, California, is enhanced susceptibility to water erosion (Pilkey et al., 1983).

Dune Water Tables

Coastal sand dunes can be important sources of water resources, for example in the Netherlands where 40×106 m^3 of water per annum are extracted for urban consumption (Van der Meulen and Jungerius, 1989) and in South Carolina where 20 well systems tap barrier island/dune complex water to supply Charleston (Kana et al., 1984). Relative sea level rise along the USA eastern seaboard, and over-pumping, can result in the incursion of sea water and displacement of potable freshwater. In South Carolina some coastal aquifers have not been used for drinking water since 1950 for this reason (Kana et al., 1984). Coastal dune systems in humid regions commonly have a domed water table (Willis et al., 1959), with sub-surface moisture in dune sands susceptible to dune erosion (Pye and Tsoar, 1990) because the depth of evaporative drying increases. This can lead to vegetation die-off and an enhanced erosion-drying feedback loop.

Marine Breaching and Sea Level Rise

The protection afforded by coastal dunes against storm surges and inundation is well recognised in scientific and planning circles. The protective role of dune systems embraces a range of complex morpho-dynamic and political issues, many of which relate to the impact of system stabilisation. In much of the USA there is a strong political rationale for the maintenance of coastal dune defences: most coastal communities can claim federal insurance for flood and storm damage (Pilkey et al., 1983); however, this does not cover damage due to erosion. This raises fundamental legal issues when the impacts of predicted sea level rise during the twenty-first

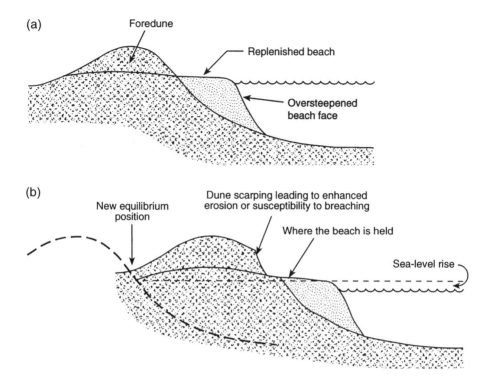

Figure 5.2 Schematic diagram showing the potential effect of beach sediment replenishment on foredune morphology and position. This example based on Pilkey *et al.* (1983), incorporates the additional effect of sea level rise

century enter the equation. In the Netherlands, where coastal dunes are a vital part of sea defences, less than a quarter of the total dune front length is currently aggrading, and almost 40 per cent is receding.

Management of dynamic primary coastal dunes to afford protection involves two main processes: beach sediment replenishment and beach/dune morphology changes. In addition to those of cost (Pikley *et al.*, 1983), problems can arise with both processes because at-a-point actions can markedly impact on system behaviour in down-drift and down-wind locations along the beach-dune system. As Psuty (1990) notes, the foredune has too frequently been viewed as a dynamic feature, but one that is spatially static, isolated in its geomorphic setting. This situation has occurred in both scientific and management contexts, though in the case of the latter Arens (1994, in Livingstone and Warren, 1996) notes how in the Netherlands management policies and programmes are developed for whole stretches of coastline.

The fundamental outcome of most attempts to stabilise foredune systems is distortion of the equilibrium beach-dune profile. Sherman and Bauer (1993) cite the work of Dolan (1972) who demonstrated that a long period of foredune and beach management and restabilisation at Cape Hatteras, North Carolina, led to over-steepening (relative to the equilibrium form) of the beach profile. This has two significant outcomes: first, the beach is eroded faster (e.g. Pikley *et al.*, 1983) (Figure 5.2) such that replenishment has to occur more frequently and at greater cost, and second, the potential for dune scarping is enhanced, in turn raising the potential for breaching during storm surges (Leatherman, 1984). The issues raised here have clear management implications in terms of whether protection in the short term is the primary goal or whether it should be protection against big events that should be the focus, because current understanding suggest that both goals can not be achieved simultaneously.

CONTINENTAL DUNE ENVIRONMENTS: MANAGEMENT ISSUES

Pressures on Marginal Semi-Arid Lands: Dune Reactivation and Vegetation Change

Images of reactivated sand dunes resulting from human pressures in semi-arid lands, often the outcome of livestock grazing (e.g. Tsoar and Møller, 1986) but also including the impacts of cultivation and recreational and military vehicle use (Marston, 1986), abound in the literature on desertification. The notion of the advancing desert/sand dune front as *the* characteristic outcome of desertification is however now well and truly debunked (e.g. Binns, 1990; Hellden, 1991; Thomas and Middleton, 1994). Nonetheless, reactivated dunes due to devegetation are a local, even regional problem in some areas. In many respects the ensuing problems and management solutions are similar to those that apply in coastal dune environments, with the emphasis on human behavioural changes since a cessation in the disturbing activity can lead to rapid recovery of natural vegetation covers (e.g. Tsoar and Møller, 1986). There are however added dimensions regarding the outcome of devegetation, relating to issues of erodibility and erosivity in the environment concerned.

The erodibility of dunes in semi-arid and related environments is a function of several factors (e.g. Tsoar and Møller, 1986) including rainfall levels, dune moisture storage, the degree and type of vegetation cover, the presence of biogenic crusts and human pressures. The last of these tends to modify the effects of the first on natural stabilising agents in the form of plants and crusts (Figure 5.3). The response of dunes to disturbance is controlled by erosivity- the energy in the environment able to effect aeolian processes to remove sediment from susceptible surfaces. In the context of continental dunes the relationship between erosivity and erodability and the resultant environmental outcome is complicated by the fact that different dune forms display different types of overall activity (Thomas, 1992).

For example, linear forms are the dune type most commonly found in partly or totally vegetated form in marginal dryland areas, dominating the ergs of the Sahel, Kalahari, Negev and interior Australia. Linear dunes can be regarded as sand-passing dune forms. Reactivation due to devegetation may result in enhanced aeolian sand transport on dune crests, where speed-up effects enhance erosivity (Wiggs *et al.*, 1995), leading to the development of steeper slopes and slip faces, but the position of the dune body tends to remain stable (e.g. Thomas, 1988) (Figure 5.4). Even where

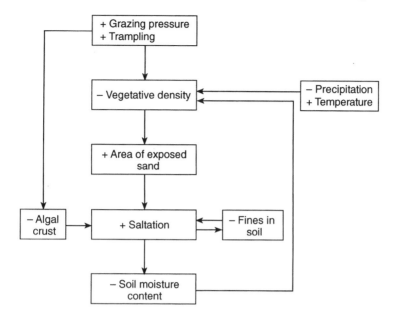

Figure 5.3   Linkages and feedbacks between vegetation destruction and dune surface processes (after Tsoar and Møller, 1986)

Figure 5.4   Grazing pressure around boreholes in the southwest Kalahari can result in devegetation and enhanced crestal activity on linear dunes (compare the bare dune crest and flanks in the foreground with the better vegetated more distant parts of the dune), but present day low wind energies mean that the effects remain very localised

devegetation has occurred at the regional scale, for example in the southern Negev Desert in the 1970s due to a substantial increase in grazing pressure (Tsoar and Møller, 1986), and morphological changes were more marked, the overall position of dunes in the landscape largely did not change and the process was naturally reversed when grazing pressures ceased. In other areas, the total array of dune types subject or susceptible to disturbance includes more mobile forms including barchan and parabolic dunes (e.g. Muhs and Maat, 1993).

The limited impacts and persistence of negative consequences of dune devegeta-tion in the above examples are partially due to the extensive, rather than intensive, nature of the disturbing land use. Blowing sand, even if localised, would be more of a hazard in situations where cultivation rather than livestock production is attempted. Certainly in the Eyre Peninsula of South Australia the problem of reactivated dunes and resultant drifting sands, and the need for management, appears to have become more acute, either actually of perceptionally, with the gradual replacement of sheep rearing by wheat cultivation in the first decades of the twentieth century (Heathcote, 1996).

The impact of devegetation on dune activity is also mitigated in some situations by relatively low wind strengths, which inhibits erosivity, relative to the conditions that are likely to have prevailed at the times of dune development (e.g. Thomas et al., 1997a). One way of assessing the combined impact of erodability and erosivity, and therefore the potential effects of human-induced devegetation, is through the use of a dune mobility or aeolian activity index (e.g. Talbot, 1984; Lancaster, 1988). These can be calibrated to show the components of the dune landscape that would be active at different index values. Bullard et al. (1997) have done this for the southwest Kalahari, showing the natural temporal variability in climatic controls on aeolian activity in the period 1960–1990 (Figure 5.5). Human actions leading to devegetation could decrease vegetation cover, thereby extending the time periods over which the erosivity component could effectively lead to sand transport prob-lems. However, disagregation of the two components of the index for individual climate stations shows that the principal constraint on sand transport on south west Kalahari linear dunes today is not vegetation cover but wind energy (Bullard et al., 1997). Sand mobility is mainly a local nuisance rather than a major problem, and the likelihood of wholesale dune reactivation through human actions in this environment is low under present climatic conditions.

The Control of Mobile Sand and Dunes

The control of mobile sand is not new: around oases in North Africa and Arabia it may have been occurring for thousands of years (Livingstone and Warren, 1996). Today a diverse array of approaches are available, some incorporating improved understanding of the dynamics and controls on sand transport, others utilising background data on the sand drift regime in the environment concerned, which can inform both the siting of constructions in the first place, thus avoiding the need for subsequent management, and the design of protective measures. Good detailed accounts of individual approaches are provided by Watson (1990) and Pye and Tsoar (1990); the focus here therefore is on the different management strategies,

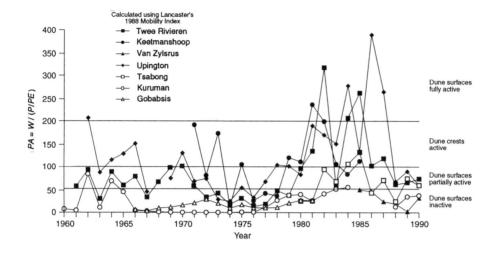

Figure 5.5 Trends in aeolian activity index values in the southwest Kalahari, 1960–1990. Natural variability in activity results from temporal changes in climatic factors that influence erodibility and erosivity (after Bullard *et al.*, 1997)

which can be both on-site and off-site, rather than on listing individual methods. Tackling the problems encountered with drifting sand falls under three headings: reducing sand supply; immobilisation of sand including enhancing deposition; and diversion or enhancement of sand transport. Techniques applicable to each are not mutually exclusive, for example sand fences can contribute to all three of these categories.

Reducing sand supply requires identification of the source area of sediment and then the application of either a stabilisation method such as mulching using an armouring layer of coarser material (Watson, 1985) or chemical applications, such as asphalt (Watson, 1990). In both cases there are risks of undercutting or destruction by sand blasting (Azizov and Atabaev, 1987), while livestock pressure can also lead to surface break-up, as reported by Harsh *et al.* (1995) in the Thar Desert. Planting in the source area is another possibility that may both bind surface sediments and reduce wind shear; sand fences may also reduce entrainment in source areas by changing velocity profiles. However, in all cases, source area stabilisation is only viable when sources are very localised, and as Livingstone and Warren (1996) note, many localised sources are only secondary stages on longer sand transport pathways from further afield.

In some situations *enhancing deposition* upwind of a construction can be used as a protection measure. If the sand transport rate is low and a moisture source is available, planting may be a viable approach and once established trees shrubs and grasses can be self-renewing (e.g. Kebin and Kaiguo, 1989). Effective planting to prevent encroachment of mobile dunes has been described for the southeastern Tengger Desert in China by Zhu *et al.* (1992). Planting can be labour intensive, but aerial seeding has also been used, with a plant first-year survival rate of 8 per cent reported for an experiment in India (Kumar and Shankarnarayan, 1988). The stabilised sand deposits have even been used to increase the area of agricultural land. In many instances fences are constructed to trap sand, but there precise location, orientation and design can have marked impacts on effectiveness (e.g. Watson, 1990). In every case fences have to be managed and renewed: a single fence line is rarely sufficient to halt sand advance for long and as sand is trapped new fences are needed above those that become buried. Furthermore, as sand is trapped, the aerodynamics of the resultant accumulation alters over time. Livingstone and Warren (1996) report artificial dunes over 40 m high that have accumulated around north African oases as a result of the maintenance and renewal of palm frond fences. If transport rates are very low, trenches to trap sand may also be used (Harsh et al., 1995).

When sand supply can not be controlled it has to be accepted as a fact of life, in which case management techniques can aim for *transport diversion or enhancement*. In both cases the goal is to avoid deposition around human constructions. Establishing fences to deflect sand transport away from or around constructions can have variable success, since the porosity of the fence will determine whether it deflects transport or encourages deposition (Watson, 1990). Enhancing transport rates through or past constructions may be used particularly if it is a road or railway that requires protection from inundation. The design of the construction may naturally enhance throughput, for example a tar road will enhance saltation rates, though kerbstones will act as depositional nuclei (Figure 5.6), while the angle of embankment slopes can also be designed to speed-up transport (Mainguet, 1991). The encroachment of mobile dune forms can be problematic and options for management can include levelling or armouring dune surfaces (Bailey, 1907, in Livingstone and Warren, 1996) so that the form naturally disperses and sand saltates faster though the construction.

RESEARCH ADVANCES: KEY ISSUES IN THE DUNE MANAGEMENT AGENDA

Dune management occurs in a range of environments for a range of purposes. In many instances the most effective route to management is likely to be through inducing changes in human behaviour, but increasingly the emphasis is on finding solutions that allow human activities to be integrated into dynamic aeolian environments. In coastal contexts engineering has a long history of contributing to dune management, but in continental dunefields management efforts are today better informed largely through the effective extrapolation from and application of the outcomes of research into aeolian processes and dryland geomorphology. As the impacts of predicted global warming will affect both coastal and continental dune systems, the role of science within dune management is likely to contribute to future developments to an even greater extent. In this section a number of important areas where geomorphic developments can contribute to the understanding or application of effective management are considered.

Figure 5.6 Sand saltating over a road surface in Kuwait during a sand storm, July 1997. The slightly embanked roadway and hard tarmac surface results in high saltation rates, preventing sand accumulation. In the foreground the presence of a small kerb acts as a depositional nucleus for the gradual accumulation of sand. This leads to the need for regular road clearance

Scale and Data Issues

Dune management, and dune research, is scale dependent. Research occurs at time scales ranging from second-to-second measurement of variations in aeolian flux and wind shear to long-term studies of dune system evolution over tens of thousands of years. The spatial scale of study ranges from the microscale monitoring of grain movement to the investigation of patterns of forms and processes across whole dune systems. Sherman (1995) notes, in the context of coastal dune systems but with equal relevance to continental systems, that at present it is difficult to integrate the findings from studies at these different scales, and this has implications for effective long-term dune management. Management tends to occur at two distinct scales (Livingstone and Warren, 1996): the small scale where it is 'a matter of aeolian

Table 5.1 Parameters controlling dune system behaviour and their inclusion in empirical studies (after Sherman, 1995)

Attributes measured fairly well and/or frequently	Attributes that are measured infrequently or poorly
Wind velocity	*Usually important*
Wind direction	Spatial variations in the wind field
Particle size and sorting	Wind velocity profile
Vegetation type	Sediment transport rate (including variability)
	Sediment moisture content
	Occasionally or locally important
	Sediment composition
	Sediment salt content
	Changes in surface conditions
	Bedform geometry
	Vegetation geometry
	Temperature gradients
	Evaporation rates
	Surface obstructions

geomorphology and plant ecology' (p. 154) and the dunefield scale where a wide range of components relating to sediment supply, climate and human use need to be integrated.

The biggest problems presently exist in the expansion of the findings of reductionist aeolian process studies to scales that are relevant to management issues, and one reason for this is that sediment transport studies tend not to measure all the parameters that are essential for a complete and effective understanding of system dynamics (Table 5.1). The major reason behind these data gaps appears to be linked to issues of resourcing, but Sherman (1995) argues that surrogate data can be drawn together from diverse sources to help overcome the problem. The precision may not be as great as with directly collected field data, but it may be appropriate for management goals and the opportunity to parametise a wider range of the components identified in Table 5.1 may more than compensate for this. The examples that follow all include evidence of the integration and development of data from a range of sources to contribute to a fuller understanding of dune system behaviour.

Beach–Dune Interactions

The links between marine sediment supply and coastal dune construction are well recognised, especially when the millennia-scale evolution of systems is considered (e.g. Orme, 1990). At the level of processes studies, however, the marine-dominated beach system and the aeolian-dominated dune system have usually been treated as independent systems (Sherman and Baeur, 1993). Many researchers (e.g. Short and Hesp, 1982; Pye, 1990; Psuty, 1992) have now demonstrated how these systems interact at a range of time scales, and these interactions have marked management implications, for example in determining whether it is viable to attempt to stabilise mobile dune surfaces, protect structures, or restore foredunes through beach

Table 5.2 Linkages between beach and dune sediment budgets and morphological outcomes (after Sherman and Bauer, 1993)

Beach budget	Dune budget	Dune morphology
Positive	Positive	Beach or dune ridges
Positive	Steady state	Various
Positive	Negative	Blowouts and deflation hollows
Steady state	Positive	Vertical dune growth
Steady state	Steady state	Various
Steady state	Negative	Blowouts and deflation hollows
Negative	Positive	Dune growth and onshore migration
Negative	Steady state	Various
Negative	Negative	Dune erosion and washover

sediment replenishment schemes, or for establishing which part of the integrated system management actions are most sensibly applied to.

Psuty (1992) and Sherman and Bauer (1993) have provided summaries of the linkages between beach and dune sediment budgets and resultant foredune morphologies, applicable at month-to-decadal time scales (Table 5.2). A number of assumptions underpin the scheme, including a limited direct role for vegetation, but it does indicate the manner in which coastal dune system morphologies might change in response to natural or anthropogenic modifications to sediment budgets. Using data gained from investigations in northwest England and other coastal dune locations, Pye (1990) has produced a schematic model that takes further the linkages in Table 5.2 to include both a temporal dimension and the role of vegetation and wind energy (Figure 5.7).

Temporal Variability in Dune Systems and the Impact of Future Climate Changes

Controls on dune activity operate variably at a range of timescales. It has already been noted that temperate environment coastal dune systems appear to respond to a cycle of major disturbances (Gares and Nordstrom, 1991), and that in semi-arid areas the regular occurrence of droughts can alter the balance of factors controlling erodibility and erosivity towards conditions favouring enhanced aeolian activity (Bullard et al., 1997). Predicted global warming may impact on environments with dune systems not only to increase the level of natural aeolian activity but to enhance their susceptibility to human impacts. Modelled climate changes predict both higher temperatures and a greater drought frequency in many semi-arid areas, as well as reduced soil moisture levels (e.g. Williams and Balling, 1995), while sea level rise could markedly alter the sediment budget regimes in beach-dune systems (Van der Meulen, 1990).

Identifying dune system responses to such changes is a major research and management challenge. In coastal systems some lessons may be learned from the responses and outcomes of human actions in dune systems on subsiding coastlines, for example in the eastern USA. The situation for continental dune systems is more

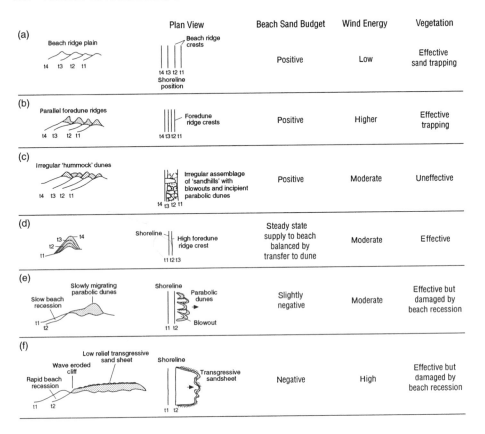

Figure 5.7 Schematic diagram showing the relationships between sediment budget, wind regime and vegetation and resultant coastal dune morphology changes over times t1–t4 (developed from Pye, 1990)

complex in many ways since the principal impact on activity changes is climate itself, and the resolution of predictions for many semi-arid areas in GCMs is poor. Where data are available at suitable resolution, dune behaviour changes may be predicted. Muhs and Maat (1993) and Stetler and Gaylord (1996) have attempted to model future changes in the area of mobile sands in the US interior. Muhs and Maat (1993) used data from two $CO_2 \times 2$ GCMs and Lancaster's (1988) dune mobility index to predict the response of currently stable sand seas in the US High Plains. Currently most dunefields and sand sheets in the Great Plains are inactive, with only localised crestal activity. For 25 out of 40 localities for which data were available, marked increases in activity are predicted, with only interdune areas remaining stable. From a management perspective Muhs and Maat (1993) also suggest that recent land use changes, from pastoralism to centre-pivot agriculture, enhance susceptibility to aeolian sand dune mobilisation; however, aquifer depletion may in fact lead to a significant reduction in irrigated agriculture in the next 30 years (Thomas et al., 1997b).

Stetler and Gaylord (1996) have integrated a Regional Climate Model with a number of different aeolian-climate indicators for a small dune area in Washington State to determine trends in unvegetated dune sand and dune mobility under a climate with mean annual temperature rise of 4°C. They predict a 400 per cent increase in mobile sand. These studies indicate two different approaches, at different spatial scales, for the prediction of aeolian activity changes. Both suggest that continental dune systems will be sensitive to climate changes over short time scales.

Conflicts Between Short-Term and Long-Term Dune Management Goals

A key issue in dune management at a time of climatic change uncertainty is to attempt to resolve the short-term goals of specific actions with the longer-term likelihood of major changes. In parts of the UK, decisions have already been taken not to continue protective management on some stretches of coastline. It is inevitable that short- to medium-term losses will result from such actions, but this may be necessary to delimit future impacts of rising sea levels and to avoid storm surge catastrophes. This is well illustrated in a North American context. For the Cape Hatteras area of North Carolina, deKimpe et al. (1991) have modelled the changes that will accrue from a formal decision to abandon the foredune maintenance that Dolan (1972) showed to enhance storm surge damage. By the year 2000 over 20 per cent of the dune system is predicted to have been breached or eroded, and in the longer term, assuming the current rate of sea level rise, approximately 70 per cent of the system will have been lost in the transformation to a new equilibrium state. Ultimately, around the world's coastlines, there are major and difficult decisions to be made about dune management, but the resolution of conflicts of interests lies squarely in the realm of politicians and planners, not engineers and geomorphologists, whose role will be to predict the outcomes of a range of possible management scenarios.

CONCLUSIONS

Both continental and coastal dune systems are important resources for a range of human activities. Appropriate, scientifically informed, management decisions that address issues at the natural environment-human action interface can be practised at a range of spatial and temporal scales. The scientific data available to afford effective management is not always available, but the growth in both micro-scale process studies and investigations of the longer term evolution and dynamics of dune systems is contributing to better-informed management strategies. However, and without even considering the social, political and economic challenges to management, there are three key areas that introduce uncertainty into effective action: integrating the array of scientific data – often collected at different scales – needed for a rigorous understanding of dune system behaviour; matching the scale of management to the scale of system dynamics; and predicting the impact of climate forcing on system behaviour in the decades to come.

REFERENCES

Azizov, A. and Atabaev, B., 1987. Influence of windborne sand on erosion resistance of soils treated with various preparations. *Problems of Desert Development*, **2**, 5–90.

Binns, T., 1990. Is desertification a myth? *Geography*, **75**, 106–113.

Bullard, J.E., Thomas, D.S.G., Livingstone, I. and Wiggs, G.F.S., 1997. Dunefield activity and interactions with climatic variability in the southwest Kalahari Desert. *Earth Surface Processes and Landforms*, **22**, 165–174.

Camp, P.D., 1989. Dune mining at Richards Bay – Zululand. In J.D. Ward, M.K. Seely and A. McLachlan *Dunes '89 Meeting: Abstract and Programme*. Koch Namib Research Foundation, Swakopmund, p. 67.

Carter, R.W.G., Hesp, P.A. and Nordstrom, K.F., 1990a. Erosional landforms in coastal dunes. In *Coastal Dunes*, K.F. Nordstrom, N. Psuty and R.W.G. Carter (Eds). John Wiley, Chichester, pp. 217–250.

Carter, R.W.G., Nordstrom, K.F. and Psuty, N., 1990b. The study of coastal dunes. In *Coastal Dunes*, K.F. Nordstrom, N. Psuty and R.W.G. Carter (Eds). John Wiley, Chichester, pp. 1–16.

Clark, S.S., 1975. The effect of sand mining on the coastal heath vegetation in New South Wales. *Proceedings of the Ecological Society of Australia*, **9**, 1–16.

Davies, J.L., 1973. *Geographical Variation in Coastal Development*. Hafner, New York.

de Kimpe, N.M., Dolan, R. and Hayde, B.P., 1991. Predicted dune recession on the outer banks of North Carolina. *Journal of Coastal Research*, **7**, 53–54.

Dolan, R., 1972. Barrier dune system along the outer banks of North Carolina: a reappraisal. *Science*, **176**, 286–288.

Doody, J.P., 1989. Management for nature conservation. *Proceedings of the Royal Society of Edinburgh*, **96B**, 247–265.

Eastwood, D.A. and Carter, R.W.G., 1981. The Irish dune consumer. *Journal of Leisure Research*, **13**, 273–281.

Gares, P.A. and Nordstrom, K.F., 1991. Coastal dune blowouts: dynamics and management implications. *Proceedings, Coastal Zone '91. American Society of Civil Engineers*, pp. 2851–2862.

Goldsmith, V., 1978. Coastal dunes. In *Coastal Sedimentary Environments*, R.A. Davis Jr (Ed.). Springer-Verlag, New York, pp. 171–225.

Goudie, A.S., 1994. Deserts in a warmer world. In *Environmental Change in Drylands*, A.C. Millington and K. Pye (Eds). John Wiley, Chichester, pp. 1–24.

Greeley, E. and Iversen, J.D., 1985. *Wind as a Geological Process on Earth, Mars, Venus and Titan*. Cambridge University Press, Cambridge.

Hack, J.T., 1941. The dunes of western Navajo country. *Geographical Review*, **31**, 240–362.

Hansom, J.D., 1988. *Coasts*. Cambridge University Press, Cambridge.

Harsh, L.N., Tewari, J.C. and Kumar, S., 1995. Sand dune fixation and afforestation in arid regions. In *Land Degradation and Desertification in Asia and the Pacific Region*, A.K. Sen and A. Kar (Eds). Scientific Publishers, Jodhpur, pp. 171–184.

Heathcote, L., 1996. Settlement advance and retreat: a century of experience on the Eyre Peninsula of South Australia. In *Climate Variability, Climate Change and Social Vulnerability in the Semi-arid Tropics*, J.C. Ribot, R. Magalhaes and S.S. Panagides (Eds). Cambridge University Press, Cambridge, pp. 109–122.

Hellden, U., 1991. Desertification – time for an assessment? *Ambio*, **20**, 372–383.

Hicks, D., 1983. Landscape evolution in consolidated coastal dune sands. *Zeitschrift für Geomorphologie*, Supplementband 45, 245–250.

Illenberger, W.K. and Rust, I.C., 1988. A sand budget for the Alexandria coastal dunefield, South Africa. *Sedimentology*, **35**, 513–521.

Jennings, J.N., 1964. The question of coastal dunes in tropical humid environments. *Zeitschrift für Geomorphologie*, **8**, 150–154.

Jungerius, P.D. and de Jong, J.H., 1989. Variability of water repellence in the dunes along the Dutch coast. *Catena*, **16**, 491–497.

Jungerius, P.D. and Van der Meulen, F., 1989. The development of dune blowouts, as measured with erosion pins and sequential air photographs. *Catena*, **16**, 369–376.

Kana, T.W., Michel, J., Hayes, M.O. and Jensen, J.R., 1984. The physical impact of sea level rise in the area of Charleston, South Carolina. In *Greenhouse Effect and Sea Level Rise: A Challenge for This Generation*, M.C. Barth, J.G. Titus and W.D. Ruckelshaus (Eds). Van Nostrand Reinhold, New York, pp. 105–150.

Kebin, Z. and Kaiguo, Z., 1989. Afforestation for sand fixation in China. *Journal of Arid Environments*, **16**, 3–10.

Koechlin, J., 1997. Ecological conditions and degradation factors in the Sahel. In *Societies and Nature in the Sahel*, C. Raynault (Ed.). Routledge, London, pp. 12–36.

Kumar, S. and Shankarnarayan, K.A., 1988. Aerial seeding on sand dunes: seedling survival and growth. *Journal of Tropical Forestry*, **4**, 124–134.

Lancaster, N., 1988. Development of linear dunes in the south-western Kalahari. *Journal of Arid Environments*, **14**, 233–244.

Leatherman, S.P., 1984. Coastal geomorphic responses to sea level rise: Galveston Bay, Texas. In *Greenhouse Effect and Sea Level Rise: A Challenge for This Generation*, M.C. Barth, J.G. Titus and W.D. Ruckelshaus (Eds). Van Nostrand Reinhold, New York, pp. 151–178.

Livingstone, I. and Warren, A., 1996. *Aeolian Geomorphology: An Introduction*. Longman, London.

Lyners, M.M., Weide, D.L. and Von Till Warren, E., 1980. *Impacts: Damage to Cultural Resources in the California Desert*. US Bureau of Land Management, Riverside CA.

Maarel, van de E., 1979. Environmental management of coastal dunes in the Netherlands. In *Ecological Processes in Coastal Environments*, R.L. Jefferies and A.J. Davy (Eds). Balkema, Rotterdam, pp. 145–158.

Mainguet, M., 1991. *Desertification: Natural Background and Human Mismanagement*. Springer-Verlag, Berlin.

Marston, R.A., 1986. Maneuver-caused wind erosion impacts, South Central New Mexico. In *Aeolian Geomorphology*, W.G. Nickling (Ed.). Allan and Unwin, Boston, pp. 273–290.

Muhs, D.R. and Maat, P.B., 1993. The potential response of eolian sands to greenhouse warming and precipitation reduction on the Great Plains of the USA. *Journal of Arid Environments*, **25**, 351–361.

Nordstrom, K.F., 1988. Dune grading along the region coast, USA: a changing environmental policy. *Applied Geography*, **8**, 101–116.

Nordstrom, K.F., Carter, R.W.G. and Psuty, N.P., 1990. Directions for coastal dune research. In *Coastal Dunes*, N.F. Nordstrom, N. Psuty and R.W.G. Carter (Eds). John Wiley, Chichester, pp. 381–387.

Nordstrom, K.F., 1994. Beaches and dunes of human-altered coasts. *Progress in Physical Geography*, **18**, 497–516.

Orme, A.R., 1990. The instability of Holocene coastal dunes: the case of the Morro dunes, California. In *Coastal Dunes*, N.F. Nordstrom, N. Psuty and R.W.G. Carter (Eds). John Wiley, Chichester, pp. 315–336.

Pilkey, O.H. Sr, Pilkey, W.D., Pilkey, O.H. Jr and Neal, W.J., 1983. *Coastal Design: A Guide for Builders, Planners and Home Owners*. Van Nostrand Reinhold, New York.

Polis, A., 1989. Destruction and conservation of North American sand dunes. In *Dunes '89 Meeting: Abstract and Programme*, J.D. Ward, M.K. Seely and McLachlan (Eds). Koch Namib Research Foundation, Swakopmund, p. 71.

Psuty, N.P., 1989. An application of science to management problems in dunes along the Atlantic coast of the USA. In Coastal Sand Dunes, C.H. Gimmingham, W. Ritchie, B. Willetts and A.J. Willis (Eds). *Proceedings of the Royal Society of Edinburgh*, **B96**, 289–307.

Psuty, N.P., 1990. Foredune mobility and stability, Fire Island, New York. In *Coastal Dunes*, N.F. Nordstrom, N. Psuty and R.W.G. Carter (Eds). John Wiley, Chichester, pp. 159–176.

Psuty, N.P., 1992. Spatial variability in coastal foredune development. In *Coastal Dunes: Geomorphology, Ecology and Management for Conservation*, R.W.G. Carter, T.G.F. Curtis and M.J. Sheehy-Skeffingham (Eds). Balkema, Rotterdam, pp. 3–13.

Pye, K., 1982. Morphological development of coastal dunes in a humid tropical environment, Cape Bedford and Cape Flattery, North Queensland. *Geografiska Annaler*, **A64**, 213–227.

Pye, K., 1990. Physical and human influences on coastal dune development between the Ribble and Mersey estuaries, northwest England. In *Coastal Dunes*, K.F. Nordstrom, N. Psuty and R.W.G. Carter (Eds). John Wiley, Chichester, pp. 339–359.

Pye, K. and Tsoar, H., 1990. *Aeolian Sand and Sand Dunes*. Unwin Hyman, London.

Raynault, C., 1997. The demographic issue in the western Sahel: from the global to the local scale. In *Societies and Nature in the Sahel*, C. Raynault (Ed.). Routledge, London, pp. 37–55.

Reynolds, W.J., 1987. Coastal structures and long term shore migration. In *Coastal Zone 87*. American Society of Civil Engineers, New York, pp. 414–426.

Ritche, W. and Penland, S., 1990. Aeolian sand bodies of the south Louisiana coast. In *Coastal Dunes*, K.F. Nordstrom, N. Psuty and R.W.G. Carter (Eds). John Wiley, Chichester, pp. 105–127.

Sherman, D.J., 1995. Problems of scale in the modeling and interpretation of coastal dunes. *Marine Geology*, **124**, 339–349.

Sherman, D.J., 1997. Coastal dunes of California. In *California's Changing Coastline*, R. Flick (Ed.). University of California Press, Berkeley.

Sherman, D.J. and Bauer, B.O., 1993. Dynamics of beach-dune systems. *Progress in Physical Geography*, **17**, 413–447.

Short, A.D. and Hesp, P.A., 1982. Wave, beach and dune interactions in southeastern Australia. *Marine Geology*, **48**, 259–284.

Sorenen, R.M., Weisman, N. and Lennon, G.P., 1984. Control of erosion, inundation, and salinity intrusion cased by sea level rise. In *Greenhouse Effect and Sea Level Rise: A Challenge for This Generation*, M.C. Barth, J.G. Titus and W.D. Ruckelshaus (Eds). Van Nostrand Reinhold, New York, pp. 179–214.

Stetler, L.D. and Gaylord, D.R., 1996. Evaluating eolian-climatic interactions using a regional climate model from Hanford, Washington (USA). *Geomorphology*, **17**, 99–113.

Stokes, S., Thomas, D.S.G. and Washington, R., 1997. Multiple episodes of aridity in Africa south of the equator since the last interglacial cycle. *Nature*, **388**, 154–158.

Talbot, M.R., 1984. Late Pleistocene rainfall and dune building in the Sahel. *Palaeoecology of Africa*, **16**, 203–214.

Thomas, D.S.G., 1988. The geomorphological role of vegetation in the dune systems of the Kalahari. In *Geomorphological Studies in Southern Africa*, G.F. Darclis and B.P. Moon (Eds). Balkema, Rotterdam, pp. 145–158.

Thomas, D.S.G., 1992. Desert dune activity, concepts and significance. *Journal of Arid Environments*, **22**, 31–38.

Thomas, D.S.G., 1997a. Sand seas and aeolian bedforms. In *Arid Zone Geomorphology: Process, Form and Change in Drylands*, 2nd Edn. John Wiley, Chichester.

Thomas, D.S.G., 1997b. Reconstructing ancient arid environments. In *Arid Zone Geomorphology: Process, Form and Change in Drylands*, 2nd Edn. John Wiley, Chichester.

Thomas, D.S.G. and Middleton, N.J., 1995. *Desertification: Exploding the Myth*. John Wiley, Chichester.

Thomas, D.S.G. and Shaw, P.A., 1991. *The Kalahari Environment*. Cambridge University Press, Cambridge.

Thomas, D.S.G., Stokes, S. and Shaw, P.A., 1997a. Holocene aeolian activity in the southwestern Kalahari Desert, southern Africa: significance and relationships to late Pleistocene dune building events. *The Holocene*, 7, 273–281.

Thomas, D.S.G., Squires, V. and Glenn, E., 1997b. The North American dust bowl and desertification: economic and environmental interactions. In *World Atlas of Desertification*, 2nd Edn, N.J. Middleton and D.S.G. Thomas (Eds). UNEP/Edward Arnold, London, pp. 149–154.

Tsoar, H. and Møller, J.T., 1986. The role of vegetation in the formation of linear sand dunes. In *Aeolian Geomorphology*, W.G. Nickling (Ed.). Allen and Unwin, Boston, pp. 75–95.

Van der Meulen, F., 1990. European dunes: consequences of climate change and sea level rise. In *Dunes of the European Coasts*, Th.W.M. Bakker, P.D. Jungerius and J.A. Klijn (Eds). *Catena* Supplement 18, 209–223.

Van der Meulen, F., 1990. European dunes: consequences of climate change and sea level rise. In *Coastal Dunes*, K.F. Nordstrum, N. Psuty and R.W.G. Carter (Eds). John Wiley, Chichester, pp. 209–223.

Van der Meulen, F. and Jungerius, P.D., 1989. Landscape development in Dutch coastal dunes: the breakdown and restoration of geomorphological and hydrological processes. *Proceedings of the Royal Society of Edinburgh*, **96B**, 219–229.

Wasson, R.J., Rajaguru, S.N., Misra, V.N. *et al.*, 1983. Geomorphology, Late Quaternary stratigraphy and palaeoclimatology of the Thar Desert. *Zeitschrift für Geomorphologie* Supplementband 45, 117–152.

Watson, A., 1985. The control of wind blown sand and moving dunes: a review of methods of sand control in deserts with observations from Saudi Arabia. *Quarterly Journal of Engineering Geology*, **18**, 237–252.

Watson, A., 1990. The control of blowing sand and mobile dunes. In *Techniques for Desert Reclamation*, A.S. Goudie (Ed.) John Wiley, Chichester, pp. 35–85.

Wiggs, G.F.S., Livingstone, I., Thomas D.S.G. and Bullard, J.E., 1994. The effect of vegetation removal on airflow structure and dune mobility in the southwest Kalahari. *Land Degradation and Rehabilitation*, **5**, 13–24.

Wiggs, G.F.S., Livingstone, I., Thomas D.S.G. and Bullard, J.E., 1995. Dune mobility and vegetation cover in the southwest Kalahari Desert. *Earth Surface Processes and Landforms*, **20**, 515–530.

Williams, M.A.J. and Balling, R.R. Jr, 1995. *Interactions of Desertification and Climate*. Edward Arnold, London.

Willis, A.J., Folks, B.F., Hope-Samson, J.F. and Yemm, E.W., 1959. Braughton Burrows: the dune system and its vegetation. Part 1. *Journal of Ecology*, **47**, 1–24.

Wilson, I.G., 1973. Ergs. *Sedimentary Geology*, **10**, 77–106.

Wolfe, S.A. and Nickling, W.G., 1993. The protective role of sparse vegetation in wind erosion. *Progress in Physical Geography*, **17**, 50–68.

Zann, L.P., 1995. *Our Sea, Our Future: Major Findings of the State of the Marine Environment Report for Australia*. Department of the Environment, Townsville.

Zhu Zhenda, Wand Xizhang, Wu Wiei *et al.*, 1992. China: desertification mapping and desert reclamation. In *UNEP* (ed. Middleton N.J. and Thomas, D.S.G.) *World Atlas of Desertification*, 1st Edn, N.J. Middleton and D.S.G. Thomas (Eds). Edward Arnold, London, pp. 46–49.

6 Physics of Aeolian Movement Emphasising Changing of the Aerodynamic Roughness Height by Saltating Grains (the Owen Effect)

DALE A. GILLETTE
Atmospheric Sciences Modeling Division, Air Resources Laboratory, USA

Experimental and theoretical developments that led to Owen's work describing the change of the roughness height during saltation episodes are reviewed. Raupach's formula summarises Owen's theoretical work and extends it to a wider range of surface conditions. Data from an experiment at Owens Lake verified Raupach's formula. The Owen effect is important in explaining the increase of sand flux with distance and in explaining the intensity of dust sources because an increase in friction velocity caused by roughness height increase is nonlinearly related to an increase of sediment flux.

INTRODUCTION

In the last few years developments in the analysis of aeolian movement have been made that must be regarded as great advancements. A few of them (but not all) will be outlined and one of them will be emphasised.

Particle–Particle Interactions

Despite the great progress made by consideration of the fluid forces on particles to explain a large part of aeolian movement, the problem of particle–particle interaction has only recently come into its own. The obvious need for this is seen in the fact that at the end of its trajectory, a saltating particle hits the earth. At that point it: bounces, causes other sand particles to enter into saltating or 'reptating' motion, blasts fine particles that were part of its aggregate nature or that were lying in the zone of its 'crater'. Some or all of these possiblilities occur for each saltating particle. Work on particle–particle interactions is exemplified by Anderson and Haff (1988). McEwan *et al.* (1992) and McEwan and Willetts (1993) have extended models of aeolian particle movements with particle–particle interaction.

Aeolian Environments, Sediments and Landforms. Edited by A. S. Goudie, I. Livingstone and S. Stokes.
Published 1999 by John Wiley & Sons, Ltd.

Momentum Absorption by Non-erodible Elements on the Surface

The absorption of momentum by inert or non-erodible elements on the surface was brought forward for many of us by the pioneering work of Marshall (1971). This work showed that the surface could conceptually be thought of as an erodible flat surface and fixed non-erodible structures seated on the flat surface. The partitioning of how much wind stress was absorbed by the non-erodible elements and how much was left over for erosion of the surface was the great contribution of this paper. Following this paper Gillette and Stockton (1989) used the same approach to express how the threshold friction velocity was increased by the presence of non-erodible hemispheres of varying diameters (artificial gravel and lag deposits). Since the approach was successful for solid particles, Musick and Gillette (1990) applied the same technique to natural vegetation found in desert locations and found that the parameter frontal area divided by floor area gave similar results for both vegetation and gravel/lag deposits. Raupach et al. (1993) generalised this work in a three-parameter formula. Wolfe and Nickling (1996) have advanced this work using the formulation of Raupach et al. (1993). Marticorena and Bergametti (1995) used an alternate approach employing aerodynamic roughness height rather than frontal area of non-erodible elements divided by floor area to express increase of threshold friction velocity. This approach was found to be very useful for modelling dust emissions from the Sahara desert (Marticorena et al., 1997).

Three-dimensional Flow Modelling Around Objects, Especially Dunes

Three-dimensional flow modeling properly is discussed in Chapter 2 and will not be covered here.

Detailed Trajectory Studies of Individual Particles in Aeolian Movement

Studies of trajectories of individual particles in aeolian movement has gone beyond the idealised models of Bagnold (1941) and Owen (1964) largely by the studies of real particle motions in wind tunnels using high speed photography. Such studies have shown the existence of 'reptation' (sand sized particles emitted into the stream at a shallow angle) (Rice et al., 1995), breakage of crusted sediments (Rice et al., 1996a), and particle–particle collisions (Rice et al., 1996b).

Studies of Sandblasting and its Effects on Fine Particle Emissions

Chepil and Woodruff (1963) were two of the first investigators who realised the importance of particle–particle emissions of fine particles and called this phenomenon 'sand blasting'. Gillette (1981) measured vertical fluxes of particles smaller than 10 μm in diameter (F_a) along with the horizontal flux of particles of all sizes (q_{tot}). Since a large portion of q_{tot} is the horizontal saltation flux, q, the ratio F_a/q_{tot} provided experimental data on the sandblasting process. Shao et al. (1993) constructed a model showing that the ratio of F_a/q varied with the mass of the sandblasting particle and inversely as the binding energy of the fine particles emitted. Applying the theory to the data led to a tentative conclusion that the binding energies vary, with the largest binding energies belonging to pure clay aggregates and the smallest binding energies to sandy particles. The theory of Alfaro et al. (1997) used binding energies that vary with fine particle size; such a theory predicts change of fine particle size distribution with wind speed.

Mechanics of Flow Influenced by Aeolian Movement

I would like to discuss in detail the development of the mechanics of flow influenced by aeolian movement by using the rest of this chapter for this subject. I have called this the 'Owen effect' after the important theoretical work of Paul Robert Owen. I would like to present a bit of historical background on the Owen effect as it affects our very young science. Since I have limited space, I would like to recommend the paper of McEwan and Willetts (1993) for further developments of the Owen effect. The story that follows is an illustration that acceptance and application of important fundamental discoveries may take several decades.

IDENTIFICATION OF CHANGE OF AERODYNAMIC ROUGHNESS HEIGHT WITH AEOLIAN MOVEMENT: THE CONTRIBUTIONS OF RALPH BAGNOLD

During some wind events particles with a broad size-range are transported by the wind. These particles have ballistic trajectories including the so-called 'saltation' phenomenon in which the particles proceed mainly in a series of hops. One of Bagnold's greatest contributions to the science of sand movement is the clarification that the movement of sand is coupled to wind stress. He specified that, in fully developed saltation, the wind stress is totally carried at the surface by saltating particles. The wind stress may be written ρu_*^2, where u_* is the friction velocity (the characteristic velocity of the system) and ρ is the density of air. Bagnold (1941) called the horizontal mass flux of saltating sand grains q and specified that the saltating grains hop a distance λ. He observed that the horizontal speed V of the saltating particle divided by the hop length λ has the dimensions of t^{-1} where t is time. The ratio of u_* to g (the acceleration of gravity) also has the dimensions of t^{-1}. Bagnold assumed the ratios V/λ and g/u_* were related by a constant of proportionality:

$$\Gamma V/\lambda = g/u_*$$ (1)

He then obtained his frequently quoted equation that relates the flux of sediment (q) to the cube of the friction velocity.

$$q = \frac{\Gamma \rho}{g} u_*^3$$ (2)

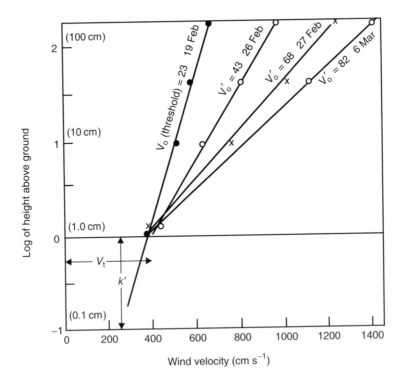

Figure 6.1 Bagnold's empirical solution for the change of aerodynamic roughness height z_0 by saltating grains (after Bagnold, 1946)

This equation has been derived in many ways and with more refined assumptions, but it remains one of the strongest structures of the aeolian scientific framework.

Bagnold also discovered that not only does the wind affect sand particles by moving them but also movement of sand particles changes the nature of the wind. Bagnold's explanation for these observations was a largely empirical one. His measurements of boundary layer wind were interpreted to show a 'focus point' at threshold velocity V_t at a height k' to which 'wind velocity rays converge during sand-driving'. Thus, for any strength of wind during a saltation episode, the wind profile up to height k' was said always to be the same; above k', the slope of the wind profile changed with the strength of the wind. Above that focus point at height k', the wind increased logarithmically with height (see Figure 6.1). In Figure 6.1, k' is equal to about 1 cm and V_t is equal to about 380 cm s^{-1}.

P.R. Owen's Theory

Eighteen years later, Paul Robert Owen (1964) described his interest in the physics of aeolian movement as follows: 'A vivid and detailed account of the phenomenon is given by Bagnold (1941) in his book *The Physics of Blown Sands and Desert Dunes*; indeed, it is largely to this delightful work that I owe my interest in the subject. But there are two questions which the book imperfectly answers. What is the effect of the saltation on the airflow at large distance from the surface? What determines the concentration of particles engaging in the saltation?'

Owen put forward two hypotheses upon which his theory was built: '(i) the saltation layer behaves, so far as the flow outside it is concerned, as an aerodynamic roughness whose height is proportional to the thickness of the layer; (ii) the concentration of particles within the saltation layer is governed by the condition that the shearing stress borne by the fluid falls, as the surface is approached, to a value just sufficient to ensure that the surface grains are in a mobile state'. Physically, the process that Owen envisioned is represented in Figure 6.2.

Before saltation begins (Figure 6.2(a)), the fluid transports momentum by turbulent eddy transfer. During saltation (Figure 6.2(b)), the sand grains interact with the fluid and transfer a part of the wind momentum to the ground. This occurs because the saltating particles strike the ground in a parabolic trajectory (implicit in the definition of saltation). These particles carry momentum absorbed at heights near the tops of their trajectories. This momentum transport is more efficient than that by air eddy transport. Consequently, momentum from greater heights in the air layer is required to replace the particle-transported momentum. This sequence of momentum transport is highly similar to that caused by an increase of aerodynamic roughness height of the surface (z_0). The result is the formation of an internal boundary layer shown in Figures 6.2(b) and (c). The increased u_* caused by saltating particles grows upward in the wind profile toward the height of the pre-existing boundary layer height. The Owen theory specifies an increased z_0 that will match the effect on the wind profile of the saltating particles. Owen (1964) expressed the profile as

$$u(z)/u_* = 2.5 \ln(2gz/u_*^2) + 9.7$$

so that z_0 is identified with the dimensional group – velocity squared divided by gravitational acceleration:

$$z_0 = \frac{u_*^2}{\gamma g} \tag{3}$$

where γ is 95. Indeed, the form of the eqn (3) is identical to Charnock's formula (eqn (4)) as quoted by Hicks (1972) for aerodynamic roughness height observations at sea related to friction velocity:

$$z_0 = \frac{\alpha u_*^2}{g} \tag{4}$$

The value of α in eqn (4) of 0.016 is very close to $1/\gamma = 0.0105$.

M.R. Raupach's Formula for the Owen Effect

Raupach (1991) incorporated the theory of Owen (1964) to provide an expression to compute the aerodynamic roughness height of smooth or rough surfaces when a

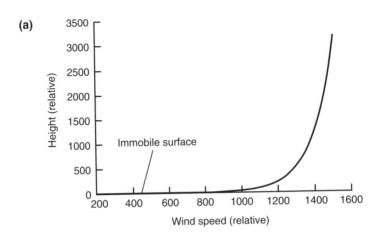

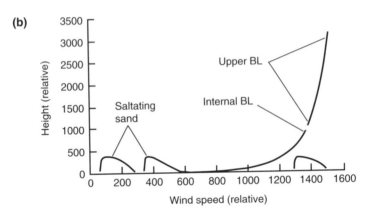

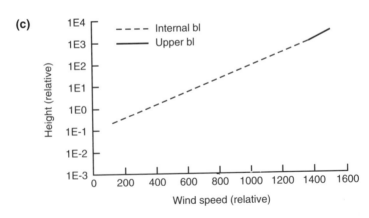

Figure 6.2 (a) Schematic plot of a wind profile without saltating particles. (b) Schematic plot of a wind profile with saltating particles. (c) Schematic plot of a wind profile with saltating particles where the wind speed is plotted logarithmically

saltation layer is developed. Unlike Owen's formula, however, the expression for the aerodynamic roughness height, when saltation occurred z_{0salt} is applicable to smooth or rough surfaces. The expression for z_{0salt} is equal to aerodynamic roughness height for wind speed below the threshold of particle movement, z_{0NS}. For this condition, it is roughly equal to Owen's aerodynamic roughness height for saltation on smooth surfaces. The main advantage of Raupach's formula is that it provides values for the aerodynamic roughness height for situations that Owen's formula could not, i.e. for rough surfaces that have particles moving in saltation. The Raupach formula provides z_{0salt} for a wider range of potentially eroding surfaces than Owen's theory does.

Eqn (5a) relates wind speed to friction wind velocity as

$$U(z) = \frac{u_*}{k} \ln \left(\frac{z}{z_0} \right) \tag{5a}$$

where $U(z)$ is wind speed at height z, u_* is friction velocity, k is von Karman's constant (0.4) and z_0 is aerodynamic roughness height (Kaimal and Finnigan, 1994). For conditions of no erosion, we may express the change of wind speed with height as eqn (5b)

$$U(z) = \frac{u_*}{k} \ln \left(\frac{z}{z_{0NS}} \right) \tag{5b}$$

where z_{0NS} is aerodynamic roughness height for *non-saltating* conditions.

For conditions when wind erosion is occurring, the aerodynamic roughness height during saltation z_{0salt} is given by the Raupach formula as:

$$z_{0salt} = \left[A \frac{u_*^2}{2g} \right]^{1-R} z_{0NS}^R \tag{6}$$

Here A is a constant of order 1; $R = u_{*t}/u_*$, where u_{*t} is the threshold friction velocity for wind erosion; and g is the acceleration of gravity. Thus, three quantities are required for the evaluation of the Raupach formula: the threshold friction velocity u_{*t}, the friction velocity u_*, and the aerodynamic roughness height z_{0NS}.

Verification of Raupach's Formula

The observational studies that verify the Raupach formula need to be field observations because wind tunnels were criticised by Raupach (1991) as having insufficient length to develop equilibrium saltation layers. Field measurements by Rasmussen and Mikkellsen (1991) described the aerodynamic roughness height of beach surfaces during wind erosion. Unfortunately z_0 values for the initial noneroding surfaces were not measured. Gillette *et al.* (1997) presented data for more than a year taken from three simultaneously operating towers towers on erodible surface, each

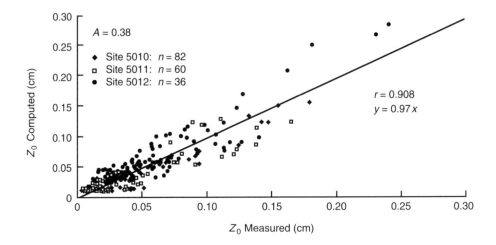

Figure 6.3 Plot of the saltation roughness heights z_0 computed using Raupach's formula [eqn (6)] vs the apparent roughness height z_0 measured for wind friction velocity (after Gillette et al., 1998)

having a different surface roughness. A sufficient volume of data was obtained so that conditions near neutral stratification (−0.03 < Ri < 0.03) and constant wind direction could be maintained.

The threshold friction velocities of Gillette et al. (1997) were the lowest observed friction velocities at which particle movements were measured (by Sensit instrument responses). Aerodynamic roughness height z_0 was computed for the data for each 20-min averaging period using eqn (5a). Aerodynamic roughness heights z_{0NS} for noneroding conditions were averages of the z_0 values for nonsaltating conditions. For assurance that cases where saltation occurred were excluded, an upper limit for u_* equal to 38 cm s^{-1} was selected which was below the lowest threshold velocity. The lower limit for u_* was 20 cm s^{-1} because low gradients for low wind speeds led to unacceptable errors in friction velocity estimations. For the three sites, arithmetic mean aerodynamic roughness heights and standard deviations in parentheses were calculated to be 0.01 (±0.009), 0.01 (±0.009) and 0.025 (±0.0l) cm. These roughness heights were in relative qualitative agreement with observations of the surface for the sites.

The values for u_*, z_{0NS}, and u_{*t} were used to calculate z_{0salt} in eqn (6) to compare with the z_0 values calculated from eqn (5a) using the experimental data for $u_* > u_{*t}$. Figure 6.3 shows a comparison of the measured z_0 and the computed z_{0salt} values when the A parameter of eqn (6) is set to 0.38. The value of A was set to 0.38 as the best fit to the data. Since Raupach had originally predicted the value of A to be 0.2 to 0.3, the Raupach formula is very consistent with observations. The slope of the regression is close to 1 (0.97), suggesting good agreement between measured and computed aerodynamic roughness heights for a range of z_0 between 0.01 and 0.3 cm. The dispersion is relatively low, and r^2 is 0.82 for 278 pairs of data.

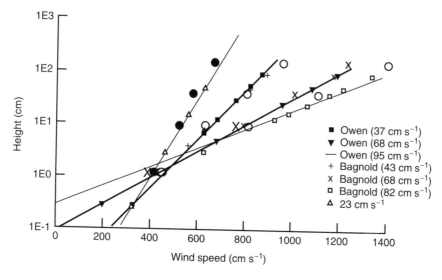

Figure 6.4   Comparison of Bagnold and Owen wind profiles

Comparison of the Bagnold's Solution with Raupach's Formula of the Owen Effect

A comparison of the Bagnold solution of the form of wind velocity profiles and the Raupach formula is given in Figure 6.4. The fittings for the data of Figure 6.4 are calculated by two methods: Raupach's formula is used by taking arbitrary values for u_*, calculating z_0 using eqn (6) and then using eqn (5a) to calculate $U(z)$. The interpretation of Bagnold (1941) shows his fitting of experimental data to the focus point at 1 cm and 380 cm s^{-1}. The parameters are set at $u_{*t} = 23$ cm s^{-1}, $A = 0.38$, and $z_{0NS} = 0.001$. Friction velocities for the Owen effect profiles were chosen to be to 37, 68, and 95 cm s^{-1}. The friction velocities for the Bagnold profiles were 43, 68 and 82 cm s^{-1}. As seen in Figure 6.4, Owen Effect profiles are practically identical to the Bagnold profiles for height greater than 1 cm. This comparison clearly shows that the Owen theory reproduces the experimental results as well as Bagnold's does, but it eliminates three constants needed to specify a wind profile. If Occam's razor were used, Owen's theory would be chosen since it has the smaller number of independent assumptions.

Consequences of the Owen Effect

Increase of Friction Velocity with Saltation Versus Without Saltation

The Owen effect is most easily detected as an increase of the ratio of friction velocity u_* to mean wind U. (See Figure 6.5, where u_* is plotted vs U at 1-m height for one of the three locations at Owens Lake described in the last section.) The dash–dot curve is a regression of the data. Without the Owen effect, the points would fall around a

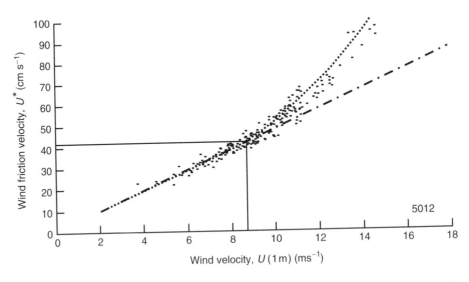

Figure 6.5 Plot of friction velocity vs mean wind at 1 m. Increase of the friction velocity above the linear relationship with the mean wind at 1 m is a result of the Owen effect (after Gillette et al., 1998)

straight line (corresponding to a constant drag coefficient). The curved portion of the curve shows the Owen effect where saltating particles effectively increase the aerodynamic roughness height z_0 of the surface.

Above the erosion threshold, the saltation wind friction velocity can be written in the form

$$\Delta u_* = u_{*salt} - u_{*NS} \tag{7}$$

Here Δu_* is a function of $(U - U_t)^2$

$$\Delta u_* = \chi(U - U_t)^2 \tag{8}$$

Gillette *et al.* (1998) showed that χ decreases slowly from $z_{0NS} = 0.005$ to 0.02 cm and then increases more rapidly from $z_{0NS} = 0.02$ to 0.1 cm. The values range from 0.00333 to 0.00282, which represents a maximal difference of 15 per cent. Consequently, this coefficient can be considered constant over the whole range of z_0 and equal to 0.003.

Thus, the increase of wind friction velocity caused by saltation of soil particles can be simply described by eqn (8) where χ is equal to 0.3. This increase of wind friction velocity results in an increase of sediment flux since it is roughly related to sediment flux as in eqn (1), of Bagnold. This expression gives results similar to the interactive computation using Raupach's formula of the saltation aerodynamic roughness height z_0. Because it provides a direct relation between the saltation wind friction velocity and parameters that can be easily measured (the wind velocity and initial roughness height in nonsaltating conditions), it is a convenient tool for modelling of the Owen effect.

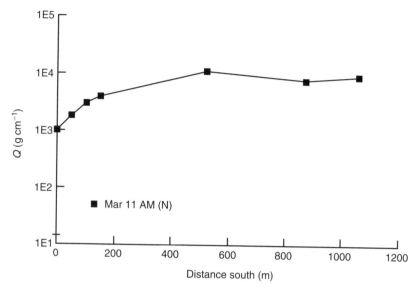

Figure 6.6 Integrated soil mass fluxes versus distance south of the zero point for $Q = \int q\,dt$ vs distance south of a point on the floor of Owens Lake. Data were obtained 11 March 1993

Explanation of the Fetch Effect

The fetch effect is an increase of sand flux with distance downwind. Field studies of wind erosion at Owens Lake (Gillette *et al.*, 1996) have provided statistically significant measurements of increases of friction velocities with distance caused by increase of aerodynamic roughness height. Two sets of Q (equal to $\int qdt$) are shown in Figure 6.6. These data represent north winds (denoted by N in Figure 6.6) that were experienced on 11 March 1993. Before 0500 on March 11, the entire area was crusted and no wind erosion was recorded. After 0500, the crust broke and was pulverised back to a distance of 503 m north of the first measurment station. During this storm winds (almost due North) were parallel to the direction of the measurement stations. The distance between each of the first four measurement stations was 50 m downwind and the distance x shown on Figure 6.6 of the first station was arbitrarily set to 0 m. The steepest gradients of Q vs x occurred immediately following the breakup of the crust when the surface was in its most aggregated condition. For x greater than 150 m, very little gradient of Q with x is seen. Therefore, the data for $0\,\mathrm{m} < x < 150\,\mathrm{m}$ were of primary interest for this analysis.

Figure 6.7 shows the estimated aerodynamic roughness heights for $x < 520$ m along with the 90th percentile confidence interval for the 150 m downwind location. Only the upper limit of the confidence interval is shown, the lower limit being zero. During a time interval from about 0930 to 1100 LST, z_0 is higher at every measurement location than z_0 at 0800 or 1200 LST. For all five locations the aerodynamic

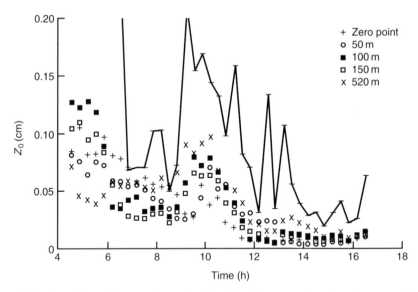

Figure 6.7 Aerodynamic roughness height for five sampling locations vs time on 11 March 1993 at Owens Lake. The solid line indicates the 90th percentile confidence interval upper limit for the 150 m location. The lower limit of the 90th percentile confidence interval is zero (after Gillette *et al.*, 1996)

roughness height increased during higher wind friction velocities compared with aerodynamic roughness heights for lower wind friction velocities. Furthermore, although not significant, the data suggest larger z_0 for downstream locations compared with upstream locations, in agreement with Owen's theory.

Figure 6.8 shows the friction velocities for $x < 520$ m during the storm of 11 March 1993 and the 90th percentile confidence intervals for the 150 m value. At the most intense period of the storm (shortly before 1000 LST when all the sampling locations showed an increased aerodynamic roughness height), the friction velocity at 100 and 150 m downwind appear higher than those at the zero point and at 50 m. To distinguish a 90 per cent significant difference between the friction velocities at the zero point and 150 m locations, the difference would need to exceed the sum of 90% confidence intervals for both locations. Indeed, individual pairs of u_* measurements showed significant differences for three 20-min intervals during the time interval of maximum wind speeds (0900–1100 LST). One of these three 20-min intervals was analysed for the time interval centred at 0930 for friction velocity. The data showed that friction velocity was significantly larger at 150 m compared with that at 0 m and suggests that friction velocity was larger at 100 m compared with that 0 or 50 m (but not significantly). A significant friction velocity increase with downwind distance (fetch) coupled with a significant z_0 increase with friction velocity time (1000 compared with 0800 or 1200 LST) provides support for Owen's theory of the fetch effect. The friction velocity data do not show a significant difference after 1200 LST when wind speeds were lower and aerodynamic roughness heights were more-or-less uniform for all the measurement locations. Data taken before 0800 shows the effect

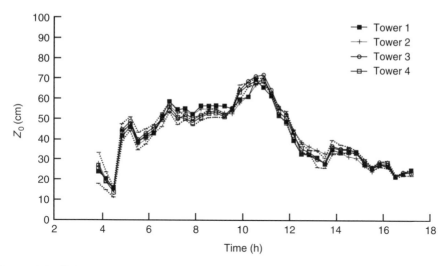

Figure 6.8 Friction velocities for five sampling locations versus time on 11 March 1993. The solid lines indicate the 90th percentile confidence interval for the 150 m location (after Gillette *et al.*, 1996)

of the rough crusted surface and its destruction in the first few hours of the storm. The most crusted (farthest North) have larger aerodynamic roughness heights and friction velocities than at the less crusted 520 m location.

Because both z_0 and u_* increase with distance downwind, and mass flux is a strong function of the u_* [see eqn (2)] the Owen effect offers a strong explanation for the fetch effect.

CONCLUSIONS

The story of the Owen effect told above shows a leisurely advance in our knowledge of aeolian processes including striking insights by Bagnold and Owen, generalisations of formulae by Raupach, and experimental verification by Gillette *et al.* The time scale of this unfoldment is about 50 years.

REFERENCES

Alfaro, S., Gaudichet, A., Gomes, L. and Maille, M., 1997. Modeling the size distribution of a soil aerosol produced by sandblasting. *Journal of Geophysical Research*, **102**, 11239–11249.

Anderson, R.D. and Haff, P.K. 1988. Wind modification and bed response during saltation of sand in air. *Acta Mechanica, Suppl.*, 21–51.

Bagnold, R.A., 1941. *The Physics of Blown Sand and Desert Dunes*. Methuen & Co., London.

Chepil, W. and Woodruff, N. 1963. The physics of wind erosion and its control. In *Advances in Agronomy*, A. Norman (Ed.). Academic Press, New York.

Gillette, D.A., 1981. Production of dust that may be carried great distances. In *Desert Dust: Origin, Characteristics and Effect on Man*, T. Pewé (Ed.). Geological Society of America, Boulder, Colorado, pp. 11–26.

Gillette, D.A. and Stockton, P., 1989. The effect of non-erodible particles on wind erosion of erodible surfaces. *Journal of Geophysical Research*, **94**, 12885–12893.

Gillette, D.A., Herbert, G., Stockton, P.H. and Owen, P.R., 1996. Causes of the fetch effect in wind erosion. *Earth Surface Processes and Landforms*, **21**, 641–659.

Gillette, D.A., Marticorena, B. and Bergametti, G., 1998. Change in the aerodynamic roughness height by saltating grains: Experimental assessment, test of theory and operational parameterization. *Journal of Geophysical Research*, **103**, 6203–6209.

Gillette, D., Hardebeck, E. and Parker, J., 1997. Large-scale variability of wind erosion mass flux rates at Owens Lake: The role of roughness change, particle limitation, change of threshold friction velocity, and the Owen effect. *Journal of Geophysical Research*, **25**, 25989–25998.

Hicks, B.B., 1972. Some evaluations of drag and bulk transfer coefficients over water bodies of different sizes. *Boundary Layer Meteorology*, **3**, 201–213.

Kaimal, J.C. and Finnigan, J.J., 1994. *Atmospheric Boundary Layer Flows, Their Structure and Measurement*. Oxford University Press, New York, pp. 14–19.

Marshall, J., 1971. Drag measurements in roughness arrays of varying density and distribution. *Agricultural Meteorology*, **8**, 269–292.

Marticorena, B. and Bergametti, G., 1995. Modeling the atmospheric dust cycle, I., Design of a soil derived dust emission scheme. *Journal of Geophysical Research*, **100**, 16415–16430.

Marticorena, B., Bergametti, G., Aumont, B. *et al.*, 1987. Modeling the atmospheric dust cycle, 2, Simulation of Saharan sources. *Journal of Geophysical Research*, **102**, 4387–4404.

McEwan, I.K. and Willetts, B.B., 1993. Adaptation of the near-surface wind to the delelopment of sand transport. *Journal of Fluid Mechanics*, **252**, 99–115.

McEwan, I., Willetts, B. and Rice, M., 1992. The grain/bed collision in sand transport by wind. *Sedimentology*, **39**, 971–981.

Musick, H.B. and Gillette, D.A., 1990. Field evaluation of relationships between a vegetation structural parameter and sheltering against wind erosion. *Land Degradation and Rehabilitation*, **2**, 87–94.

Owen, P.R., 1964. Saltation of uniform sand grains in air. *Journal of Fluid Mechanics*, **20**, 225–242.

Rasmussen, K. and Mikkellsen, H., 1991. Wind tunnel observations of aeolian transport rates. *Acta Mechanics, I* (Suppl.), 135–144.

Raupach, M.R., 1991. Saltation layers, vegetation canopies and roughness lengths. *Acta Mechanica, I* (Suppl.), 83–96.

Raupach, M.R., Gillette, D. and Leys, J., 1993. The effect of roughness elements on wind erosion threshold. *Journal of Geophysical Research*, **98**, 3023–3029.

Rice, M., Willetts, B. and McEwan, I., 1995. An experimental study of multiple grain-size ejecta produced by collisions of saltating grains with a flat bed. *Sedimentology*, **42**, 695–706.

Rice, M., Willetts, B. and McEwan, I., 1996a. Wind erosion of crusted soil sediments. *Earth Surface Processes and Landforms*, **21**, 279–293.

Rice, M., Willetts, B. and McEwan, I., 1996b. Observations of collisions of saltating grains with a granular bed from high-speed cine-film. *Sedimentology*, **43**, 21–31.

Shao, Y., Raupach, M. and Findlater, P., 1993. Effect of saltation bombardment on the entrainment of dust by wind. *Journal of Geophysical Research*, **98**, 12719–12726.

Wolfe and Nickling (1996). Shear stress partitioning in sparse vegetated desert canopies. *Earth Surface Processes and Landforms*, **21**, 607–619.

7 Wind Erosion on Agricultural Land

JOHN LEYS

Centre for Natural Resources, Department of Land and Water
Conservation, Gunnedah, Australia

Wind erosion on agricultural land is an important issue because of on- and off-site effects. On-site effects such as soil degradation, capital infrastructure destruction and reduction in the agricultural productivity all impact on the landholder, while off-site effects such as air and water pollution impact on the rural and urban communities beyond the farm boundary. However, quantifying the magnitude, extent and impact of wind erosion is extremely difficult on agricultural land because of the interaction between soil texture, vegetation, soil moisture, slope and land management activities. Recent research in Australia has used numerical models and a geographic information system to predict the location, magnitude and temporal nature of erosion. Other research has quantified soil and vegetation factors that control wind erosion in southeastern Australia and identified threshold levels for soil texture, clay content, dry aggregation and vegetation cover required for effective erosion protection. To demonstrate the impact of wind erosion on the soil, field wind tunnel and field scale experiments have indicated how wind erosion subtly winnows the finer fractions from the soil leaving the coarser lag material behind. Off-site, research into the emission of dust and particle-associated pesticide has shown how dust is a pathway for environmental contaminants from agricultural operations to sensitive riverine environments. Despite these recent advancements, significant areas of fundamental research are required to cope with the spatial and temporal components of wind erosion processes.

INTRODUCTION

This chapter will discuss topics related to wind erosion, defined here as the movement of sediment by wind and the resultant wearing down or abrasion of the underlying surface. There are a considerable number of books and chapters which discuss and review wind erosion (Bagnold, 1941; Greeley and Iversen, 1985; Cooke et al., 1993; McTainsh and Leys, 1993; Nickling, 1994). These publications review the topic in a detailed and concise manner, so rather than repeat what has already been done, this chapter will focus on recent developments in wind erosion, particularly those from the southern hemisphere. The majority of publications on wind erosion are from authors living and working in the northern hemisphere, so this chapter will help balance the information available by presenting research and concepts from Australia, a country with a wide range of wind erosion issues, although not always at the scale and intensity of other places in the world (Pye, 1987; Middleton, 1997).

Other chapters in this book discuss the physics and processes of sediment initiation, transport and deposition and the bedforms and landforms caused by the

Aeolian Environments, Sediments and Landforms. Edited by A.S. Goudie, I. Livingstone and S. Stokes.
© 1999 John Wiley & Sons, Ltd.

Figure 7.1 Dust storm over Mildura on the Murray River, Australia. Dust storms move huge quantities of soil and can carry contaminates like nutrients, herbicides and pesticides associated with the dust

erosion and deposition of aeolian sediments. This chapter will focus on wind erosion processes, control methods, and the on- and off-site effects of wind erosion associated with agricultural land. The discussion of control factors will focus on those factors that land managers can manipulate. It will not cover the human factor which is such an important variable in the control of wind erosion (Baidu-Forson and Napier, 1998).

Wind erosion is a natural erosion process which is responsible for evolution of many landscapes. On desert and coastal landscapes wind erosion is a natural feature; however, agricultural lands are predominantly stable and wind erosion is considered as undesirable because of the negative impacts on agricultural production and associated environmental pollution (Figure 7.1). The intensity and magnitude of wind erosion is largely increased by human activity such as cultivation (Clausnitzer and Singer, 1996) vehicle movement on unsealed roads (Nicholson *et al.*, 1989) and inappropriate land management systems (Leys, 1990).

This chapter is divided into four sections. The first section discusses issues that make the study of wind erosion on agricultural land separate from those on coastal or desert dunes. The second section discusses some of the major methods to control wind erosion and the factors that land managers can manipulate to control erosion. The third section describes some of the on-site impacts of not controlling wind erosion and the fourth section describes one of the more important off-site impacts of wind erosion. The chapter concludes with a discussion of some of the challenges and research directions that face agricultural wind erosion workers.

WIND EROSION ON AGRICULTURAL LANDS

This section discusses the extent of wind erosion and the modelling efforts that have been undertaken to describe the processes that control erosion, and how such modelling will help in the management of wind erosion control.

Extent of Wind Erosion

Generally, the incidence of wind erosion increases as the rainfall decreases (Pye, 1987) (McTainsh and Leys, 1993) and the texture of the soil becomes increasingly sandy (Leys *et al.*, 1996). Semi-arid lands used for both cropping and grazing are the most prone to erosion because of the high evaporation to rainfall ratio, which results in a dry soil surface and low vegetation production compared with more humid areas. The use of fallowing methods that keep the ground weed free for 10–12 months prior to sowing with the aim of storing additional soil moisture and mineralising soil nitrogen (Sims, 1977) is a major cause of wind erosion (Leys, 1990). In recent years in Australia, there have been large areas of grazing land converted to dryland cereal production due to increased economic returns from cereal grain production compared to livestock (Leys *et al.*, 1994). As world populations continue to grow, many of the semi-arid grazing lands of the world will come under increased agricultural pressure to produce more food and fibre, and wind erosion is likely to increase unless land management systems that maintain soil cover and encourage soil health are implemented. As a consequence, there will be an increased need to demonstrate to land managers and administrators the magnitude of the problem, the impact that wind erosion has both on- and off-site and which land management practices are most effective in reducing wind erosion. To do this, it will be necessary to map the areas which experience wind erosion, quantify the magnitude of the erosion and suggest remedial measures.

The extent of wind erosion is difficult to determine and it is even more difficult to describe the distribution on agricultural land. Maps of dust sources (Pye, 1987, Fig. 4.1) tend to be interpretative maps. Such maps tend to be based on the assumption that if the area is: (1) semi-arid or arid; and (2) the area or region is know to have dust storms on a semi-regular basis; then the area is a dust source region.

Some maps that are based on desert land forms, such as those in Cooke *et al.* (1993, Fig. 16.1) and use the location of dune fields and other aeolian landforms to infer where wind erosion might occur. Other maps used dust storm frequency (McTainsh *et al.*, 1990) to infer where wind erosion is most active. However, such approaches for describing the distribution of wind erosion are not totally appropriate. In the case of the map McTainsh *et al.* (1990) (Figure 7.2) the map fails to identify the dust source areas that lie in the agricultural areas of the 'wheat/sheep' belt which runs down the southeastern coast line from Brisbane to Adelaide. The wheat/sheep belt is a wind erosion region because the predominant agricultural systems fail to cope with the regular dry periods that this region experiences. Therefore, once vegetation cover is removed and the soils are either over-cultivated or excessively trampled by stock, they blow and create major dust storms. So the reason such areas do not get mapped is because they are 'management induced' erosion areas. In fact, many arid areas of Australia produce much less dust than higher rainfall agricultural areas. The reasons for why an agricultural area begins to produce dust are varied, and it is this complexity that makes the understanding and controlling wind erosion a challenge.

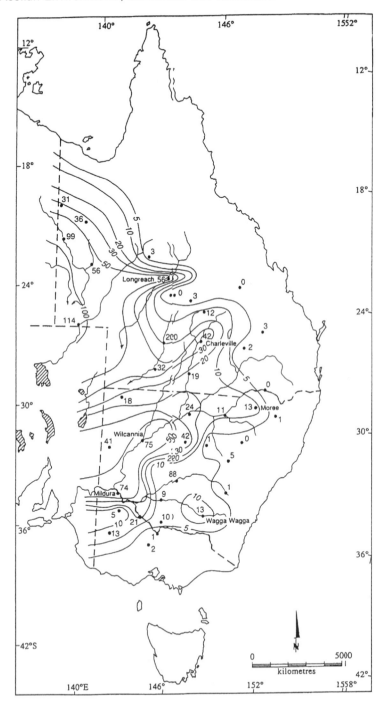

Figure 7.2 Map of dust emission zones in Australia (after McTainsh *et al.*, 1990).
Reproduced by permission of CSIRO Publishing

Modelling

The model development and calibration process enables the limits of knowledge on wind erosion processes to be examined and creates the opportunity to investigate the impacts of wind erosion and determine the spatial extent and magnitude of the process.

Wind erosion processes on agricultural land are generally more complex than for desert or coastal sand dunes because of: (a) the wider soil texture range; (b) the temporal and spatial variation in soil properties, such as aggregation, crusts and soil moisture content; (c) temporal and spatial variation in vegetation, such as growth and decomposition of annual crops and pastures; and (d) the irregular modification of the surface by land management practices such as tillage and grazing. However, there are also similarities with desert and coastal sands and it is for this reason recent research into the aeolian processes on agricultural land has taken an inter-disciplinary approach using conceptual and numerical models from physicists, engineers and geomorphologists, each of whom have each contributed different outlooks to the problem of sediment entrainment, transport and deposition.

Quantitative wind erosion studies have their origins in the classic work of Bagnold (1941) whose detailed experiments and theories influences most aspects of aeolian research. The extensive works of W.S. Chepil (summarised in Chepil and Woodruff, 1963) and colleagues from the United States Department of Agriculture, Wind Erosion Research Unit also provide a valuable contribution. However, this research was largely empirical, thereby making it relatively site specific. These empirical studies culminated in the wind erosion equation (WEQ) (Woodruff and Siddoway, 1965).

The original WEQ used long (annual) averages to estimate annual average soil loss. For estimates over shorter periods, the WEQ has been modified (Bondy et al., 1980; Cole et al., 1983). More recent revisions of the WEQ have led to the Revised Wind Erosion Equation (RWEQ) which includes input parameters such as planting date, tillage method and amount of residue from the previous crop, and a weather generator is then used to predict future erosion (Comis and Gerrietts, 1994). The empirical nature of the WEQ limits its transferability from the Central Great Plains of the USA, for which it was originally developed. Also, the complex interactions between the variables controlling wind erosion are not fully accounted for in the empirical WEQ and RWEQ. While the WEQ is probably the most widely used wind erosion equation, it is acknowledged that it has major limitations (Nickling, 1985). As a result, the Wind Erosion Research Unit is developing a more physically based model called the Wind Erosion Prediction System (WEPS). WEPS includes sub-models for weather generation, crop growth, decomposition, soil, hydrology, tillage and erosion (Hagen, 1991; Comis and Gerrietts, 1994); however, at the time of writing, the model had not been released.

Several other workers have developed empirical, semi-empirical and numerical models to address agricultural wind erosion issues. The WEQ has been modified to estimate the PM_{10} emissions from the Northwest Columbia Plateau (Stetler and Saxton, 1996), and to estimate changes in agricultural production (Houldsworth, 1995). However, the limitations of these empirical models have resulted in other researchers pursuing process-based numerical models.

Physical models of sediment mass transport (q) are based on the cubic relationship between friction velocity (u_*) as identified by Bagnold (1941)

$$qg/\rho u_*^3 \tag{1}$$

where g is acceleration due to gravity and ρ is air density. Equation (1) has formed the basis for many different sediment transport equations that have been derived to fit experimental data, some of which are tabulated in Greeley and Iversen (1985, p. 100). Others returned to the physical processes of saltating grains by modelling the saltation subprocesses of grain trajectories, wind velocity profile modifications and grain bed impacts (Anderson and Haff, 1988; Werner, 1990; Ungar and Haff, 1987; Anderson and Haff, 1991) and how such process emit dust from the bed (Shao *et al.*, 1993; Berkofsky and McEwan, 1994). Analytical expressions of how the sediment flux varies downwind, have also been developed and calibrated (Stout, 1990). However, these sediment transport equations are limited to bare, level, loose sands. While these conditions occur on desert dunes and beach foreshores, they occur far less on agricultural areas.

To overcome this limitation, researchers have added other factors that account for the conditions in an agricultural field. Examples of such models include the Texas Tech Wind Erosion Equation (Gregory, 1986; Gregory, 1991) which includes factors such as: the threshold friction velocity; the fraction of wind energy transferred to the soil surface or the stress partition; the length of the eroding surface, to account for avalanching effects; the erodibility of the soil, as described by the crushing energy (Hagen *et al.*, 1988); and an abrasion factor of the soil (Gregory, 1986). A full discussion can be found in Gregory *et al.* (1994). Another physical model was that of Gillette and Passi (1988), which was used to estimate dust emissions caused by wind erosion across the USA (Gillette and Passi, 1988). This model used a theoretical function between wind speed and dust flux developed by Owen. The threshold wind speeds were determined from a large data base of measured threshold friction velocities and were used to account for the effects of soil texture, live and dead vegetation, surface roughness and soils moisture. Surface roughness was included by using an empirical relationship (Armbrust *et al.*, 1964). The effect of field length was acknowledged as an input to the model but was ignored in the analysis. The model also includes a calibration constant.

There are other models: such as the Guelph Wind Erosion Assessment and Management Model (GWEAM) (Nickling and Satterfield-Hill, 1992), the Soil-Derived Dust Emission Scheme (SDDEM) (Marticorena and Bergametti, 1995) and the Wind Erosion Assessment Model (WEAM) (Shao *et al.*, 1996). These last two models are excellent examples of the synthesis of existing theory, numerical models and experimental data published in the aeolian research community literature.

The Wind Erosion Assessment Model (WEAM) is a physically based numerical model using a combination of established and recently developed wind erosion theory. The model is fully described in Shao *et al.* (1996). In brief, the model aims to account for the combined effect of the four main interacting processes that govern wind erosion: climate, soil, vegetation and land use (Shao and Leys, 1998). From the physics viewpoint, wind erosion is a result of two opposing forces: the capacity of the wind to initiate and sustain erosion, and the ability of the soil to resist

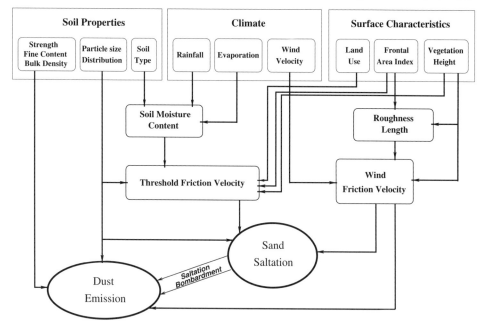

Figure 7.3   Schematic diagram of the Wind Erosion Assessment Model (after Shao *et al.*, 1994). Reproduced by permission of the Australian Association of Natural Resource Management

erosion. The physical quantity used to cause and sustain erosion is the friction velocity u_*, which represents the wind shear or drag on the soil. The opposing quantity offered by the soil is the threshold friction velocity u_{*t}, which defines the minimum friction velocity required for erosion to occur. u_* is determined by wind flow conditions and the surface roughness, while u_{*t} is determined by surface factors such as soil texture, aggregation and moisture. This is the fundamental concept underpinning WEAM and is shown diagrammatically in Figure 7.3.

In WEAM, saltation is quantified using the Owen sand transport equation (Owen, 1964); dust flux is quantified using Australian wind tunnel results and theory (Shao *et al.*, 1993); the threshold friction velocity of the particle sizes present are quantified using wind tunnel results and theory (Greeley and Iversen, 1985); the influence of soil moisture content is characterised from an empirical relationship derived from Australian wind tunnel results (Shao *et al.*, 1996); the influence of soil vegetative cover is based on the on stress partition theory (Raupach, 1992) and previously published wind tunnel results (Raupach *et al.*, 1993). WEAM has not been developed to the point whereby submodels for crop growth, decomposition, hydrology, and tillage have been incorporated, however a sub-model for weather conditions has been completed which provides friction velocities, rainfall, and soil moisture (Shao *et al.*, 1996). The real advantage of WEAM over existing models is its coupling to a GIS which enables the spatial and temporal representation of the input parameters and resultant horizontal and vertical fluxes to be displayed. Shao and Leslie (1997) have demonstrated how the model can be applied by modelling dust emission for the Australian continent and describing a large dust storm event that occurred in central Australia in February 1996.

In summary, modelling wind erosion on agricultural land is extremely complex. It has all the complexity of the aeolian processes operating on bare, level, loose sandy surfaces, plus the additional factors of soil type, soil aggregation, soil crusts, non-erodible and erodible surface roughness, topographic variation, wind breaks, crop growth and residue decomposition. Add to this list the temporal and spatial variation of the processes, the effects of land management such as cultivation and grazing, and the full scope of the problem becomes apparent. Despite this, new process-based models, which combine the best available knowledge from a range of disciplines with GIS, are demonstrating what can be achieved. One of the major advantages for pursuing such modelling is to answer fundamental questions about the erosion process, thereby creating an understanding of how to control erosion.

WIND EROSION CONTROL

From an agricultural view point, there are several strategies and erosion control factors that can be manipulated by the land manager to reduce erosion. This section describes some, but not all, of these strategies and factors, and reports threshold levels that provide effective erosion protection derived from Australian field wind tunnel studies.

Avoidance Strategy

The simplest way to control wind erosion on agricultural land is to avoid farming or grazing soil landforms that are predisposed to erosion. This is the under pinning concept to land capability, which is defined as 'the ability of land to accept a type and intensity of use permanently or for specific periods under specific management and without permanent damage' (Houghton and Charman, 1986). In Australia, eight land classes are used to rank land from highest to lowest capability (Table 7.1). Each land class is suitable for different crops and or different management.

The avoidance strategy has been applied in the USA and is one part of the rational behind deciding what land is to be excluded from farming under the Conservation Reserve Program (Ervin, 1989).

The avoidance strategy has also been applied to the development of Crown lands in Australia (i.e. Crown land is land owned by the government and leased to individuals and companies for agriculture, mining and urban development). In the case of grazing and agricultural leases in the State of New South Wales (NSW), clearing and cultivation of land is not permitted on soils with a sandier surface texture than a loamy sand and some sandy loam soils (generally those with less than 13 per cent soil clay content). The delineation of this threshold soil texture is based on five independent methods.

1.   Visual interpretation of maximum soil fluxes measured at 75 km h^{-1} (Leys, 1991) using a portable field wind tunnel (Raupach and Leys, 1990).

Table 7.1 Land capability classification for rural land

Land suitable for regular cultivation

Class I	No special soil conservation measures required
Class II	Simple soil conservation measures required, such as adequate crop rotation and stubble retention farming practices
Class III	Intensive soil conservation measures required, such as wind breaks or strip cropping with management practices in Class II

Land suitable for grazing and occasional cultivation

Class IV	Simple soil conservation measures required, such as stock control and occasional cropping with stubble retention farming practices
Class V	Intensive soil conservation measures required, such as improved pastures or perennial crops with management practices in Class IV

Land suitable for grazing only

Class VI	Judicious soil conservation management is required to ensure adequate ground cover and maintenance of perennial cover

Land unsuitable for general rural production

Class VII	Land best protected by green timber because of the erosion hazard
Class VIII	Should not be cleared, grazed or logged, but utilised for activities compatible with nature conservation

2. A threshold clay content of 13 per cent is required to maintain the sediment flux to a low level ($5 \, g \, m^{-1} s^{-1}$) for bare, level, cultivated soils (Leys *et al.*, 1996).

3. The rapid increase in the erodibility index (I) used for the Wind Erodibility Groups. For the soil textures: sandy loam, loamy sand and sand, I increases from 193 to 300 to 695 $t \, ha^{-1}$ respectively (Soil Conservation Service, 1988).

4. One of the assumptions of the Owen sand transport equation (Owen, 1964), is that the constant c should be in the order of unity. However, when Owen's equation is fitted to the sediment flux data for nine soils presented in Leys (1991) ranging in texture from sand to clay, the predicted values of c for the soils with higher clay content than sandy loam have values less than 0.2. This indicates that these soils have low saltation levels and correspondingly low erosion rates. Secondly, when clay content and predicted c values are plotted against each other for those soils with a c value greater than 0.2, then c equals zero, a condition indicating that saltation ceases to be the dominant sediment transport mechanism, when clay content is about 14 per cent (Figure 7.4).

5. The threshold friction velocities for loose, level soils with sandy loam or higher clay content are generally greater than $0.3 \, m \, s^{-1}$ (Leys, 1991). In the aeolian country of southeastern Australia, less than 20 per cent of winds exceed this friction velocity.

The threshold soil texture is based on the assumption that soils with clay contents greater than 13 per cent, have sufficient inherent stability to control erosion even when bare and cultivated in average years. This is not to say that these soils will not pose an erosion hazard, for if their aggregation is broken down by excessive cultivation or by extended drought and stock trampling, then they will also erode.

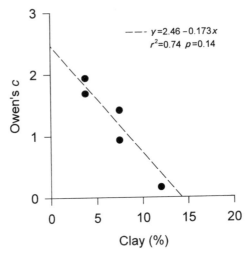

Figure 7.4 Relationship between Owen's constant c and clay content for soils with a c > 0.2 as described by Leys (1991)

While the avoidance strategy is a good method for reducing the risk of erosion, it does not alleviate the wind erosion problems associated with agricultural land because inappropriate land management methods can predispose even the most inherently stable of soils to the risk of erosion. Therefore there are a number of erosion control factors that land managers can manipulate to control erosion. Two of these factors, dry aggregation and soil cover, will be discussed below.

Dry Aggregation

The percentage of dry aggregation greater than 0.84 mm has long been recognised as one of the major factors which controls soil erodibility (Chepil, 1950, 1953, 1954; Zobeck, 1991; Fryrear et al., 1994; Larney et al., 1995). Dry aggregation is used in the determination of soil erodibility (I) for use in the Wind Erosion Equation (see Table 1, exhibit 505.61(b), Soil Conservation Service, 1988). Based on the dry aggregation greater than 0.84 mm, an index value for I in tons/acre is given. Percentage dry aggregation is the inverse of erodible fraction (EF) which is used in the Revised Wind Erosion Equation to describe the soil erodibility component for the model (Fryrear et al., 1994).

Chepil (1942) reports that the percentage of dry aggregates in six size classes (<0.42; 0.42–0.83; 0.83–2.0; 2.0–6.4; 6.4–12.7; and >12.7 mm) can be used to estimate soil erodibility. The practicalities of gathering this large amount of sieving data was recognised by Chepil (1953, p. 479) and again later in Chepil (1958, p. 33) when he reports a method for predicting the soil erodibility of a soil from the percentage dry aggregation greater than 0.84 mm.

Recently, the empirical relationship between the percentage mass of dry aggregation greater than 0.85 mm as measured by gentle hand sieving (%DA) and erosion

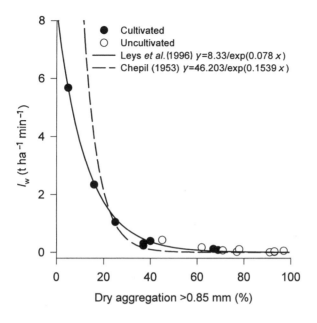

Figure 7.5   Relationship between erosion rate I_w and dry aggregation level >0.85 mm for Australian (solid line) and US (dashed line) soils (after Leys *et al.*, 1996). Reproduced by permission of CSIRO Publishing

rate measured at 65 km h^{-1} (I_w) was determined for Australian conditions (Leys *et al.*, 1996) and the results are shown for nine soils in the bare cultivated and uncultivated condition in Figure 7.5.

I_w decreases rapidly over the range 0 to 40 %DA and becomes negligible greater than 40 per cent. There is a highly significant relationship between I_w and %DA ($r^2 = 0.99$, $P < 0.01$) and there is good agreement between this relationship and that of Chepil (1953). Differences do occur when %DA is less than 20 per cent. Chepil's model predicts markedly higher soil erosion rates in this range (e.g. for a %DA of 5 per cent, Chepil's model returns an I_w of 21.4 t ha^{-1} min^{-1} while the Australian model predicts 5.7 t ha^{-1} min^{-1}). It is likely that the differences are a function of the wind tunnel testing methods, i.e. Chepil used a laboratory wind tunnel and trays of soils while Leys *et al.* (1996) used a portable field wind tunnel (Raupach and Leys, 1990) over natural surfaces. A full discussion of differences and reasons is presented in Leys *et al.* (1996).

To determine a threshold level of dry aggregation required to provide effective wind erosion control, a methodology was devised using the portable field wind tunnel (Raupach and Leys, 1990). Based on years of experience with the tunnel, comparative observations between the surface and the sediment flux, and the clay and dry aggregation relationships against sediment flux presented in Figure 7.5, an erosion control target of 5 g m^{-1} s^{-1} (or 0.7 t ha^{-1} min^{-1}) was set. Using the regression equation in Figure 7.5, the threshold dry aggregation level required to provide effective wind erosion control (i.e. keep $I_w < 0.7$ t ha^{-1} min^{-1}) is 31 per cent.

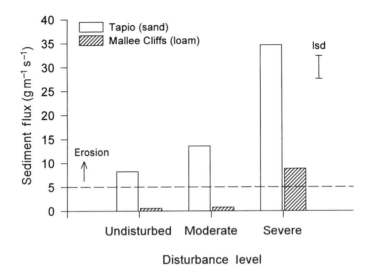

Figure 7.6 Sediment flux of a sand and loam textured soil with different crust disturbance levels (after Leys and Eldridge, 1998). Reproduced by permission of John Wiley & Sons Ltd

There are a number of studies that identify the influence of tillage under different crop rotations (Siddoway, 1963; Armbrust, 1982), soil moisture levels (Wagner *et al.*, 1992; Larney and Bullock, 1994), and soil densities (Lyles and Woodruff, 1961) and identify the optimal conditions for optimising aggregation for erosion control. In rangelands, dry aggregation is closely related to biological crust levels (Leys and Eldridge, 1998). The level of aggregation is also related to the level of disturbance of the crusted surface. For a loamy soil with a good level of biological crust cover (50 per cent) Leys and Eldridge (1998) report that the soil had a %DA of 60 per cent when undisturbed, 45 per cent when moderately disturbed and 39 per cent when severely disturbed. On a loamy sand soil with 32 per cent biological crust cover the %DA level was 38 per cent when undisturbed, 25 per cent when moderately disturbed and 15 per cent when severely disturbed. The influence of the increased aggregation levels and biological crust cover is seen in Figure 7.6.

Dry aggregation is therefore one land management option for erosion control. Aggregation can be maximised in cropped soils by working the soil at optimal moisture contents and reducing the number and aggressive nature of the cultivations. Aggregation can be maximised in rangeland soils by minimising disturbance by stock, which in turn maximises biological crust cover.

Vegetation Cover

The effectiveness of vegetation cover to reduce wind erosion by absorbing the force of the wind has been shown by many authors (Chepil, 1944; Siddoway *et al.*, 1965; Lyles and Allison, 1981; Fryear, 1985; Findlater *et al.*, 1990). This research found an empirical exponential relationship between cover and erosion rates. Much of the US research was undertaken in laboratory wind tunnels with simulated cover using wooden dowels (Fryrear, 1985); however, Australian research was undertaken on prostrate lupin stubble (Findlater *et al.*, 1990) and wheat stubble (Leys, 1991).

The research into the role of cover in controlling erosion can be divided between the empirical and the process approach. While it is acknowledged that the current modeling effort focuses on drag partitioning (Marshall, 1971; Raupach, 1992; Raupach *et al.*, 1993; Wyatt and Nickling, 1997) as the process for explaining the effectiveness of cover in controlling erosion, for land managers, the need for a simpler index of protection is required. One of the simpler indices is prostrate soil cover, for which Fryrear (1985) devised an empirical relationship. Although the relationship had been initially tested for lupin (Findlater *et al.*, 1990) and wheat stubble (Leys, 1991), there was a need to test it over a wider range of soils and prostrate cover types.

The empirical exponential relationship between cover and erosion rates was tested with the portable wind tunnel for two soil textures, sand and loamy sand, in the Murray Mallee of South Australia (Leys, 1998) and the results are presented in Figure 7.7. Based on this research, a threshold cover level of 46 ± 1 per cent for the sand and 27 ± 1 per cent for the loamy sand is required to provide adequate erosion protection. The differences in threshold cover required for each soil texture relates to the aggregation differences of the sand and loamy sand soils.

As discussed above in the dry aggregation section, dry aggregation is an important factor in controlling wind erosion. Bisal and Ferguson (1970) recognised this when they proposed an empirical relationship between threshold wind velocity and soil cover and non-erodible soil aggregates. The concept that the level of vegetation cover decreases as the level of aggregation increases can be used to explain the differences in the cover level required for the two soil textures studied. The method was to analyse the data for the sand site ($n=120$) and the loamy sand site ($n=70$) presented in Figure 7.7, using a multiple variable regression including soil cover and dry aggregation (Minitab, 1994).

The results from the regression analysis suggest that when the levels of %DA is set to zero (i.e. no aggregation is present) then the sandy soil would require 53 per cent and the loamy sand soil 54 per cent soil cover to provide effective protection (i.e. maintain the erosion rate to $\leq 5\ \mathrm{g\,m^{-1}\,s^{-1}}$) for a 75 $\mathrm{km\,h^{-1}}$ wind measured at 10 m height. The outcome of these findings is that soil cover levels should be maintained above 50 per cent, as this will ensure erosion protection in dry periods when dry aggregation decreases due to cultivation and stock trampling.

ON-SITE EFFECTS OF WIND EROSION

There has been considerable research into the effects of wind erosion on agricultural land. In general, erosion has adverse effects on agricultural production by: sandblasting emerging crops (Armbrust, 1984; Bird *et al.*, 1992); removal of topsoil (Leys and McTainsh, 1996); reshaping the surface thereby making it difficult to traverse with wide agricultural implements; reducing soil clay and silt content

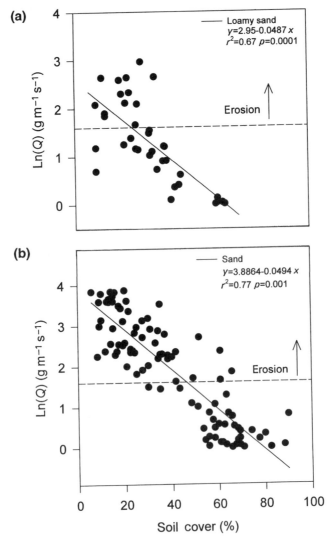

Figure 7.7 Relationship between sediment flux and soil cover for two soil textures. The horizontal dashed line represents the erosion control target ($5 \mathrm{~g~m^{-1}~s^{-1}}$)

(Zobeck and Fryrear, 1986); reducing organic matter (Daniel and Langham, 1936; Lyles and Tatarko, 1986); burying or undermining infrastructure such as fences and roads (Pye, 1987); burying adjacent land with sand drift thereby limiting the drifted lands production in the short term (Gaynor and MacTavish, 1981); silting up of waterways and irrigation channels, reducing visibility (Hagen and Skidmore, 1976); depositing unwanted dust and associated contaminates off-farm; and raising airborne particulate levels (Saxton, 1996), with particle sizes less than 10 μm (PM_{10}), which can have adverse health effects (Choudhury *et al.*, 1997).

Winnowing

One of the on-site effects listed above is winnowing of the surface and the resultant change in surface texture and this has not been adequately described due to the difficulties in measuring the winnowing effect over short periods. Recently the effects have been quantified under field conditions in Australia (Leys and McTainsh, 1999). This study showed how the particle-size distribution (PSD) of the surface of a cultivated sandy soil and the eroded sediments, changed during natural and simulated wind erosion events. To identify the changes in particle size, Folk's (1971) quantum concept was applied to discriminate the component log-normal distributions (or quantum) within the parent soil, surface grain layer and the eroded sediment particle-size distributions. Figure 7.8 shows the particle-size distribution and log-normal distributions and component quantum of a sandy soil before and after 30 min exposure to a 75 km h^{-1} wind. The particle-size distribution (left-hand side of figure) indicates little differences between the two surfaces. However, the quantum analysis (right-hand side of figure) indicates that particle-size populations (quantum) have changed from three quanta at 90, 180 and 300 μm before the erosion event, to two quanta at 90 and 300 μm after the erosion event. The 300 μm quantum has increased from 60 per cent before the erosion event to 86 per cent after 30 minutes of erosion. The 180 μm has been removed from the surface as eroded sediment. MIX analysis of the eroded sediments over the 30 min showed that 78–100 per cent of the sediment was in the 180 μm quantum.

Their study shows that changes in particle size of the surface are not simply due to the formation of lags bought about by the selective removal of fines, but are also influenced by deposition and *en masse* removal. Changes in surface particle size were also dependent upon: parent soil particle size; micro-topographic factors, such as how cultivation ridges and furrows responded to erosion, and the effects of non-erodible roughness elements on deposition; and wind conditions, in particular wind velocity, duration of events and fetch.

These research results raise significant questions about the validity of wind erosion models which ascribe a time-independent erodibility function to a soil based upon a single particle-size distribution. The removal of sediment from the soil surface changes the soil surface microtopography and the particle-size distribution of the soil surface. These temporal changes then influence the sediment flux rate for the surface (Figure 7.9). Generally the sediment flux rate decreases with time because: (i) the stress partitioning between the exposed non-erodible elements and the erodible sediment changes with time (Chepil and Woodruff, 1963; Lyles *et al.*, 1974; Nickling and McKenna Neuman, 1995); and/or (ii) 'armouring' of the surface occurs as the particle-size distribution of the surface becomes coarser (Leys and McTainsh, 1998), with corresponding higher threshold friction velocities.

OFF-SITE EFFECTS OF WIND EROSION

The above discussion describes the on-site effect of wind erosion and the degradation of the soil; however, assessments of the condition of the environment have seen

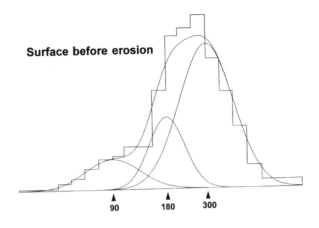

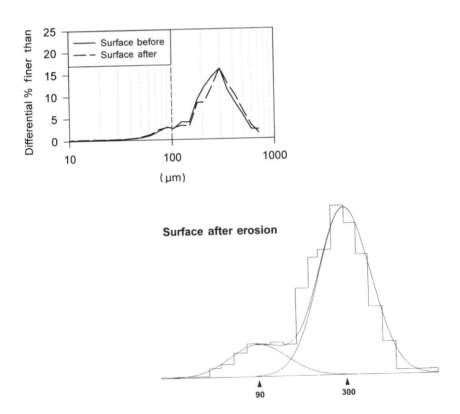

Figure 7.8 Particle-size distribution before and after erosion (left-hand side) and MIX analysis of the same distributions (right-hand side) showing the winnowing effect (i.e. removal of 180 μm quantum) on a sandy soil 30 min after exposure to a 75 km h^{-1} wind

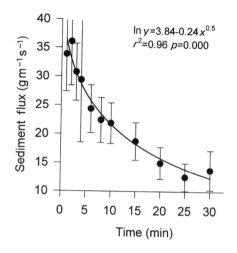

$$\ln y = 3.84 - 0.24\,x^{0.5}$$
$$r^2 = 0.96 \quad p = 0.000$$

Figure 7.9 Decrease in sediment flux rate with time on a bare sandy surface

agricultural areas identified as a major source of dust (Chow and Watson, 1992), with 34 per cent of emissions in 1990 in the USA from unsealed roads, 16 per cent from cultivation and 9 per cent from erosion. As a result, there has been increased interest in the impact of windblown dust emanating from agricultural land on off-site air quality (Saxton, 1995, 1996; Stetler and Saxton, 1995, 1996) and human health (Clausnitzer and Singer, 1996). Dust storms often lead to dust concentrations of particulates less than 10 μm diameter (PM_{10}) which exceed air quality health standards. These particulates may travel significant distances from source. In Australia, urban areas such as Brisbane (Knight *et al.*, 1995) and Melbourne (Raupach *et al.*, 1994) have experienced dust storms whose source material originated hundreds to thousands of kilometres upwind. The deposition of dust in itself is a problem, however, recent studies have also identified herbicides and pesticides which are associated with the dust bring new problems.

Although most of the suspended dust emitted by agricultural operations is deposited within less than 100 m of its source (Larney *et al.*, 1998), a portion of it may be transported long distances by wind (Chow and Watson, 1992) and may also carry herbicides and pesticides. Gaynor and MacTavish (1981) reported that approximately 43 per cent of simazine was removed from a treated area by wind erosion on a sandy soil in southwestern Ontario, Canada. The simazine was deposited downwind at quantities sufficiently high to be phytotoxic to susceptible crops or to impair the quality of adjacent irrigation ponds or waterways. Glotfelty *et al.* (1989) reported movement of atrazine, simazine and alachlor by wind erosion, while Larney *et al.* (1996) found 2,4-D enrichment in windblown sediment compared with the source soil.

In recent years, endosulfan has been detected in waterways of the cotton-growing region of southeastern Australia (Arthington, 1995; Cooper, 1996). This has implications for the long-term sustainability of both the cotton industry (the major user of endosulfan) and Australia's natural environment. Dust transport has been identified as one of four possible pathways by which the pesticide endosulfan entered the riverine environment, the others being spray drift, waterborne runoff and vapour transport initiated by volatilisation (Raupach and Briggs, 1996). A recent study in Australia quantified the dust and associated endosulfan emission caused by wind and on-farm activities and the deposition of the dust both on- and off-farm (Leys et al., 1998; Larney et al., 1998; Leys et al., 1999).

Dust Emission

A portable wind tunnel was used to determine the threshold friction velocities, sediment flux rates and the pesticide source strength for an unsealed farm road and a cultivated field (Leys et al., 1999). The study demonstrates that road surfaces represent an important source of wind erosion on cotton farms. The wind erosion potential, as measured by sediment flux in a portable field wind tunnel, of an unsealed road was 53 times greater than that of an adjacent cultivated cotton field in northern New South Wales, whereas endosulfan emissions were 1.6 times greater for the cotton field than the road because the endosulfan concentrations on the road dust were only 1 per cent that of the field dust. The threshold wind velocity necessary to initiate erosion (U_t), friction velocity (u_*), threshold friction velocity (u_{*t}) and the aerodynamic roughness length (z_0) of the road surface were all significantly lower than the field surface. These finding suggest that dust emissions were greatest from roads compared with the adjacent cotton fields, but the endosulfan emission, nine days after application, were higher from the field because of the higher source strength of endosulfan on the field soil. If roads are accidentally over-sprayed or have chemical drift deposited on them, they could then become a significant source of endosulfan emission.

Having identified that the roads are major sources for wind erosion, the next step was to verify if anthropogenic activities, such as vehicle movement and cultivation, were major sources of dust and endosulfan emissions (Leys et al., 1998). A vehicle travelling at 80 km h^{-1} on an unsealed road was a greater source of TSP emission (3.7 g m^{-1} travelled) than an 8 m wide inter-row cultivator travelling at 8 km h^{-1} (1.7 g m^{-1}). However, the particle size distribution of the TSP from inter-row cultivation was finer (mode of 19–22 μm) than that from vehicular traffic on unsealed roads (mode of 32 μm) and hence may be transported further. Endosulfan source strength from inter-row cultivation was 3.6 μg m^{-1} of travel (or 0.45 μg m^{-2}) which was only 6.0 × 10^{-4} per cent of that applied, 4 days after endosulfan application. This was slightly higher than the endosulfan source strength from vehicular traffic on an unsealed road (3.1 μg m^{-1} of travel), only 2 days after spraying. On unsealed roads, particle-associated endosulfan mass fractions declined rapidly with time due to volatilisation and photodegradation and a decrease in endosulfan-enriched source sediment due to removal by repeated vehicle passes. These findings suggest that anthropogenic activities are a regular source of dust on-farm during dry weather. Endosulfan source strength will be highest shortly after spraying because the levels of endosulfan decline rapidly over the following days. Therefore to reduce endosulfan emissions, strategic closure of contaminated roads or minimisation of dust emissions can successfully decrease off-site movement. If emissions should occur, the next question to be answered is what levels are being deposited off-farm.

Dust Deposition

The objectives of the dust deposition project (Larney *et al.*, 1998) were to examine: (1) dust deposition from vehicular traffic on an unsealed road on a cotton farm, and around a cotton field in the 11–65 h period after endosulfan application; (2) endosulfan deposition in the 11–65 h period after endosulfan application, and over a 3-month period at on-farm and off-farm (non-target) locations in the cotton-growing region of northern New South Wales. Dust deposition rates from vehicular traffic varied from 0.013 $g m^{-2}$ per vehicle at 1 m, and 0.002 $g m^{-2}$ per vehicle at 100 m from an unsealed road. Dust deposition, which was caused my vehicle movement on unsealed farm roads around the sprayed field, in the 11–65 h period after endosulfan spraying varied from 0.30 $g m^{-2}$ at 10 m to 0.14 $g m^{-2}$ at 1000 m from the field. The highest endosulfan deposition value in this post-spraying period was 95 $\mu g m^{-2}$ at 5 m and 13.3 $\mu g m^{-2}$ at 1000 m from the field. The results indicate that vapour as well as dust deposition were measured in this research. Endosulfan concentrations in the dust alone are estimated to be in the order of 1–10 $\mu g g^{-1}$ of dust.

Over a 3-month monitoring period (December 1996–March 1997), the average daily deposition rate of endosulfan was 0.16 $\mu g m^{-2}$ day^{-1} for the off-farm sites compared with 0.35 $\mu g m^{-2}$ day^{-1} for the on-farm site. Converting the off-farm deposition rate to water concentrations for a 1 m depth of non-flowing water gives a concentration of 1.6×10^{-4} $\mu g l^{-1}$ which is well below the ANZECC guideline of 0.01 $\mu g l^{-1}$ and that of measured levels of endosulfan in rivers of the study area in 1995/96 (Cooper, 1996).

These studies highlight the importance of dust movement off-farm and illustrate the impact of agricultural activities on the environment. However, they also allow land managers to mitigate the impact of dust emissions by implementing best management practices which can minimise the movement of dust and pesticides. The adoption of best management practices is currently under way in the study area (Williams, 1997).

CONCLUSIONS

This chapter has discussed issues pertaining to wind erosion on agricultural land. The processes operating on agricultural land are generally more complex than sand dunes or beaches because of the interaction of soil texture, vegetation, soil moisture slope and land management activities. Despite this, there have been substantial advances in our understanding and ability to model wind erosion on agricultural land. From this modelling effort and previous research undertaken to control wind erosion, there are several options for land managers to control erosion. The impetus for controlling erosion is to reduce on- and off-site impacts of wind erosion. On-site impacts affect the land manager directly, while off-site impacts associated with dust and associated contaminants affect the environment and other communities. Because of the impacts that wind erosion has, there are still substantial areas of research to be undertaken to answer the questions that are posed to the wind erosion community.

The following three areas for additional research are suggested because they are fundamental to the erosion process.

1. A sediment transport model could be developed for supply limited situations. All current models are based on a transport limited model, yet for agricultural areas, the erosion processes are generally supply limited.
2. Feedback mechanisms that adjust for the evolving surface conditions of the eroded surface.
3. Development of the time-dependent aspects of the factors that control wind erosion, such as aggregation, crust cover and vegetation cover.

With these fundamental processes accounted for, then there is a greater chance of understanding and predicting the impacts of wind erosion on man and the environment.

ACKNOWLEDGMENTS

Much of the research presented here is a result of collaborative projects with many of my fellow wind erosion researchers. I express my sincere appreciation to these people, in particular Grant McTainsh, Michael Raupach, Yaping Shao, Francis Larney and Jochen Müller, because without their assistance, this work would not have been achievable. Similarly, I wish to acknowledge the funding I have received from the Department of Land and Water Conservation, Rural Credits Development Fund, Land and Water Resources Research and Development Corporation, Murray Darling Basin Commission and the Cotton Research and Development Corporation.

REFERENCES

Anderson, R.S. and Haff, P.K., 1988. Simulation of eolian saltation. *Science*, **241**, 820–823.
Anderson, R.S. and Haff, P.K., 1991. Wind modifications and bed response during saltation of sand in air. *Acta Mechanica Supplement*, **1**, 21–51.
Armbrust, D.V., 1984. Wind and sandblast injury to fieldcrops: effects of plantage. *Agronomy Journal*, **76**, 991–993.
Armbrust, D.V., Chepil, W.S. and Siddoway, F.H., 1964. Effects of ridges on erosion of soil by wind. *Soil Science Society Proceedings*, **28**, 557–560.
Armbrust, D.V., Dickerson, J.D., Skidmore, E.L. and Russ, O.G., 1982. Dry soil aggregation as influenced by crop and tillage. *Soil Science Society of America Journal*, **46**, 390–393.
Arthington, A., 1995. 'State of the rivers in cotton growing areas', LWRRDC Occasional Paper Series No. 02/95 (Land and Water Resources Research and Development Corporation, Canberra, ACT).
Bagnold, R.A., 1941. *The Physics of Blown Sand and Desert Dunes*. Chapman and Hall, London.

Baidu-Forson, J. and Napier, T.L., 1998. Wind erosion control within Niger. *Journal of Soil and Water Conservation*, **53**, 120–124.

Berkofsky, L. and McEwan, I., 1994. The prediction of dust erosion by wind: an interactive model. *Boundary-Layer Meteorology*, **97**, 385–406.

Bird, P.R., Bicknell, D., Bulman, P.A. *et al.*, 1992. The role of shelter in Australia for protecting soils, plants and livestock. *Agroforestry Systems*, **20**, 59–86.

Bisal, F. and Ferguson, W.S., 1970. Effect of non-erodible aggregates and wheat stubble on initiation of soil drifting. *Canadian Journal of Soil Science*, **50**, 31–34.

Bondy, E., Lyles, L. and Hayes, W.A., 1980. Computing soil erosion by periods using wind energy distribution. *Journal of Soil and Water Conservation*, **35**, 173–176.

Chepil, W.S., 1944. Utilization of crop residues for wind erosion control. *Scientific Agriculture*, **24**, 307–319.

Chepil, W.S., 1950. Properties of soil which influence wind erosion: II. Dry aggregate structure as an index of erodibility. *Soil Science*, **69**, 403–414.

Chepil, W.S., 1953. Factors that influence clod structure and erodibility by wind. I. Soil texture. *Soil Science*, **75**, 473–483.

Chepil, W.S., 1954. Seasonal fluctuations in soil structure and erodibility of soil by wind. *Soil Science Society of America Proceedings*, **18**, 13–16.

Chepil, W.S. and Woodruff, N.P., 1963. The physics of wind erosion and its control. *Advances in Agronomy*, **15**, 211–302.

Choudhury, A.H., Gordian, M.E. and Morris, S.S., 1997. Associations between respiratory illness and PM10 air pollution. *Archives of Environmental Health*, **52**, 113–117.

Chow, J.C. and Watson, J.G., 1992. Fugitive emissions add to air pollution. *Environmental Protection*, **3**, 26–31.

Clausnitzer, H. and Singer, M.J., 1996. Respirable dust production from agricultural operations in the Sacramento Valley, California. *Journal of Environmental Quality*, **25**, 877–884.

Cole, G.W., Lyles, L. and Hagen, L.J., 1983. A simulation model of daily wind erosion soil loss. *Transactions of the ASAE*, **26**, 1758–1765.

Comis, D. and Gerrietts, M., 1994. Stemming wind erosion. *Agricultural Research*, **42**, 8–15.

Cooke, R.U., Warren, A. and Goudie, A.S., 1993. *Desert Geomorphology*. UCL Press, London.

Cooper, B., 1996. Central and North West Regions Water Quality Program: 1995/96 Report on Pesticides Monitoring. Technical Report. No. TS 96.048. NSW Department of Land and Water Conservation, Technical Services Directorate, Parramatta, NSW.

Daniel, H.A. and Langham, W.H., 1936. The effect of wind erosion and cultivation on the total nitrogen and organic matter content of soil in the Southern High Plains. *Journal of the American Society of Agronomy*, **28**, 587–596.

Ervin, C., 1989. Implementing conservation title. *Journal of Soil and Water Conservation*, **44**, 367–370.

Findlater, P.A., Carter, D.J. and Scott, W.D., 1990. A model to predict the effects of prostrate ground cover on wind erosion. *Australian Journal of Soil Research*, **28**, 609–622.

Fryrear, D.W., 1985. Soil cover and wind erosion. *Transactions of the ASAE*, **28**, 781–784.

Fryrear, D.W., Krammes, C.A., Williamson, D.L. and Zobeck, T.M., 1994. Computing the wind erodible fraction of soils. *Journal of Soil and Water Conservation*, **49**, 183–188.

Gaynor, J.D. and MacTavish, D.C., 1981. Movement of granular simazine by wind erosion. *Horticultural Science*, **16**, 756–757.

Gillette, D.A. and Passi, R., 1988. Modelling dust emission caused by wind erosion. *Journal of Geophysical Research*, **93**, 14233–14243.

Glotfelty, D.E., Leech, M.M., Jersey, J. and Taylor, A.W., 1989. Volatilization and wind erosion of soil surface applied atrazine, simazine, alachlor and toxaphene. *Journal of Agriculture and Food Chemistry*, **37**, 546–551.

Greeley, R. and Iversen, J.D., 1985. *Wind as a Geological Process on Earth, Mars, Venus and Titan*. Cambridge University Press, Cambridge.

Gregory, J.M., 1991. Wind erosion: prediction and control procedures. Report prepared for

US Army Corps of Engineers, Waterways Experimental Station, Vicksburg, Mississippi. Texas Tech University, Lubbock, Texas.

Gregory, J.M. and Borrelli J., 1986. The Texas tech. wind erosion equation. *American Society of Agricultural Engineers*, 86-2528, 1–13.

Gregory, J.M., Tock, R.W. and Wilson, G.R., 1994. Atmospheric Loading Of Dust and Gases: Impact On Society. Document. No. 69-5626-3-481. Prepared for the Soil Conservation Service by the Secretary of Agriculture.

Hagen, L.J. and Skidmore, E.L., 1976. Wind erosion and visibility problems. No. 76-2019. American Society of Agricultural Engineers, St Joseph, MI.

Hagen, L.J., Skidmore, E.L. and Layton, J.B., 1988. Wind erosion abrasion: Effects of aggregate moisture. *Transactions of the ASAE*, **31**, 725–728.

Houghton, P.D. and Charman, P.E.V., 1986. *Glossary of Terms Used in Soil Conservation*. Soil Conservation Service of NSW, Sydney.

Houldsworth, B., 1995. Central and North West Regions Water Quality Program: 1994/95 Report on Nutrient and General Water Quality Monitoring. Technical Report. No. TS 95.088. NSW Department of Land and Water Conservation, Technical Services Directorate, Parramatta, NSW.

Knight, A.W., McTainsh, G.H. and Simpson, R.W., 1995. Sediment loads in an Australian dust storm: implications for present and past dust processes. *Catena*, **24**, 195–213.

Larney, F.J. and Bullock, M.S., 1994. Influence of soil wetness at time of tillage and tillage implement on soil properties affecting wind erosion. *Soil and Tillage Research*, **29**, 83–95.

Larney, F.J., Bullock, M.S., McGinn, S.M. and Fryrear, D.W., 1995. Quantifying wind erosion on summer fallow in southern Alberta. *Journal of Soil and Water Conservation*, **50**, 91–95.

Larney, F.J., Bullock, M.S., Timmermans, J.G. and Cessna, A.J., 1996. Wind erosion studies in Alberta. In *Proceedings of the Soil Quality Assessment for the Prairies Workshop*. Edmonton, Alberta, Canada, January 22–24, 1996.

Larney, F.J., Leys, J.F., Müller, J. and McTainsh, G.H., 1998. Dust and endosulfan deposition in a cotton-growing area of northern New South Wales, Australia. *Journal of Environmental Quality*, **220**, 55–70.

Leys, J.F., 1990. Blow or grow? A soil conservationist's view to cropping Mallee soils. In *The Mallee Lands: A Conservation Perspective*, J.C. Noble, P.J. Joss and G.K. Jones (Eds). CSIRO, Melbourne, pp. 280–296.

Leys, J.F., 1991. The threshold friction velocities and soil flux rates of selected soils in south-west New South Wales, Australia. *Acta Mechanica* (Suppl. 2), 103–112.

Leys, J.F., 1991. Towards a better model of the effect of prostrate vegetation cover on wind erosion. *Vegetatio*, **91**, 49–58.

Leys, J.F., 1998. Wind Erosion Processes and Sediments in Southeastern Australia. Unpublished PhD thesis. Griffith University, Brisbane.

Leys, J.F., Craven, P., Murphy, S. *et al.*, 1994. Integrated resource management of the Mallee of South-Western New South Wales. *Australian Journal of Soil and Water Conservation*, **7**, 10–19.

Leys, J.F. and Eldridge, D.J., 1998. Influence of cryptogamic crust disturbance on wind erosion of sand and loam rangeland soils. *Earth Surface Processes and Landforms*, **23**, 963–974.

Leys, J.F., Koen, T. and McTainsh, G.H., 1996. The effect of dry aggregation and percentage clay on sediment flux as measured by portable wind tunnel. *Australian Journal of Soil Research*, **34**, 849–861.

Leys, J.F., Larney, F.J. and McTainsh, G.H., 1999. Sediment flux rates and threshold friction velocities of an unsealed road and cultivated field using a portable wind tunnel. *Australian Journal of Soil Research*, submitted.

Leys, J.F., Larney, F.J., Müller, J.F. *et al.*, 1998. Anthropogenic dust and endosulfan emissions on a cotton farm in northern New South Wales, Australia. *Science of the Total Environment*, **220**, 55–70.

Leys, J.F. and McTainsh, G.H., 1996. Sediment fluxes and particle grain-size characteristics of wind-eroded sediments in southeastern Australia. *Earth Surface Processes and Landforms*, **21**, 661–671.

Leys, J.F. and McTainsh, G.H., 1999. Effects of wind erosion on the surface particle-size distribution of a sandy surface. *Sedimentology*, in review.

Lyles, L. and Allison, B.E., 1981. Equivalent wind erosion protection from selected crop residues. *Transactions of the ASAE*, **24**, 405–408.

Lyles, L. and Tatarko, J., 1986. Wind erosion effects on soil texture and organic matter. *Journal of Soil and Water Conservation*, **41**, 191–193.

Lyles, L. and Woodruff, N.P., 1961. Surface soil cloddiness in relation to soil density at time of tillage. *Soil Science*, **91**, 178–182.

Marshall, J.K., 1971. Drag measurement in roughness arrays of varying density and distribution. *Agricultural Meteorology*, **8**, 269–292.

Marticorena, B. and Bergametti, G., 1995. Modeling the atmospheric dust cycle. *Journal of Geophysical Research*, **100**(D8), 16 415–16 430.

McTainsh, G.H. and Leys, J.F., 1993. Soil erosion by wind. In *Land Degradation Processes in Australia*, G.H. McTainsh and W.C. Broughton (Eds). Longman Cheshire, Melbourne, pp. 188–233.

McTainsh, G.H., Lynch, A.W. and Burgess, R.C., 1990. Wind erosion in eastern Australia. *Australian Journal of Soil Research*, **28**, 323–239.

Middleton, N.J., 1997. Desert dust. In *Arid Zone Geomorphology: Process Form and Change in Drylands*, 2nd Edn, D.S.G. Thomas (Ed.). John Wiley, London, pp. 413–436.

Minitab, 1994. *Minitab Reference Manual, Release 10 for Windows*. Minitab Inc., Pennsylvania, USA.

Nicholson, K.W., Branson, J.R., Geiss, P. and Cannell, R.J., 1989. The effects of vehicle activity on particle resuspension. *Journal of Aerosol Science*, **20**, 1425–1428.

Nickling, W.G., 1985. Recent advances in the prediction of soil loss by wind. In *4th International Conference on Soil Erosion and Conservation*, Maracay, Venezuela, November 3–8, 1985, International Soil Conservation Organisation.

Nickling, W.G., 1994. Aeolian sediment transport and deposition. In *Sediment Transport and Depositional Processes*, K. Pye (Ed.). Blackwell Science, London, pp. 293–350.

Nickling, W.G. and Satterfield-Hill, K., 1992. Wind erosion prediction: The Guelph Wind Erosion Assessment and Management Model (GWEAM). In *The International Conference on the Application of Geographical Information Systems to Soil Erosion Management*, S.H. Luk (Ed.). Taiyuan, China, pp. 95–106.

Owen, P.R., 1964. Saltation of uniform grains in air. *Journal of Fluid Mechanics*, **20**, 225–242.

Pye, K., 1987. *Aeolian Dust and Dust Deposits*. Academic Press, London.

Raupach, M.R., 1992. Drag and drag partition on rough surfaces. *Boundary Layer Meteorology*, **60**, 375–395.

Raupach, M.R. and Briggs, P., 1996. Modelling the aerial transport of endosulfan to rivers. Part 2: Transport by multiple pathways. Technical Report. No. 121. CSIRO Centre for Environmental Mechanics, Canberra, ACT.

Raupach, M.R., Gillette, D.A. and Leys, J.F., 1993. The effect of roughness elements on wind erosion threshold. *Journal of Geophysical Research*, **98**, 3023–3029.

Raupach, M.R. and Leys, J.F., 1990. Aerodynamics of a portable wind erosion tunnel for measuring soil erodibility by wind. *Australian Journal of Soil Research*, **28**, 177–191.

Raupach, M.R., McTainsh, G.H. and Leys, J.F., 1994. Estimates of dust mass in recent major dust storms. *Australian Journal of Soil and Water Conservation*, **7**, 20–24.

Saxton, K.E., 1995. An interim technical report of the Northwest Columbia Plateau Wind Erosion Quality Project. Miscellaneous Publication. No. MISC0182. Washington State University, College of Agriculture and Home Economics, Pullman, WA.

Saxton, K.E., 1996. Wind erosion and its impact on off-site air quality in the Columbia Plateau – an integrated research plan. *Transactions of the ASAE*, **38**, 1031–1038.

Shao, Y. and Leslie, L.M., 1997. Wind erosion prediction over the Australian continent. *Journal of Geophysical Research*, **102**(D25), 30 091–30 105.

Shao, Y. and Leys, J.F., 1998. Wind erosion assessment and prediction using dynamic models and field experiments. In *Climate Prediction for Agricultural and Resource Management*. Bureau of Resource Sciences, Canberra.

Shao, Y., Raupach, M.R. and Findlater, P.A., 1993. Effect of saltation bombardment on the entrainment of dust by wind. *Journal of Geophysical Research*, **98**, 20 719–20 726.

Shao, Y., Raupach, M.R. and Leys, J.F., 1996. A model for predicting aeolian sand drift and dust entrainment on scales from paddock to region. *Australian Journal of Soil Research*, **34**, 309–342.

Shao, Y., Raupach, M.R. and Short, D., 1994. Preliminary assessment of wind erosion patterns in the Murray-Darling Basin. *Australian Journal of Soil and Water Conservation*, **7**, 46–51.

Siddoway, F.H., 1963. Effects of cropping and tillage methods on dry aggregate soil structure. *Soil Science Society Proceedings*, **27**, 452–454.

Siddoway, F.H., Chepil, W.S. and Armbrust, D.V., 1965. Effect of kind, amount and placement of residue on wind erosion control. *Transactions of the ASAE*, **8**, 327–331.

Sims, H.J., 1977. Cultivation and fallowing practices. In *Soil Factors in Crop Production in a Semi-arid Environment*, J.S. Russel and E.L. Greacen (Eds). University of Queensland Press, St Lucia, Queensland.

Soil Conservation Service, 1988. *National Agronomy Manual. Part 502*. United States Department of Agriculture, Washington DC.

Stetler, L.D. and Saxton, K.E., 1995. Fugitive Dust (PM10) Emissions and soil mass relations for agricultural fields in Washington State. In *AWMA Speciality Conference on Particulate Matter: Health and Regulatory Issues*. Pittsburgh, PA.

Stetler, L.D. and Saxton, K.E., 1996. Wind erosion and PM10 emissions from agricultural fields on the Columbia Plateau. *Earth Surface Processes and Landforms*, **21**, 673–685.

Stout, J.E., 1990. Wind erosion within a simple field. *Transactions of the ASAE*, **33**, 1597–1600.

Ungar, J.E. and Haff, P.K., 1987. Steady state saltation in air. *Sedimentology*, **34**, 289–299.

Wagner, L.E., Ambe, N.M. and Barnes, P., 1992. Tillage-induced soil aggregate status as influenced by water content. *Transactions of the ASAE*, **35**, 499–504.

Werner, B.T., 1990. A steady state model of wind-blown sand transport. *Journal of Geology*, **98**, 1–17.

Williams, A., 1997. *Australian Cotton Industry Best Management Practices Manual*. Cotton Research and Development Corporation, Narrabri, NSW.

Woodruff, N.P. and Siddoway, F.H., 1965. A wind erosion equation. *Soil Science of America Proceedings*, **29**, 602–608.

Wyatt, V.E. and Nickling, W.G., 1997. Drag and shear stress partitioning in sparse desert creosote communities. *Canadian Journal of Earth Sciences*, **34**, 1486–1498.

Zobeck, T.M. and Fryrear, D.W., 1986. Chemical and physical characteristics of windblown sediment. II. Chemical characteristics and total soil and nutrient discharge. *Transactions of the ASAE*, **29**, 1037–1041.

8 Wind Erosional Landforms: Yardangs and Pans

ANDREW S. GOUDIE

School of Geography, University of Oxford, UK

There are two main wind erosional landforms – yardangs and pans. Recent studies of yardangs have revealed some of their morphometric characteristics and given some indication of their rate of formation. It is now clear that they result from a combination of processes (abrasion, deflation, gully development, mass movements, salt weathering etc.) and evolve through a series of stages with different morphologies. There is also an increasing amount of morphometric information for pans and their associated lunettes, the history of which is now being revealed by stratigraphic and sedimentological studies. The origin of pans is still a matter of controversy and most are the product of a complex set of factors. There have been very few field studies of wind flows and sediment movements around either pans or yardangs. Such data could feed into models of the interactions between wind flow characteristics and surface topography, and collection of these data provides a major research opportunity.

INTRODUCTION

Wind erosion, as Chapter 7 indicated, is a potent cause of change on agricultural land. However, wind erosion is no less an important factor in the development of certain landforms, including closed depressions (pans), streamlined hillocks (yardangs), deflated surfaces such as stone pavements, and micro-forms (miscellaneous ventifacts, dreikanter etc.) (Goudie, 1989).

Good recent reviews of the diversity of wind erosional landforms, both large and small, are provided by Laity (1994), Livingstone and Warren (1996) and Breed *et al.* (1997). In this chapter recent views on two of the larger and more widespread types of wind erosion forms, namely yardangs and pans, will be discussed.

YARDANGS

Introduction

Yardang is a Turkmen word that was introduced by Sven Hedin as a term for wind-abraded ridges of cohesive material (Hedin, 1903). Since his time, yardangs (Figure 8.1) have been identified in a large number of arid areas, and developed on a large range of materials (Table 8.1). They have also been recognised in the stratigraphic

Aeolian Environments, Sediments and Landforms. Edited by A.S. Goudie, I. Livingstone and S. Stokes.
© 1999 John Wiley & Sons, Ltd.

Figure 8.1   Two large, orientated yardangs rising up above a wind-planed surface developed on dolomitic limestone in Bahrain

record (Tewes and Loope, 1992; Jones and Blackley, 1993). They show a considerable range in sizes from small centimetre-scale ridges (micro-yardangs), through forms that are some metres in height and length (meso-yardangs) to features that may be tens of metres high and some kilometres long (mega-yardangs or ridges) (Cooke *et al.*, 1993, pp. 296–297). They can also be classified according to their morphology, and Halimov and Fezer (1989) described eight different types from the Qaidam Basin in Central Asia and related them to a cycle of development and obliteration. General reviews of the yardang literature have been provided by McCauley *et al.* (1977a, b). There has been considerable debate in this literature as to the relative importance of deflation, aeolian abrasion, fluvial incision and mass movements in moulding yardang forms, but there are relatively few data available on yardang morphometry and on rates of yardang formation.

Morphometry

Some data for the Qaidam Basin in Central Asia were provided by Halimov and Fezer (1989) (Table 8.2). They found that the ratios of their length, width and height were 10:2:1. At Rogers Lake in California, Ward and Greeley (1984) found a 1:4 width to length ratio. At Kharga Depression in Egypt, surveying of 50 yardangs with differential (kinematic) GPS revealed volume, length, width, height ratios of

Table 8.1 Studies of yardangs

Source	Location	Lithology
Blackwelder (1934)	California	Lunette material
Breed *et al.* (1997)	Western Desert,	Lake beds, spring beds, Eocene
Grolier *et al.* (1980)	Egypt	limestone, Dakhla shale, Nubian sandstone, Palaeocene Tarawan Chalk
Brookes (1986)	Dakhla, Egypt	Tarawan formation limestones, Quaternary lake beds
Capot-Rey (1957)	S. Central Africa	Cretaceous and Cambrian clays
Clarke *et al.* (1996)	Mojave Desert, California	Holocene playa dunes
Doornkamp *et al.* (1980)	Bahrain	Dolomite
Gabriel (1938)	Lut, Iran	Silty clays and sands
Gabriel (1938)	Khash Desert, Afghanistan	Clay
Goudie (1979)	Rajasthan, India	Eocene limestones
Halimov and Fezer (1989)	Qaidam Depression, China	Clastic sediments and evaporites
Hedin (1903)	Lop Nor	Lake beds
Lancaster (1984)	Namib	Limestones, dolomites, silcretes
Mainguet (1968, 1972)	Borkou, Sahara	Palaeozoic and Lower Mesozoic sandstones
McCauley *et al.* (1977a,b)	S. Central Peru	Upper Oligocene to Miocene age siltstones
McCauley *et al.* (1977a,b)	Northern Peru	Weakly to moderately consolidated Upper Eocene to Palaeocene sediments (shales and sandstones)
Riser (1985)	Mali	Holocene lacustrine beds
G.L. Wells (pers. comm.)	High Andes, Chile, Argentina	Ignimbrites
Ward and Greeley (1984)	Rogers Lake	Lunette material

Table 8.2 Morphology of different types of yardangs in Qaidam Basin, Central Asia (from Halimov and Fezer, 1989)

Type	Structural dip (°)	Height (m)	Width (m)	Length (m)	Width of passages(m)
Mesa	0–10	10–15	100–2	1000–2	Varying
Saw tooth crest	15–15	3–1	5–10	30–10	
Cone	45	20	30	30	2
Pyramid	Flat	4–15	6–15		10–50
Very long ridges		>30	>50	100–5000	200–350
Hogbacks		10–30	15–6	100–35	
Whalebacks		3	5	15	100
Low streamlined whalebacks		3–0.5	5–1	30–5	100–500

Table 8.3   Morphometric features of yardangs from the
Kharga Depression, Egypt

	Range (m)	Mean (m)	Median (m)
Width	1.4–36.3	6.57	5.05
Length	4.04–1189.53	28.62	20.59
Height	0.49–8.7	2.62	2.30
Volume (m^3)	2.6–5131	572.3	160.3

18.7:9.9:2.7:1 (Goudie *et al.*, 1999). This study also provides data on the relationships between the different morphometric variables, demonstrating how strongly
correlated they are (Table 8.3).

The available morphometric data presented here are relatively sparse and are all
for eroded Quaternary lacustrine and hydro-aeolian sediments. There is plainly a
need for further studies, not least of large features developed in more resistant
bedrocks, where the ratios and relationships of the different variables are likely to
be very different.

The orientations of yardangs and other types of ridge and swale system tend to be
closely clustered and seem to conform to the direction of the strong, supposedly
unidirectional wind regimes in which they develop. However, there are few meteorological data on most yardang fields, or on the pattern of wind flow around individual
yardangs. The installation of arrays of anemometers and sediment traps connected
to data-loggers could be a productive area for further research, and could be used to
test the aerodynamic model of wind flow around yardangs that was developed on the
basis of interpretation of erosional markings by Whitney (1983).

Rates of Yardang Formation

Data on rates of yardang formation are few. The most widely used method to
estimate rates of incision is to date the Quaternary sediments in which they have
often developed. This can be achieved by radiocarbon or luminescence techniques.
There are, for example, various studies in the Sahara (e.g. Riser, 1985) which show
that lake and swamp deposits dating back to the mid-Holocene have been deflated
to produce yardangs that are often as much as 5 to 10 m high. Assuming that this
deflation has taken place over the last 4000 or so years, a rate of formation of up to
2.5 m per 1000 years can be interpolated. Likewise, if the 8 m high Lop Nor
yardangs in China have been eroded since the fourth century AD, their rate of
incision would be about 5 m per 1000 years (McCauley *et al.*, 1977a, pp. 11–20).
The yardangs at Rogers Lake in the Mojave Desert appear to be eroding at their
headward ends at a rate of about 2 m per 1000 years, while lateral erosion, caused
by a combination of abrasion and deflation, appears to be taking place at a rate of
about 0.5 m per 1000 years (Laity, 1994, p. 531).

Once again, however, there are no data for bedrock yardangs, and the installation
of micro-erosion meters (or similar devices for monitoring surface lowering) would
not only provide data on gross rates of formation, but would also offer the

Figure 8.2 A small yardang in the Kharga Oasis, Egypt, displaying the typically undercut prow

possibility of identifying variations in rates of change over different parts of a yardang. For yardangs developed in softer Quaternary sediments, arrays of simple erosion pins could provide comparable data.

Plainly, given the rates at which yardangs have developed in the late Holocene, they are features that are both young and ephemeral. However, mega-yardangs or ridges developed in bedrock may be old and persistent features that have been shaped over hundreds of thousands of years by Trade Winds that may have had higher velocities and higher rates of deflation and abrasion during glacial episodes (Rea, 1994).

Processes Acting on Yardangs

There is general agreement that wind abrasion and deflation are the prime processes causing yardang development. That abrasion is important is indicated by polished, fluted and sand-blasted slopes, and the undercutting of the steep windward face and lateral slopes (Figure 8.2). Abrasion is probably the dominant process in hard bedrock yardangs. However, deflation may be important in the evolution of soft sediment yardangs, but whether in such circumstances it is more important than abrasion is the subject of debate (Laity, 1994). In any event, abrasion is only likely to be the dominant process on the lower portions of yardangs, where saltating and creeping sand grains and granules can be effective.

There are also subsidiary processes that act on yardangs. These include fluvial erosion and gully formation, which may provide an avenue along which wind erosion may occur. Gully action may also occur on yardangs developed on soft sediments when they are subjected to heavy rain showers. Excessive fluvial erosion,

Figure 8.3   Large yardangs developed in old lake beds to the lee of the limestone escarpment in the Dakhla Oasis, Egypt. Note the desiccation cracks

however, would obliterate yardangs. Mass movements may also be significant when yardang slopes are oversteepened, while exudation of salts from old lake and swamp sediments may create puffy surfaces that are especially prone to deflation. Some yardangs have their forms modified by desiccation cracks (Figure 8.3) that developed in drying lake sediments, and the cracks may also be the sites of slope failure when oversteepening takes place.

An Evolutionary Model of Yardang Development

Table 8.4 shows an ideal model of yardang evolution in unconsolidated Quaternary sediments, and builds upon that of Halimov and Fezer (1989). It starts with the formation of susceptible deposits under humid conditions. Desiccation takes place and the sediments then suffer from initial incision by wind and/or by fluvial gullying. This produces a landscape of high ridges and mesas which are cut through to the base of the sediments by narrow corridors. Abrasion progressively leads to the widening of the corridors and the retreat of the noses of the ridges. This is a time when oversteepening leads to extensive failure of slopes, especially along cracks produced by desiccation and contraction.

The ridges gradually become transformed into cones, pyramids, saw-tooth forms, hogbacks etc. Then, after relief is reduced to less than about 2 m, the whole surface is subjected to abrasion to give a classic, simple, aerodynamic form, with secondary vortices, etc. Finally, a new low-level surface will be formed once the last vestiges of the yardangs have been removed.

Table 8.4 An evolutionary model of yardang development

Formation of lake bed, swamp deposits or cultivation deposits under humid
conditions in mid-Holocene
↓
Desiccation
↓
Incision of beds by wind and/or water towards base of sediment
↓
Formation of high ridges and mesas
↓
Abrasion leads to widening of corridors and plains between ridges, Oversteepening
causes mass movement. Salt weathering encourages deflation
↓
Ridges become transformed into cones, pyramids, saw tooth forms, hogbacks etc.
↓
After relief is reduced to less than 2 m, the whole surface is subjected to abrasion
to give simple aerodynamic form
↓
Gradual removal of sediment until plain surface formed

PANS

Introduction

In many parts of the world's drylands there are closed topographic depressions –
pans – many of which have a characteristic morphology that has been likened to a
clam, a heart or a pork chop. Recent analyses of available maps and remote sensing
imagery have demonstrated the great areal extent and wide distribution of these
features (Goudie and Wells, 1995). They are especially well developed in southern
Africa, on the High Plains of the USA, in the Pampas of Argentina, Manchuria,
western and southern Australia, and the West Siberian steppes and Kazakhstan.

One of the prime controls of pan distribution is the availability of susceptible
surfaces. In South Africa they are preferentially developed on the fine-grained Ecca
Shales and on the sandy Kalahari Beds. In the USA pans are especially well
displayed on the Ogallala sediments, especially where the calcrete caprock is thin or
lacking (Wood *et al.*, 1992). Examples are provided in Table 8.5. Other pans occur
in particular types of topographic situation. For example, palaeolacustrine pans are
widespread in many of the desiccated and deflated pluvial lake basins of the western
USA. Palaeodrainage pans occur where deflation has disrupted former river
courses, as in Western Australia (Figure 8.4) or in the case of some of the larger
basins in the Texan High Plains. Some of the South African pans have developed
where drainage systems have been dismembered or altered by tectonic deformation,
but most disrupted drainage systems are probably the result of climatic deteriora-
tion. A further class of pan is the interdunal type, formed by deflation of interdune
swales and the noses of parabolic dunes. Some of the aligned pans of the High
Plains of the USA are the relict interdunal basins of a formerly more extensive
Pleistocene sand ridge desert. Finally, there are coastal surface pans caused by

Table 8.5   Favoured materials for pan development

Location	Material
South Africa	Ecca and Dwyka shales and sandstones
Botswana	Kalahari beds (sands, calcretes etc.)
Zimbabwe, Zambia and Zaire	Kalahari sands
Namibia (Aminuis)	Karoo sediments
Tunisia	Neogene clays and sands, Cretaceous marls
Yorke Peninsula, S. Australia	Permian clays and sands, coastal sediments
Egypt (SW desert)	Nubian sandstone
Stirling area, W. Australia	Tertiary marine sediments, Middle Proterozoic sedimentaries
Mongolia	Sandstones, coarse granites
Altiplano of Chile, Bolivia and Argentina	Ignimbrite sheets and volcanoclastic sediments
High Plains, USA	Sediments (including calcrete), shales (Pierre, Carlile and Steele) and sandstone (Mesaverde and Fox Hills)

Figure 8.4 A group of characteristically shaped pans developed by deflation along an ancient drainage line in Western Australia

deflation of coastal sediments. Examples of these occur in the Cape Agulhas area of South Africa, the Pampa Deprimada in Argentina, and on the eastern seaboard of the USA in the shape of the controversial Carolina Bays.

Morphometry

One of the most important developments in pan studies in recent years has been the accumulation of quantitative data on pan morphology. Such studies plot pan densities (see, for example, Goudie and Wells, 1995, figs 2, 5, 11, 12, and table 2), and pan geometric characteristics, including the frequency distributions of pan area and long axis lengths (see Goudie and Wells, 1995, fig. 20; and Sabin and Holliday, 1995, fig. 8).

Of particular note has been the consideration of the quantitative relationships between pans and the lunette dunes that occur on some of their lee sides. For example, only 1100 of the 25 000 pans on the Southern High Plains of the USA have lunettes (Sabin and Holliday, 1995). Playa variables that seem to be strongly associated with the presence of these lunettes are area and depth, but we are still far from knowing why some pans have lunettes and others do not. Not all large or deep pans have lunettes, though in general few small pans have them.

The Stratigraphy of Pans and Lunettes

Analysis and dating of sediments on the floors of pans and in neighbouring lunettes (Figure 8.5) have done much to increase our knowledge of the age and history of pans. For example, Holliday et al. (1996) undertook a study of the lithostratigraphy and chronostratigraphy of a range of basins in the Southern High Plains, and found that many of them had a complex and lengthy history. Among their conclusions were the following (p. 953):

> . . . some basins have a prolonged history as depressions, persisting in more or less the same location as the High Plains surface aggraded by eolian addition (Blackwater Draw Formation) throughout the Pleistocene. Sizes of the basins varied through time as they were encroached upon by the Blackwater Draw Formation, enlarged by fluvial, lake margin, and aeolian erosion, were filled and re-exposed, or were buried. Some basins are newly formed on the High Plains surface and have no apparent predecessors.

Lunette sediments have been equally productive, and Holliday (1997, p. 54) proposed the following sequence of events in the High Plains:

> The lunettes began forming as low-carbonate sand dunes in the late Pleistocene as playa basins were formed or deepened by wind erosion. The erosion repeatedly alternated with stability. The environment probably was cool and dry, but one or more cool and wet intervals 25 000–8000 yr B.P. resulted in a rise in the water table and deposition of lacustrine carbonate in the deepest basins. There may have been short departures towards warmer (and probably towards drier) conditions throughout this time. Episodically dry conditions 15 000–8000 yr B.P. resulted in deflation of the carbonate and further dune construction by repeated accretion of calcareous sandy

Figure 8.5 One of the world's largest lunettes, developed to the lee of the Sebkha el Kelbia (Tunisia) and formed of deflated lake sediments

loam or loamy sand. The low carbonate sand was deposited during widespread drought and deflation 8000–5000 yr B.P. The dunes have been largely stable in the late Holocene.

This is a model that is broadly supported by a study of a lunette in the Central Great Plains of Kansas by Arbogast (1996). Comparable studies of lunette stratigraphy and dates are also being undertaken in southern Africa (Lawson, 1998).

The Origin of Pans

The origin of pans is something that has intrigued geomorphologists for over a century. Hypotheses have included deflation, excavation by animals, and karstic and pseudo-karstic solution. Arguments on this issue are recurrent and recent years have seen some important contributions to the debate (e.g. Grobler, 1989; Verhagen, 1991; Wood et al., 1992; Gustavson et al., 1995). What is becoming increasingly clear, however, is that a range of processes has been involved in the initiation and maintenance of pans and that no one hypothesis can explain all facets of their often long histories and their variable sizes and morphologies.

An attempt at presenting an integrated model of pan development was made by Goudie and Wells (1995) and this has now been developed further (Figure 8.6). Its essentials are as follows. First, pans occur preferentially in areas of relatively low effective precipitation. This predisposing condition of low precipitation means that vegetation cover is sparse and that deflational activity can occur. Moreover, once a small initial depression has formed, and the water in it has evaporated to give a saline environment, the growth of vegetation is further retarded. This further encourages

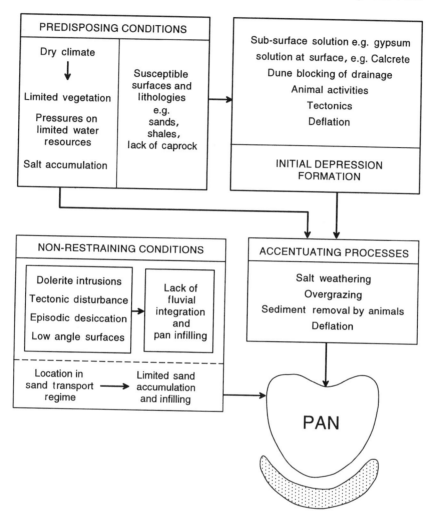

Figure 8.6 A model of pan formation

deflation. The role of deflation in the removal of material from a depression may be augmented by animals (Verhagen, 1991), who tend to concentrate at pans because of the availability of water, salt licks and a lack of cover for predators. Trampling and overgrazing expose the soil to deflation and the animals would also physically remove material on their skins and in their bladders. Aridity also promotes salt accumulation so that salt weathering (Figure 8.7) could attack the bedrock in which the pan might be located (Goudie and Viles, 1997). It is also important that any initial depression, once formed and by whatever means, should not be obliterated by the action of integrated or effective fluvial systems. Among the factors that can cause a lack of fluvial integration are low angle slopes, episodic desiccation and dune encroachment, the presence of dolerite intrusions and tectonic disturbance. This model of pan

Figure 8.7 Salt weathering of a boulder on the margins of Death Valley playa, California, USA. This produces fine material which can then be removed by deflation

formation is remarkably similar to that independently developed for the High Plains by Gustavson *et al.* (1995, pp. 38–39):

> Whereas the formation of playa basins involved many processes they most likely started with runoff collecting in small topographic depressions on the High Plains surface. Some playa basins began when subsidence was induced by salt dissolution, probably as early as late Ogallala time. Some playa basins probably began as solution pans on the caprock caliche as early as the late Pliocene. Others, which are underlain by parts of the Blackwater Draw Formation, are younger and may have been initiated by conditions such as differential compaction, animal wallows, or blowouts where vegetation was missing. Runoff that ponded in these depressions killed vegetation or hindered plant growth, thus facilitating deflation of surface sediment when the pond waters evaporated. As the first small basin expanded, fluvial erosion and lacustrine sedimentation became more important. Centripetal drainage enlarged the basin by eroding the basin margin and carrying sediment to the basin floor. Periodic flooding continued to keep the center of the playa basin poorly vegetated to relatively clear of vegetation. Wind deflated dry sediment from the playa center, and this deflation may have accelerated after large herds of passing bison pulverized dried surface soils;. sediments from these basins were carried downwind . . .

CONCLUSIONS

Recent developments in the study of both yardangs and pans have shown various similarities. First, we now have a relatively good idea of the distribution of both types of landform. This is largely because of the availability of remote sensing imagery. The picture is, however, not as complete as that for dunes (see McKee, 1979, for a comparison). Secondly, recent years have seen the accumulation of morphometric data both for yardangs and pans, but further work is required on the form of large yardang ridges developed in bedrock. Thirdly, our understanding of the age and history of these erosional forms has been substantially improved by the analysis of the sediments with which they are associated. Fourthly, it is clear that while the origin of both yardangs and pans is still the subject of debate, it is certain that they are the result of a complex mix of processes that has varied through time. Finally, there have been remarkably few field studies that have attempted to monitor wind flow and sediment movement on and around the landforms. This contrasts with what is now being achieved by those studying aeolian depositional landforms. Such data could feed into models of the interactions between wind flow characteristics and surface topography, and collection of these data provides a major research opportunity.

REFERENCES

Arbogast, A.F., 1996. Late Quaternary evolution of a lunette in the central Great Plains: Wilson Ridge, Kansas. *Physical Geography*, **17**, 354–370.

Blackwelder, E., 1934. Yardangs. *Bulletin Geological Society of America*, **45**, 159–166.

Breed, C.S., McCauley, J.F., Whitney, M.I. *et al.*, 1997. Wind erosion in drylands. In *Arid Zone Geomorphology*, D.S.G. Thomas (Ed.). Wiley, Chichester, pp. 437–464.

Brookes, I.A., 1986. Quaternary geology and geomorphology of Dakleh oasis and environs, south central Egypt: reconnaissance findings. *Discussion Paper 32*, Dept. of Geography, York University, Toronto.

Capot-Rey, R., 1957. Le vent et le modelé éolien au Bourkou. *Travaux, Institut Recherche Sahariennes*, **15**, 149–157.

Clarke, M.L., Wintle, A.G. and Lancaster, N., 1996. Infra-red stimulated luminescence dating of sands from the Cronese Basins, Mojave Desert. *Geomorphology*, **17**, 199–205.

Cooke, R., Warren, A. and Goudie, A., 1993. *Desert Geomorphology*. UCL Press, London.

Doornkamp, J.C., Brunsden, D. and Jones, D.K.C., 1980 *Geology, Geomorphology and Pedology of Bahrain*. GeoAbstracts, Norwich.

Gabriel, A., 1938. The southern Lut and Iranian Baluchistan. *Geographical Journal*, **92**, 193–208.

Goudie, A.S., Stokes, S., Cook, J. *et al.*, 1999. Yardang landforms from Kharga Oasis, south-western Egypt. *Zeitschrift für Geomorphologie, Supplementband*, **116**, 1–16.

Goudie, A.S., 1979. Arid geomorphology. *Progress in Physical Geography*, **3**, 421–426.

Goudie, A.S., 1989. Wind erosion in deserts. *Proceedings of the Geologists' Association*, **100**, 83–92.

Goudie, A.S. and Viles, H.A., 1997. *Salt Weathering Hazards*. John Wiley, Chichester.

Goudie, A.S. and Wells, G.L., 1995. The nature, distribution and formation of pans in arid zones. *Earth-Science Reviews*, **38**, 1–69.

Grobler, N.J., Loock, J.C. and Behownek, N.J., 1989. Development of pans in palaeo-drainage in the north-western Orange Free State. *Palaeoecology of Africa*, **19**, 87–96.

Grolier, M.J., McCauley, J.F., Breed, C.S. and Embabi, N.S., 1980. Yardangs of the Western Desert. *Geographical Journal*, **146**, 86–87.

Gustavson, T.C., Holliday, V.T. and Hovorka, S.D., 1995. Origin and development of playa basins, sources of recharge to the ogallala Aquifer, Southern High Plains, Texas and New

Mexico. *Bureau of Ecoomic Geology, University of Texas, Report of Investigations*, No. 229, 44 pp.

Halimov, M. and Fezer, F., 1989. Eight yardang types in Central Asia. *Zeitschrift für Geomorphologie*, **33**, 205–217.

Hedin, S., 1903. *Central Asia and Tibet*. Scribners, New York.

Holliday, V.T., 1997. Origin and evolution of lunettes on the High Plains of Texas and New Mexico. *Quaternary Research*, **47**, 54–69.

Holliday, V.T., Hovorka, S.D. and Gustavson, T.C., 1996. Lithostratigraphy and geochronology of fills in small playa basins on the southern High Plains, United States. *Bulletin Geological Society of America*, **108**, 953–965.

Jones, L.S. and Blackey, R.C., 1993. Erosional remnants and adjacent unconformities along an eolian-marine boundary of the Page Sandstone and Carmel Formation, Middle Jurassic, South-Central Utah. *Journal of Sedimentary Petrology*, **63**, 852–859.

Laity, J.E., 1994. Landforms of aeolian erosion. In *Geomorphology of Desert Environments*, A.D. Abrahams and A.J. Parsons (Eds). Chapman and Hall, London, pp. 506–535.

Lancaster, N., 1984. Characteristics and occurrence of wind erosion features in the Namib Desert. *Earth Surface Processes and Landforms*, **9**, 469–478.

Lawson, 1998. Personal communication.

Livingstone, I. and Warren A., 1996. *Aeolian Geomorphology*. Longman, Harlow.

Mainguet, M., 1968. Le Bourkou. Aspects d'un modèle éolien. *Annales de Géographie*, **77**, 296–322.

Mainguet, M., 1972. *Le Modelé des Grés*. IGN, Paris.

McCauley, J.F., Grolier, M.J. and Breed, C.S., 1977a. Yardangs of Peru and other desert regions. *USGS Interagency Report: Astrogeology*, **81**, 177.

McCauley, J.F., Grolier, M.J. and Breed, C.S., 1977b. Yardangs. In *Geomorphology in Arid Regions*, D.O. Doehring (Ed.). Proceedings of the 8th Annual Geomorphology Symposium, pp. 233–269.

McKee, E.D. (Ed.), 1979. Global Sand Seas. *USGS Professional Paper*, 1052.

Rea, D.K., 1994. The palaeoclimatic record provided by eolian deposition in the deep sea: the geologic history of wind. *Reviews of Geophysics*, **32**, 159–195.

Riser, J., 1985. Le rôle du vent au cours des derniers millénaires dans le bassin Saharien D'Araouane (Mali). *Bulletin de l'Association de Géographes Français*, **62**, 311–317.

Sabin, Ty.J. and Holliday, V.T., 1995. Playas and lunettes on the Southern High Plains: Morphometric and spatial relationships. *Annals of the Association of American Geographers*, **85**, 286–305.

Tewes, D.W. and Loope, D.B., 1992. Palaeo-yardangs: wind scoured desert landforms at the Permo-Triassic unconformity. *Sedimentology*, **39**, 251–261.

Verhagen, B.T., 1991. On the nature and genesis of pans – A review and an ecological model. *Palaeoecology of Africa*, **21**, 179–194.

Ward, A.W. and Greeley, R., 1984. Evolution of the yardangs at Rogers Lake, California. *Bulletin Geological Society of America*, **95**, 829–837.

Whitney, M.I., 1983. Eolian features shaped by aerodynamic and vorticity processes. In *Eolian Sediments and Processes*, M.E. Brookfield and T.S. Ahlbrandt (Eds). Elsevier, Amsterdam, pp. 223–245.

Wood, W.W., Sanford, W.E. and Reeves, C.C., 1992. Large lake basins of the southern High Plains: groundwater control of their origin? *Geology*, **20**, 535–538.

9 Dust Transport and Deposition

Faculty of Environmental Sciences, Griffith University, Australia

There is a long history of reporting dramatic dust transport and deposition events, but a more structured and systematic approach to aeolian dust research has developed only in the last 15 years. Spatial and temporal patterns of dust transport have been well described at a broadscale using meteorological records of dust storm frequency. Dust sampling and other direct measures of dust concentrations have also advanced our knowledge. The most exciting recent developments are in tracking dust plumes using air parcel trajectory analyses and remote sensing. Dust deposition has generally been seen to be easy to study, hence it often has not been systematically studied. Deposition measurements have been made by a variety of, often crude, methods with the result that the data are variable in quality. Systematic studies of dust deposition have been slow to arrive, and even today there are few studies of complete dust transport systems. A conceptual model of dust transport and deposition is proposed which demonstrates some of the spatial and temporal complexities of dust transport systems, and data from a variety of studies are viewed within this framework. Understanding of the roles of dust deposition in soil formation (particularly desert loess), and dust-derived nutrient contributions to terrestrial, lacustrine and marine ecosystems, has been slow to emerge. At this point in time, we appear to be at the start of a period of more systematic dust deposition research, which should demonstrate to the scientific community the major role which dust plays in environmental processes.

INTRODUCTION

There is a long history in the research literature of studies of dramatic dust transport and deposition events. One of the earliest reports is from Charles Darwin's (1845) voyage around the world in the *Beagle*. The *Beagle* was coated in dust as it sat becalmed 300 mile off the West African coast in a Harmattan dust plume. Since then there have been many short papers on dramatic dust events, such as the report by Pitty (1968) of the Saharan dust plume which passed over London in July 1968. A significant number of these reports have appeared in the journal *Nature*. While these opportunistic reports of dust events have served to demonstrate the vagaries of nature, their generally superficial treatment of dust processes and rule of thumb estimates of dust loads and deposition rates have meant that, despite a relatively long history of research attention, knowledge of dust transport and deposition processes has, until recently, remained rudimentary and out of the mainstream of aeolian research.

Aeolian Environments, Sediments and Landforms. Edited by A.S. Goudie, I. Livingstone and S. Stokes.

There has been a significant change in research focus over the past few years, towards more systematic and rigorous dust research investigations and researchers have been rewarded with a dramatic advancement in our knowledge of dust transport and deposition. Tangible evidence of this new research attention is available in the emergence of the first specialist aeolian dust text (Pye, 1987) and in the much greater attention given to aeolian dust processes in recent aeolian geomorphology texts (e.g. Livingstone and Warren, 1996) than in earlier texts. A comparison of the outcomes of two meetings of aeolian dust researchers 18 years apart also evidences the much greater research activity in recent times. The 1977 Gothenburg Workshop on Saharan Dust (Morales, 1979) was a small gathering of researchers with 16 published papers, while the 1995 conference on African Dust Across the Mediterranean (ADAM) (Guerzoni and Chester, 1996) which, although a more narrowly focused meeting, resulted in 39 published papers.

This review will summarise the state of knowledge of dust transport and deposition and will discuss past trends and future possibilities. There will be at least two perspectives which reflect the research background of the author. Greater attention is given to dust deposition than transport, and information on dust processes in Australia figures more prominently than would be the case if weight of research attention alone was the determinant.

DUST TRANSPORT

An analysis of images from the NOAA-11 polar orbiting satellite of the volcanic dust ejected into the air by the eruption of Mount Pinatubo (Aldhous, 1991) (Figure 9.1) provides a fascinating picture of the dispersion of these particulates around the globe in the low latitudes. Figure 9.1 also shows the real impact which aeolian dusts can have upon the earth's atmosphere. Possibly of greater interest to aeolian dust researchers is that the 'before', or 'blank' image, in Figure 9.1, which received only passing mention in the above study, shows the very substantial amounts of dust being emitted from the Sahara and the Arabian deserts.

It is now well established within the aeolian research community that large quantities of dust are transported great distances within the earth's atmosphere. Atmospheric dust concentrations range from $100\,000\ \mu g\ m^{-3}$ near source to $< 1\ \mu g\ m^{-3}$ over the oceans, and as the number of studies, and their level of sophistication, increase the size of dust load estimates also seem to increase. According to D'Almeida (1986) about one billion tonnes per year of Saharan dusts alone are transported in the atmosphere. Several reviews of dust transport are available, for example, Goudie (1978, 1983), Pye (1987), Prospero (1996).

Early studies of individual dust transport events, have provided generalised evidence of the magnitudes, sources and trajectories of dust transport events, but sometimes the data used were anecdotal and subjective. A case in point might be the reports of a dramatic dust storm which engulfed Melbourne, Australia on 8 February 1983 which was recorded with impressive photographs. The dust load of this storm was estimated at $140\,000\,t$ by the Victorian Department of Conservation, Forests and Lands, and the dust source area was attributed to the wind erosion-

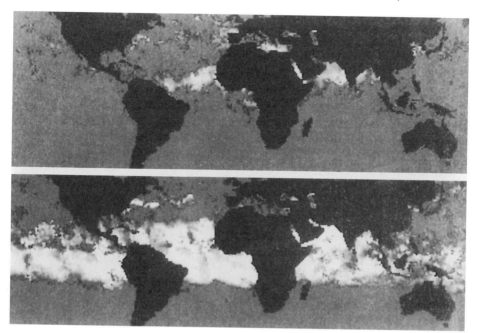

Figure 9.1   Before and after the Mt Pinatubo eruption in the Philippines: an example of the impact of dust upon the Earth's atmosphere (after Aldhous, 1991). Reproduced by permission of Macmillan Magazines

prone Mallee region of NW Victoria, around Mildura. In a subsequent, more detailed study by Raupach *et al.* (1994) the dust load was conservatively estimated to be 2 ± 1 Mt, and an analysis of the three hourly meteorological records showed that Mildura was not the source. While the Mildura area also experienced a dust storm that day, this dust did not reach Melbourne, because the winds were southerly (blowing in the opposite direction to Melbourne). The actual source area of this dust storm was western Victoria. This example highlights that, while one off studies provide useful information on dust transport events, without more detailed measurements or models, such studies may not contribute significantly to the body of knowledge of dust processes. Furthermore, with the emphasis upon, dramatic events, there is also a potential danger of creating the false impression that such events are commonplace, or conversely that if events of such a magnitude do not occur, nothing is happening.

Spatial and temporal patterns of dust transport have been described using meteorological records of dust events in several continents (summaries in Goudie, 1978, 1983; Middleton 1997). This approach has the advantage that the data are readily available from WMO stations in most countries and over long periods. For example, in Australia, 'quality' daily data are available for 72 stations for the past 31 years (Figure 9.2). This map, however also highlights a weakness of this approach; the low spatial resolution of meteorological stations. Questions also exist as to the reliability of some records. Records are most reliable for dust storms and smaller

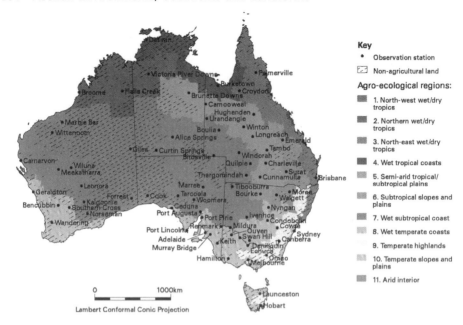

Figure 9.2 Meteorological stations with quality dust event records (1965–1996) within Agro-Ecological Regions (after McTainsh, 1998). Reproduced by permission of CSIRO Publishing

scale entrainment events, close to source, but dust haze records are less reliable. Dust entrainment events have recently been combined into a dust storm index (McTainsh, 1998). The weightings given to each of these event types within the DSI are based upon measured dust concentrations in relation to visibility.

$$DSI = (5 \times SD) + MD + (^{LDE}/_{20}) \tag{1}$$

where: DSI = dust storm index (in dust event days); SD = severe dust storm (present weather codes: 33, 34, 35 and 98)*; MD = moderate dust storm (present weather codes: 09, 30, 31 and 32)*; LDE = local dust event (present weather codes: 07 and 08)*.

Figure 9.3 shows wind erosion activity, as measured by the dust storm index, throughout Australia during 1986–1996. The effects of long-distance dust transport from inland Australia, on air pollution in eastern Australia cities has recently been inferred from relationships between dust storm occurrence and total suspended particulate measurements in Brisbane (Figure 9.4). Using empirical relationships between visibility and dust concentrations progress is being made in Australia towards estimating dust storm fluxes from these meteorological records (Reddan, 1994). The relationship between visibility and dust concentration is, however,

---

* Bureau of Meteorology (1982)

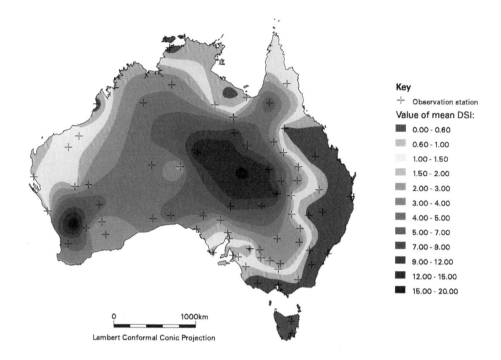

Key

+ Observation station

Value of mean DSI:

■ 0.00 - 0.60
▨ 0.60 - 1.00
□ 1.00 - 1.50
▨ 1.50 - 2.00
▨ 2.00 - 3.00
■ 3.00 - 4.00
■ 4.00 - 5.00
■ 5.00 - 7.00
■ 7.00 - 9.00
■ 9.00 - 12.00
■ 12.00 - 15.00
■ 15.00 - 20.00

0   1000km

Lambert Conformal Conic Projection

Figure 9.3 Dust events in Australia (1986–1996) based upon the Dust Storm Index (after McTainsh, 1998). Reproduced by permission of CSIRO Publishing

particle-size dependent, and as particle-size changes with transport distance, a given level of visibility impedance will represent decreasing dust concentrations with transport distance.

The increasing use of automatic instrumentation, in place of meteorological observers, in WMO networks is both good and bad news for research in dust transport and deposition. The good news is that it will increase the quantification of meteorological observations. For example, the use of instrumentation for visibility measurement will increase the reliability and resolution of visibility data, and may open up possibilities of obtaining better relationships between visibility, and dust concentrations and deposition rates. The bad news is that the observer record of dust event occurrence which has provided such a good record of broadscale dust entrainment and transport processes, may deteriorate and eventually become unuseable, as a result of the reduction in manual observations at meteorological stations,

More direct measurements of dust concentrations have been made using sun photometers and other measures of atmospheric turbidity in and around the Sahara (Jaenicke, 1979), and by direct sampling of dusts in Mali (Nickling and Gillies, 1993), Australia (Nickling *et al.*, 1999) and elsewhere. The longest quantitative record of dust transport is that of Prospero and colleagues (summary by Prospero, 1996) who have been monitoring Saharan dust at Barbados since the 1960s, and which has yielded valuable information about Saharan dust transport over the

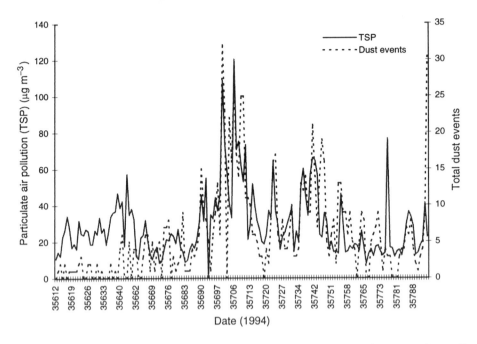

Figure 9.4 Relationship between dust event occurrence in inland eastern Australia and particulate air pollution in Brisbane (1994) (after McTainsh, 1998). Reproduced by permission of CSIRO Publishing

Atlantic ocean to the West Indies. Dust outbreaks from the West African mainland occur every 3–4 days during the northern hemisphere summer months and most of the dust in these plumes is transported westwards in the high altitude Saharan Air Layer, taking about one week to reach the Caribbean. Winter is the season of the Harmattan wind, which carries large quantities of dusts south westwards from source areas in the southern Sahara and Sahel savanna in a more southerly path over the tropical Atlantic Ocean, reaching South America (Prospero *et al.*, 1983).

AIR PARCEL TRAJECTORY MODELS AND REMOTE SENSING

Remote sensing techniques have offered great potential for measuring dust loads and tracking dust storm trajectories, but until recently this potential has not been realised. The recent, significant, advances which have been in remote sensing of dust is evidenced in the various papers presented at the ADAM conference in Sardinia (Guerzoni and Chester, 1996). For example, Dulac *et al.* (1996) describe measurements of dust optical depth using Meteosat imagery under non-cloud conditions over oceans, which offer uniform reflectance characteristics compared with continental surfaces. Their temporal trends in Saharan dust loads over the Mediterranean Sea correlate in general terms with other direct monitoring trends from Barbados

(Prospero and Nees, 1986) and Cape Verde Island, and with the occurrence of drought in dust source areas. Meteosat images are also used by Schulz *et al.* (1996) to test a 3-D dust transport model description of a dust event originating from the central western Sahara. As Figure 9.5 shows, the model and satellite image maps correlate quite well. This approach offers great promise for the more accurate identification of dust source areas, plume trajectories and dust loads.

Air parcel trajectory models provide greater precision in sourcing individual dust events than do satellite imagery. Early air parcel trajectory analyses were used for tracking radioactive dusts (Gabites, 1954) from British atomic testing in Woomera, South Australia but the routine use of computer-based air parcel trajectory analysis methods in weather forecasting did not start till the mid 1980s. Alarcon *et al.* (1996) use air parcel trajectories to model nutrient fluxes and deposition of Saharan dusts in the western Mediterranean. Swap *et al.* (1992) used wind field data from the European Centre for Medium Range Weather Forecasting (ECMWF) to track dust plumes from the Sahel region of the Sahara to the Amazon. Reiff *et al.* (1986) also used the ECMWF model to source a dust deposition event over the Netherlands in April, 1983, back to Mauritania and the Sahel region.

The spatial resolution at which dust source area(s) are defined remains at a low level despite the aid of remote sensing and air parcel trajectory models. While this may be a minor issue from a dust transport perspective, from a wind erosion perspective it is important to be able to specify a dust source area to a few square kilometres. In reality, rather than there being a single dust source area with constant source strength, there is usually a hierarchy of; dust source areas within a dust source region (Figure 9.6), and dust source regions within a desert. The dust plumes from these areas, then regions, mix downwind. The dust load within the plume and its spatial extent will depend upon the size, location and source strength of these various sources. Butler *et al.* (1996) are developing DUSTRAN; a source-based Gaussian plume dust transport model which describes the mixing and downwind dispersion of plumes from source areas of different source strengths and locations. Figure 9.7 is a three-dimensional simulation of the plume mixing that results 10 km downwind from three equal strength linear dust sources, two of which combine into one plume and a third which remains separate.

DUST DEPOSITION

Dust deposits are usually more readily observed than dust in transport, and possibly for this reason there have been even more opportunistic studies of dust deposition than dust transport. Dust deposits have been measured on houses (Péwé *et al.*, 1981, Schroeder, 1985), cars (McTainsh *et al.*, 1996), plant leaves (Gillette and Dobrowolski, 1993), snowfields (Franzen *et al.*, 1994; Wagenbach *et al.*, 1996) and on a variety of other surfaces and settings. Dust deposition in rain, which is variously referred to as blood rains and red rains (Anon, 1945) has been a particular source of fascination. Recently dust deposition research has been less opportunistic and better planned to test particular research hypotheses.

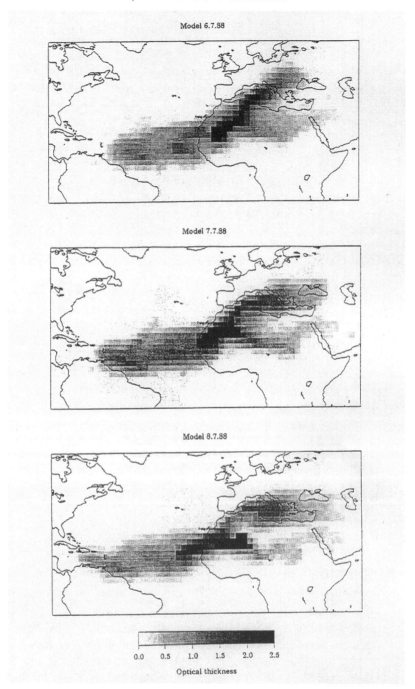

Figure 9.5   A comparison of a Saharan dust plume modelled by a tracer model and observed from Meteosat satellite imagery (after Schultz *et al.*, 1996). Reproduced by permission of Kluwer Academic Publishers

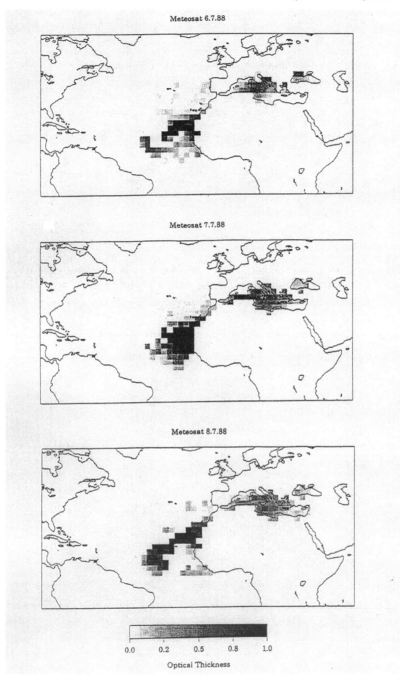

Figure 9.5   *(continued)*

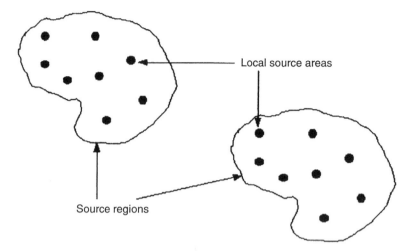

Figure 9.6   Schematic representation of multiple source areas within source regions (after Butler *et al.*, 1996). Reproduced by permission of Elsevier Science Ltd

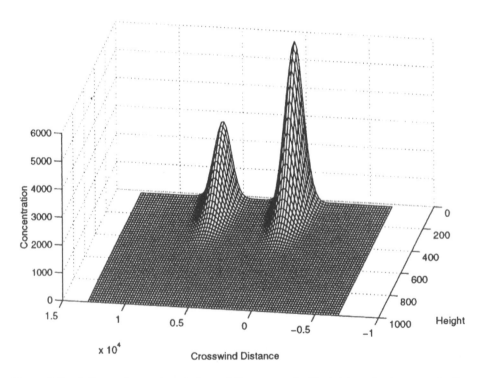

Figure 9.7   A three-dimensional view of the crosswind/height *vs* dust concentration profile resulting from the mixing of three dust sources (after Butler *et al.*, 1996). Reproduced by permission of Elsevier Science Ltd

Table 9.1 Annual dust deposition rates on land (after Middleton, 1997 and Goudie, 1995)

References	Location	Deposition rate	
		$(t\ km^{-2}\ a^{-1})$	$(mm\ 1000\ a^{-1})^a$
Sahara			
Mediterranean region			
Löye-Pilot *et al.* (1986)	Corsica	14	16
Yaalon and Ganor (1975)	Israel	22–83	25–93
Goossens (1995)	Negev	15–30	10–20
Bücher and Lucas (1975)	Pyrenees	18–23	20–26
Pye (1992)	Crete	10–100	11–112
Harmattan plume			
Maley (1980)	S Chad	109	122
McTainsh and Walker (1982)	N Nigeria	137–181	154–203
Drees *et al.* (1993)	SW Niger	200	100–150
USA			
Smith *et al.* (1970)	High Plains	65–85	73–96
Péwé *et al.* (1981)	Arizona	54	61
Gile and Grossman (1979)	New Mexico	9.3–125.8	10–141
Muhs (1983)	California	24–31	27–35
Reheis and Kihl (1995)	California/Nevada	4.3–15.7	5–18
Middle East			
Safar (1985)	Kuwait	100	112
Behairy *et al.* (1985)	W Saudi Arabia	13–109	15–122
Miscellaneous			
Inoue and Naruse (1991)	Japan	3.5–6	4–7
Tiller *et al.* (1987)	SE Australia	5–10	6–11
Kukal (1971)	Caspian Sea	39.5	44

^{a} Calculated on bulk density of dust of 0.89 g cm^3 where not derived in original reference.

Dust Deposition Rates

Dust deposition rates vary considerably through time and space. Goudie (1978) collated several published reports with absolute rates in excess of 1 000 000 tons and in excess of 300 tons km^{-1} per event. A large range of annual dust deposition rates (4.3 to 200 t km^{-2} per annum) is reported by Middleton (1997) (Table 9.1).

The large range in these annual rates may reflect a variety of factors related to: sampling techniques (which are discussed later), or sampling times and sampling locations, in addition to the magnitude of the dust transport and deposition systems being described. Calculations of annual average dust deposition rates based upon short sampling times can lead to unrepresentative results, because dust processes are highly spasmodic. For example, in a dust deposition study in northern Nigeria, McTainsh and Walker (1982) reported Harmattan dust deposition during a 12-day period (3–14 March, 1977) of 54 t km^{-2} which was 39 per cent of the total seasonal deposition, which lasted 8 months.

Reliable long-term rates are not common, and cover a wide range depending upon location in relation to dust source. Also, presenting deposition rates from a single location without taking into account distance to source(s) can make it difficult to realistically assess the significance of the data. For these reasons, probably only a small number of dust deposition rate studies have produced deposition rates which are representative of a particular region or provide an adequate temporal average. With the help of a conceptual model of dust deposition, it is, however, possible to reduce some of the complexity and ambiguity associated with interpreting dust deposition rate data from different studies.

A Conceptual Model of Dust Deposition

To realistically interpret deposition rates they need to be placed within the spatial and temporal context of the dust transport system which they represent. A comparison with fluvial systems should serve to identify the essential elements of such a system. Fluvial systems are spatially constrained by river catchment boundaries, and within them transport is channelised and therefore follows predictable paths. It is therefore a relatively simple proposition to measure the sediment discharge from a catchment at the mouth of that catchment. In contrast, aeolian dust systems have few spatial or temporal limits, as they can potentially operate anywhere and at any time, and the sediments can be transported any direction and over distances ranging from a few metres to several thousand kilometres. It is therefore much more difficult to make dust measurements at one or two locations and from these come to conclusions about dust source areas, transport loads, dust pathways or dust deposition rates.

Yaalon (1987) made some progress in resolving these spatio-temporal aspects of dust transport and deposition by classifying dust conditions into three types: long-distance dusts, medium distance dusts and local dusts, each of which has particular sources. The following model of dust deposition (Figure 9.8) extends this approach, by providing a conceptual framework, and a set of terms, to describe spatial and temporal aspects of dust systems in a downwind sequence.

Dust is initially entrained by *primary entrainment* (*Ep*) at a *primary dust source area(s)*, then is transported downwind within a plume, depositing increasingly fine dusts as *primary deposits* (*Dp*) with transport distance and at a diminishing deposition rate. If windspeeds decrease after leaving the primary dust source area, primary deposition rates will be high in the *dust deposition zone* and will decrease with transport distance. If windspeeds in the Deposition Zone subsequently increase, these primary deposits may be re-entrained by *secondary entrainment processes* (*Esa*) or human-induced activities, such as vehicle or stock movements, may re-entrain these dusts (*Esh*) converting this area into a *secondary entrainment zone*, from which the dusts will be transported downwind, depositing increasingly fine dusts as *secondary deposits* (*Ds*) with transport distance and at a diminishing deposition rate. These entrainment/deposition processes may be repeated several times and if wind directions change the dusts may be recycled within the deposition/entrainment zone. Eventually, the dusts may reach a well vegetated stable environment of deposition and make a net contribution to an ecosystem.

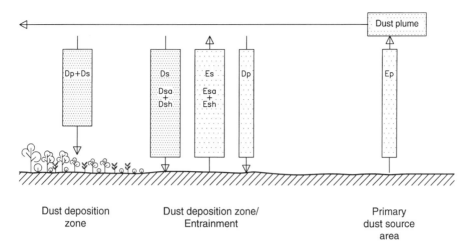

Figure 9.8 A conceptual model of dust transport and deposition

To demonstrate the utility of this approach, we shall examine data from the south east dust path in Australia (Figure 9.9). Dust storm dusts are entrained from the primary dust source areas in the floodplains of the Murray-Darling river system in north western Victoria and western New South Wales, and in the Channel Country of western Queensland (as described by Nickling *et al.*, 1999 and McTainsh *et al.*, 1999). These dust storms have high concentrations (Ep) at source, reaching 85 860 μg m^{-3} in a small dust storm at Diamantina Lakes (400 km north east of Birdsville – Figure 9.9) in 1996, and deposition rates (Dp) at source ranging from 0.003 to 0.174 t km^{-2} h^{-1}. Dust concentrations, deposition rates and the particle size of the dusts decrease with transport distance. Dust deposit particle-size characteristics at source vary considerably depending upon source area characteristics; with modes from 20–40 μm at the fine Diamantina Lakes claypan source, to 75 μm at the sandier Birdsville, on the edge of the Simpson Desert during a dust storm in 1987 (Knight *et al.*, 1995). During the same event, 600 km downwind at Charleville, within the deposition/entrainment zone, dust plume concentrations and primary deposition are lower (2130 μg m^{-3} and 0.0124 t km^{-2}, respectively) and dust deposit modal particle sizes have reduced to 15 μm. A further 700 km downwind in Brisbane, during the same event, dust concentrations were reduced to 149 μg m^{-3}.

Secondary entrainment (Es) within the deposition zone can produce high dust concentrations and deposition rates, but are usually lower than in the primary dust source area, because of higher vegetation cover levels. Background dusts within the deposition zone at Charleville have low dust concentrations (26 μg m^{-3}) and deposition rates (31.4 t km^{-2}). These suspended and deposited background dusts typically mode at 15 μm and the deposited dusts have a coarse tail to 150 μm. In summary, it should be apparent that dust concentrations, deposition rates and particle sizes can vary considerably according where and when they are sampled.

Let us now extend the timescale of this discussion into a geological time context, and to the formation of dust deposits and soils. The critical spatial relationship here

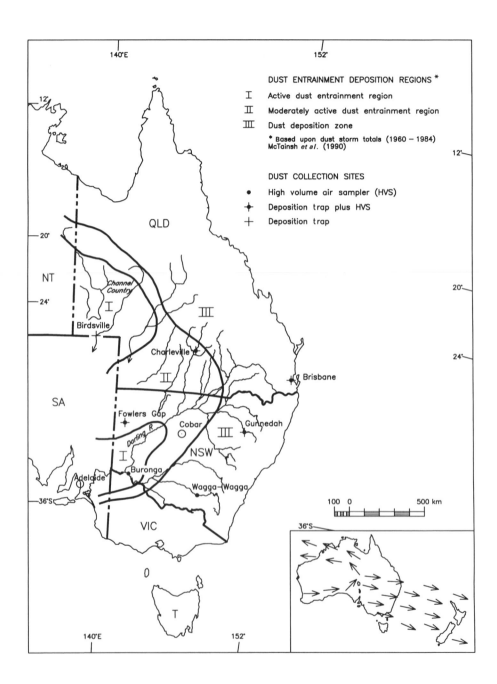

Figure 9.9 Dust entrainment and deposition regions within south east Australia

is between decreasing dust deposition rates (and hence diminishing potential to form deposits) with distance from primary dust source, and the increasing stability of the environment of deposition (and hence increasing potential to stabilise deposits) away from source. Given these opposing spatial trends, it is only at a certain critical distance from source that stable dust deposits can form. This may, in part, explain why loess deposits are seldom coarser than $60\,\mu$m or finer than $10\,\mu$m. In the east Australian context, there is another element to this spatial relationship, which is the interaction between the easterly moving south east dust path and the westerly flowing inland-draining river systems, which may result in a long-term sediment recycling system. The dusts deposited within the east Australian entrainment and deposition zones, which are not stabilised, can re-enter the rivers of the Murray-Darling catchment via runoff and return to the western floodplain primary dust source areas via the rivers. McTainsh (1985) contended that such alluvial-aeolian interrelationships may be an important element of the long-term sustainability of these sediment supply-limited dust transport systems and to the lowering of the east Australian landscape. This sediment recycling phenomenon may also operate in river systems in the Sahara, including: the Chari-Logone Rivers in the Lake Chad Basin, the Niger Inland Delta region of Mali and possibly the River Nile in Egypt.

The utility of the dust deposition conceptual model can also be seen in a Saharan dust context. The Harmattan season in West Africa lasts up to eight months each year (October to May), during which time, north easterly winds transport dusts from the Bilma-Faya Largeau area of the Lake Chad Basin, a primary dust source area, over the West African savanna dust deposition zone (Figure 9.10) producing significant deposits in Nigeria (137–181 t km^{-2} per annum, McTainsh and Walker, 1982), in Niger (200 t km^{-2} per annum, Drees et al., 1993) and elsewhere in West Africa. At the end of the Harmattan season unstable atmospheric conditions develop within the dust deposition zone as a result of the northward migration of the Inter-tropical Convergence Zone (ITCZ). Intense thunderstorms and associated downdraughts can develop, resulting in the re-entrainment of large quantities of primary dust deposits (Dp) during the Harmattan season (Figure 9.11), leading to secondary deposition downwind.

One such downdraught storm occurred in the Gao area of Mali in late April 1990. The dust from this event, which is described by Gillies et al. (1996), passed over the inland delta region of the Niger River reducing visibility to 150 m and increasing dust concentrations above 13 000 μg m^{-3} (<10 m). Other West African dust deposition studies also record high deposition rates towards the end of the Harmattan season, when the savanna dust deposition zone becomes an entrainment zone. Drees et al. (1993) found that over half of the annual dust deposition in Niger occurred during April to July, after the main Harmattan season. McTainsh and Walker (1982) recorded the highest deposition rates in the 1978–1979 season in April, in Nigeria, and Tiessen et al. (1991) recorded similarly increased deposition rates in the latter half of the Harmattan season in Ghana.

The effects of sample location, in relation to primary dust source area, upon deposition rates are also evident in the following comparison of studies of dust deposition on the northern side of the Sahara. Avila et al. (1996) report a deposition rate of 5.1 t km^{-2} per annum (over 11 years) of north Saharan dust over N.E. Spain,

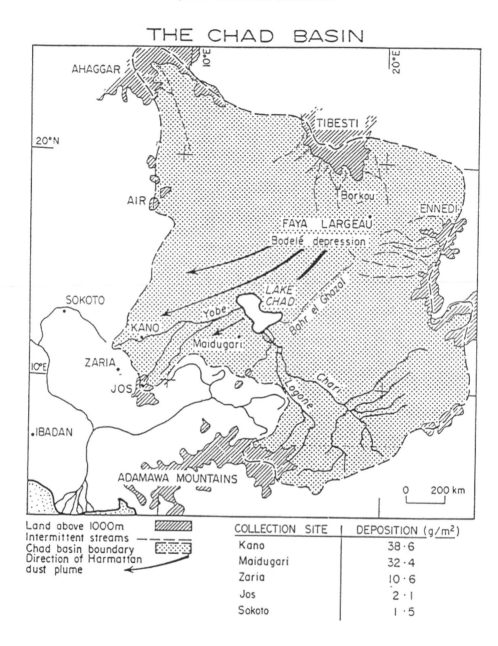

Figure 9.10 Harmattan dust entrainment and deposition regions in the southern Sahara (after McTainsh and Walker, 1982). Reproduced by permission of E. Schweizerbartsche

Figure 9.11 Local dust entrainment at Kano Northern Nigeria towards the end of the Harmattan season (during and after the event)

while closer to source, in Israel, Ganor and Foner (1996) (over 33 years) report 30–60 t km^{-2} per annum.

An alternative approach, to piecing together dust deposition and other data from independent studies in order to reconstruct dust transport and deposition systems, is to establish large dust deposition networks. Reheis and Kihl (1995) report the first regional scale and longer term dust deposition study in western USA, involving 55 dust deposition traps in southern Nevada and south eastern California over 5 years. Deposition rates ranged from 4.3 to 30 t km^{-2} per annum. These annual rates are slightly lower than those arising from the eastern Australia dust deposition network, which range from 43.8 t km^{-2} per annum at Fowlers Gap (near Broken Hill) which is within the dust entrainment zone (Figure 9.9), but not immediately at source, to 33 t km^{-2} per annum at Gunnedah 800 km downwind, within the deposition zone.

Rainfall and Dust Deposition

Up to this point in the discussion emphasis has been upon dry dust deposition, however there is recent evidence which suggests that rainfall-induced dust deposition (or wet deposition) is also an important process, particularly once dust plumes pass over humid regions and out over oceans. The role of rainfall in deposition has received increasing research attention recently. According to Loye-Pilot and Martin (1996) wet deposition is the major mode of Saharan dust deposition over the Mediterranean region. Prakasa Rao *et al.* (1992) found that wet deposition rates exceeded dry rates in an urban environment in India, and Gao *et al.* (1997) modelled wet and dry deposition in the nearshore and offshore regions in the China Sea and concluded that wet deposition contributed to 37–69 per cent of total deposition. These recent results have particular significance for studies of dust contributions to oceans (more on this later) as it means that both the particle-size characteristics and the quantities of dust deposits over oceans may be quite different as a result.

Dust Deposition Measurement Techniques

A large variety of dust deposition measurement techniques have been used over the years, and because few have been fully tested, the accuracy of these data remains unverified. Even putting aside the more unconventional opportunistic sampling techniques, such as collecting dust deposits upon cars, and table tops etc, few of the plethora of deposition measurement techniques have been rigorously tested. For example, when the dry deposition traps used by McTainsh (1980) in Nigeria were tested against the same traps containing distilled water, the dry traps were found to be 64 per cent efficient over a monthly collection period, and that efficiency decreased for longer sampling periods. A more recent study by Adams (1997) of dust deposition onto glass slides showed that their efficiency dropped off significantly after 1 week.

Amongst the variety of dust deposition traps which have been used, traps containing marbles or glass beads to stabilise the deposited dusts, appear to be the most popular. Reheis and Kihl (1995) used a cylindrical trap containing glass marbles in their long-term dust deposition study in the western USA, and in another long-term study on the Columbia Plateau, USA, Busacca *et al.* (1996) used large (0.8 m) trays

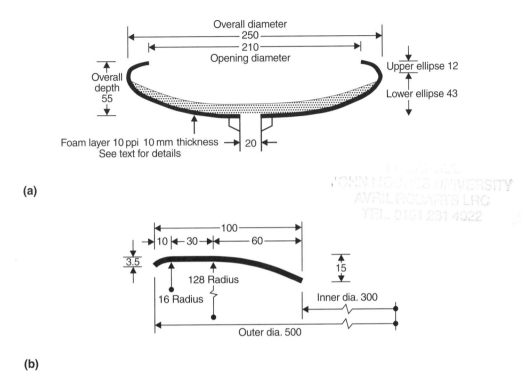

Figure 9.12   A new design dust deposition trap (after Hall *et al.*, 1994). (a) Collecting bowl for deposit gauge. (b) Deflector ring for deposit gauge. Dimensions in mm. Reproduced by permission of Elsevier Science Ltd

containing marbles. Kiefert (1995) used smaller plastic box traps (surface area 0.113 m²) containing two layers of 1 cm diameter glass beads. Few of these traps have, however, been fully tested.

A higher level of rigour has generally been applied to the development and testing of deposition measurement techniques in the particulate air pollution research area, than in aeolian geomorphology. The need to identify formal air quality standards is probably the driving force behind this greater emphasis upon tested and standard-ised dust deposition measurement techniques. The design and testing work of Hall and colleagues on dust deposition traps is a good example of this approach. The new dust deposit gauge of Hall *et al.* (1994) is the outcome of several years of testing, which started with the unconventional 'inverted frisbee' dust trap (Hall and Upton, 1988). The new dust trap (Figure 9.12) is shallow to reduce eddying; a negative effect demonstrated by Hall *et al.* (1994) in the British Standard deposit gauge, and aerodynamically shaped with an air deflecting ring which provides flat unaccelerated air flow over the gauge opening. A layer of foam is used to eliminate rainsplash and assist particle retention.

Bird strike and contamination of dust deposition traps has been an ongoing problem which has been addressed with a variety of structures. Recently, Drees and Manu (1996) demonstrated the severity of the problem, in their description of how bird urate contamination of dust traps affects the nutrient composition of dust deposits. A string bird strike preventer is used on the Hall et al. (1994) dust deposit gauge, however its effects upon the aerodynamics of the trap are not described. After trying a variety of bird deterrents to avoid contamination of wet traps, McTainsh (1980) used the simple solution which exploited the laziness of birds when drinking. A sacrificial dust trap kept full of water for the birds to drink with ease was set up adjacent to the main trap, the water level of which was a few centimetres below the lip of the trap. A novel approach was used by Stoorvogel et al. (1997) in measuring contamination in their study of Harmattan dust deposition in the rainforests of Ghana. They used Ti as a reference element, as its concentration in Harmattan dust is apparently consistent. Decreases in Ti concentrations beyond that expected for uncontaminated dusts were measured and corrections applied. This approach is worthy of further testing.

Studies of dust contributions to lakes can provide a valuable long-term record of dust deposition rates, but these are relatively rare because of the difficulties of differentiating alluvial inputs to the lakes, from aeolian inputs. The study by Busacca et al. (1996) of dust deposition into a 'pothole lake' without significant alluvial inputs, in basalt terrain on the Columbia Plateau, USA allowed these authors to discriminate the component of dust deposition due to human activities. A similar project is in progress on dust contributions to lake sediments on Fraser Island on the Queensland coast over the past 150 000 years (Longmore and McTainsh, 1997). Fraser Island, like the Columbia Plateau, offers an ideal geological setting for dust deposition studies, as the island is made up of quartz-rich coastal sands, and as such comprises of predominant quartz and a few heavy minerals, but no clay minerals. Therefore, any clays within Fraser Island lakes can be attributed to dust deposition. Studies of this kind offer great potential for understanding temporal changes in dust deposition in recent geological time.

Local environmental conditions, natural and human-induced, can also have an important influence upon dust deposition rates, but there have been few studies of these effects. Goossens and colleagues (Offer and Goossens, 1995; Goossens, 1996) have examined topographic effects upon dust deposition rates, demonstrating high upwind rates and low downwind rates as a result of 'dust shadow' effects. Boulders on the soil surface have also been shown by Goossens (1994) to increase dust deposition rates. Buildings have also been shown to strongly affect deposition rates. Tsoar and Erell (1995) show that deposition rate in Beer-Sheva in Israel was more than double that in rural areas.

Dust Inputs to Soils, Vegetation, Terrestrial Water Bodies and Oceans

Dust deposition potentially has widespread and major effects upon terrestrial and marine ecosystems, but generally speaking, understanding of the actual effects is rudimentary. There has been a debate over the existence of desert loess for the past century, and the direction of the debate has to a certain extent been influenced by dust deposition research, or the lack of it. A major reason for the persistence of the debate is that desert loess deposits are less easily identified than periglacial loesses of the mid-latitudes, and because desert loess generally occurs within geomorphically dynamic environments, their sedimentological characteristics also differ from their peri-glacial equivalents. In the wake of recent growing field and other evidence of dust processes in arid and semi-arid areas (summaries in Goudie, 1978; Middleton 1997) and the mounting evidence of large desert-derived loess deposits on the Matmata Plateau, southern Tunisia (Coudé-Gaussen and Rognon, 1988) and the Loess Plateau in China (Zhang et al., 1991), the existence of desert loess is becoming more widely accepted. For example, in a study of Saharan dust contributions to soils in Niger, the Canary Islands and Portugal, Hermann et al. (1996) report dust contributions to 'recent soil material' of between 4 and 66 per cent. Equally important, the arguments and supporting evidence as to why desert loess may not exist in some desert areas has grown (Yaalon, 1987; Tsoar and Pye, 1987) and is adding a much needed process perspective to the debate.

Yaalon (1987) examined the evidence for desert loess in and around the Sahara and identified three desert loess regions: the Negev, Israel (Yaalon and Dan, 1974), northern Nigeria (McTainsh, 1984) and the Matmata Plateau in Tunisia (Coudé-Gaussen and Rognon, 1987). These and other deposits have only recently appeared on loess maps of the world (Figure 9.13) The best studied area is the Negev, but there is sufficient information from the other regions to show that desert loess soils (unlike their peri-glacial equivalents) differ considerably in their characteristics, reflecting the influences of source area, dust transporting wind system and environment of deposition. Furthermore, Yaalon observed that as more evidence emerges, more desert loess soils are likely to be observed in and around the Sahara in the future.

Desert loess in Australia has received limited attention both within the country and internationally, and this story is yet to unfold. Pioneering work by Butler (1956) provided clear evidence of dust-derived soils on the Riverine Plain area in S.E. Australia (Figure 9.13b), but these soils were given the local name 'parna' (an aboriginal term for dusty ground), rather than being formally identified as desert loess. This naming is, in a sense, unfortunate, as it partly explains why these soils have not being given the attention in the international loess research community that they deserve.

Although dust process evidence is now available of frequent dust storm activity in western NSW and Queensland and the south east dust transport path out over the Tasman Sea and New Zealand is confirmed (McTainsh, 1989; Hesse, 1993; Kiefert and McTainsh, 1996), the evidence for desert loess in Australia is slow to emerge.

Dust Deposit Stabilisation

One of the factors which has slowed the advancement of the desert loess formation hypothesis has been the limited amount of reliable data on soil forming dust deposition rates. In this context it is important to, first, distinguish primary deposition (Figure 9.8) which represents new inputs of dust to a region from distant sources, and is as a result new soil-forming material, from secondary deposition; arising from

202   Aeolian Environments, Sediments and Landforms

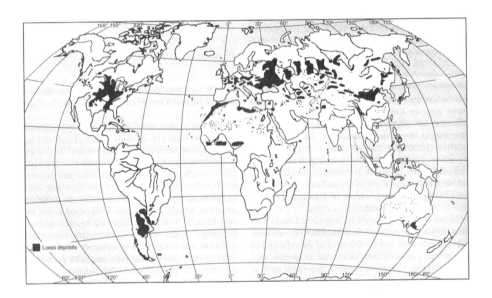

Figure 9.13 Two loess maps of the world: evidence that the tide is turning in the desert loess debate? (after Pye, 1984(a) and Livingstone and Warren, 1996 (b)). Reproduced by permission of (a) Academic Press and (b) Addison Wesley Longman Ltd

local soil entrainment (human-induced or natural) or re-entrained local dusts (Figure 9.8). The proportions of primary to secondary deposition will change according to the stability of the environment of deposition. Yaalon and Ganor (1975) estimated that 50 per cent of total deposition in Israel was locally re-entrained dusts and McTainsh (1980) came up with a similar figure of 42 per cent secondary dust deposition in Nigeria. In a study of dust deposition and particle size in Mali, McTainsh et al. (1996) conclude that the secondary dust component of total deposition, which was largely dusts entrained by vehicles and stock, was very large, though not quantified. Littmann (1997) measured dust deposition in a dunefield in Israel, and concluded that net (or primary) dust deposition was only 6 per cent, but to a certain extent this small number reflects the sampling methodology used. The deposition traps were set at 1 m and at ground level, and the local component (assumed to be $> 20 \mu$m) was determined by sieving.

There have been few studies of the effects of different surface types or vegetation upon dust stabilisation, and most of them in Israel. Danin and Ganor (1997) conclude that grasses on north facing slopes trap, and then stabilise greatest quantities of dusts. The role of biological crusts has received recent research attention and deserves more. Littmann (1997) investigated the role of dust deposition, in particular nitrogen, in crust formation on dunes in southern Israel. Littmann also concluded from particle-size analyses of the crusts that, once formed, biological crusts do not act as a dust trap (i.e. they contained only 6–8 per cent of fines), nor are nitrogen contributions in dust deposition as important as nitrogen fixing by the biological crusts. Whereas Danin et al. (1989) and Danin and Ganor (1997), also working in Israel, concluded that biological crusts do assist in trapping dust deposits. Also, Gillette and Dobrowolski (1993) found that a fine surface crust on soils in Tadzhikistan was dust-derived.

Nutrient Contributions to Soils and Terrestrial Ecosystems

As the weight of evidence in favour of the existence of desert loess soils has increased, an awareness has developed that the effects of dust deposition on soils has been even more pervasive and subtle, than merely the formation of desert loess soils in semi-arid environments. At the present time there have been few detailed studies, but even these limited results provide indications that this phenomenon will soon become accepted to be much more widespread

It is probable that dust deposition has influenced a variety of soil properties. Data on nutrient contributions in dust deposition are limited, but the best studied region is Israel. Herut and Krom (1996) working in coastal Israel estimated total deposition (wet and dry) to the north and south east Mediterranean was low (0.38 t km^{-2} per annum), but even this rate could exceed soil formation rates from weathering. This rate is considerably lower than those in more heavily populated northern Europe (~ 1 t km^{-2} per annum) (Schaug et al., 1987) and the NW Mediterranean Sea (0.51 t km^{-2} per annum), but these rates probably do not reflect primary dust deposition. At greater distances from source over the Atlantic, Pacific and Indian Oceans N deposition rates are 0.3–0.4 t km^{-2} per annum (Duce et al., 1991). Dry deposition of P in the SE Mediterranean is 0.139 t km^{-2} per annum which is about double North Atlantic and Pacific Oceans and a factor of 20 more than the South

Atlantic and Pacific Oceans. Herut and Krom (1996) estimate a wet/dry P ratio of about 0.5. Littmann's (1997) study on the edge of the Sinai-Negev sand sheet revealed a very low mean N deposition of 0.48 kg ha^{-1} per annum (with a range of 0.45 to 0.62 positively related to rainfall). The N deposition peaked at about 0.25 mg m^{-2} per day during periods of cultivation – reflecting the major source of N. Littmann makes a useful comparison between his study and others in semi-arid environments and the general positive relationship with rainfall persists. Stock and Lewis (1986) report total N deposition in coastal South Africa (rainfall 400 mm) of 2 kg ha^{-1} per year.

A study by Offer *et al.* (1996) 50 km to the east of Herut and Krom's study area, concluded that dry deposition of nitrogen, organic carbon in this region is of ecological significance. This dry region (100 mm rain) experiences active dust deposition (110–220 t km^{-2} per annum) with total N deposition ranging from 0.7 to 2 kg ha^{-1} per annum. Leys and McTainsh (1998) examined nutrient-laden dust contributions to riverine environments in sub humid regions of eastern Australia and conclude that dust contributions could be significant when rivers are not flowing – a not unusual situation in Australia.

The effects of soluble salts in dust, upon soils and terrestrial ecosystems has received little research attention, but early indications are that this may also be a significant process in particular areas. Reheis (1997) found soluble salt levels of up to 30 per cent in dust deposition downwind of Owens Lake, California and concluded that this process could have had a significant effect upon soil pH in the region. Dare Edwards and McTainsh (1992) propose that soluble salts in Australian dusts may have 'overprinted' soil profiles in eastern Australia, increasing the mobility of nutrients and particles within profiles. Schroeder (1985) describes ion concentrations in the range 4.1–14.6 per cent in dust deposits in the Sudan, which are comparable with eastern Australian dusts (2.7–9.3 per cent) (Kiefert, 1995).

In a geological time context there is some evidence that calcium and silica-rich dust deposition may have played a part in duricrust formation in arid regions. Calcretes may have been formed from deposition of calcium carbonate-rich dusts (Goudie, 1983; Khadikar *et al.*, 1998), and soluble silica in quartz and biogenic silica may have been important in silcrete formation (Summerfield, 1983) While clear evidence of these dust processes is still emerging, it is equally true in Australia at least, which has very large areas of duricrust, that hard evidence in favour of the pedogenic, ground water and other possible processes of duricrust formation is equivocal.

Dust Deposition onto Vegetation

Vegetation plays a dual role in dust deposition. Not only does it stabilise dust deposits, as discussed earlier, it can increase deposition rates by reducing wind speeds (Tsoar and Pye, 1987) and filtering dusts. Littmann (1997) found that the increased roughness produced even by leafless shrubs in an interdune depression in Israel increased deposition rates up to 1.65 times those on a nearby bare surface. According to Stoorvogel *et al.* (1997) Harmattan dust deposition in to the humid tropical forests of Ghana contributes 50 per cent of the nutrient supply to this ecosystem. Kwiecien (1997) identified significant elemental contributions in dust deposition upon leaves within a forest canopy and that reaching the ground in an industrial setting in Poland.

Filtering of dust-laden air by tree vegetation may also be an important process. In small stands of trees or where the tree spacing is wide enough for the trees to not have a windbreak effect, dust-laden air passing through trees could be filtered by their large dust trapping surface area. Such a process could occur in narrow stands of riverine vegetation in arid areas and may significantly increase dust inputs to these rivers (Leys and McTainsh, 1998).

Dust Deposition Over Oceans

There have been few long term monitoring studies of dust deposition rates over oceans, but models such as GESAMP (Duce *et al.*, 1991) which estimate dry and wet deposition from measured dust concentrations show that ocean deposition rates are orders of magnitude lower than those reported over the continents. Prospero (1996) estimates total global dust deposition to oceans of 360 Tg per annum, about half of which is deposited in the North Atlantic Ocean.

Even though total ocean deposition may be much smaller than over the continents, because all of the deposited dusts are stabilised in the ocean waters, these processes may have been responsible for significant nutrient contributions to marine ecosystems and for forming large ocean deposits.

Several authors have reported that Fe-rich dusts play an important role in sustaining marine phytoplankton populations (e.g. Duce, 1986; Lloyd, 1991). Duce and Tindale (1991) note that although alluvial fluxes of Fe are three to four times higher than atmospheric inputs, the flux of dissolved Fe from the atmosphere exceeds rivers by a factor of $\sim$3. Martin *et al.* (1990) have conducted experiments in Antartic waters to show that Fe contributions stimulate plankton productivity, which peaks in spring and summer. As this is also the dust storm season in eastern Australia (McTainsh *et al.*, 1998), Australian dusts could be source of Fe to the Southern Ocean. Australian desert soils are distinctively red, reflecting their iron content of 0.04–0.6 per cent, and dusts are further enriched (< 1.25 per cent Fe) (Kiefert, 1995).

Dust contributions to ocean deposits have been very substantial during the Quaternary and the character of the deposits provide an indicator of Quaternary environmental changes. An early relationship between dust transport systems and ocean deposits was identified in the Tropical Atlantic Ocean off the West African coast, associated with Saharan dusts (Biscaye, 1965; Parmenter and Folger, 1974; Bowles, 1975) and in the North Pacific Ocean (Blank *et al.*, 1985) in association with north Asian dusts. More recently, Hesse (1993) established a relationship between Australian dusts and Tasman Sea sediments. Several authors have also used the particle size characteristics of dust within ocean sediments as an indicator of Quaternary changes in wind strength (Parkin and Shackleton, 1973; Sarnthein *et al.*, 1981; Clemens and Prell, 1990). Advances in understanding of dust deposition processes over oceans, and in particular the role of rainfall-induced deposition may, however, be reason to rethink some of these earlier models.

CHARACTERISTICS OF SUSPENDED AND DEPOSITED DUSTS

Studies have shown that there is a relationship between dust particle size and transport distance (e.g. Tsoar and Pye, 1987), but this relationship is not clear cut. For example, evidence is growing of very large dust particles collected at locations thousands of kilometres from source. Betzer *et al.* (1988) report $> 75 \mu$m particles collected over the central Pacific Ocean 10 000 km from source. There is also a growing awareness that some dusts are mixtures of particles and aggregates, rather than being purely particulate. Evidence has been available for some time of aggregation of dusts during transport by electrostatic processes (Greeley and Leach, 1979) and Brownian motion (Friedlander, 1977). This adds another dimension to understanding dust size–transport distance relationships and also the reflectance properties of dust plumes. Arimoto *et al.* (1997) show that particle aggregation is an important process in North Pacific dusts, and studies in Australia by McTainsh *et al.* (1998) show that Australian dusts can be highly aggregated, and that aggregation levels vary considerable in time and space. Kiefert *et al.* (1996) compare the aggregation levels of Australian and Saharan dusts and suggest that the latter may be less aggregated, but given the small number of samples analysed, this suggestion requires further testing. On the one hand this adds another level of complexity, to the relationship between dust size and transport distance, but on the other may provide a sedimentological finger print for tracing dust source areas.

Interesting possibilities are also emerging for characterising dust particle-size distributions in much more detail than has been possible to date, as a result of improvements in laboratory particle-sizing instrumentation (e.g. McTainsh *et al.*, 1997) and in the statistical characterisation of multiple population particle-size distributions (Barndorff-Neilsen, 1991; Chatenet *et al.*, 1996; Leys and McTainsh, 1998).

CONCLUSIONS

It should be apparent from this review that research into dust transport and depositon is at an exciting phase of development. Although dust studies have been in the scientific literature for a long period, dust research has now passed from the early phase, when dust studies were published as environmental curiosities, to the current phase when a wide range of sophisticated new technologies are being applied to a large range of aspects of dust transport and deposition. The future of dust transport and deposition research looks bright.

REFERENCES

Adams, S.J., 1997. Dust deposition and measurement: a modified approach. *Environmental Technology*, **18**, 345–350.
Alarcon, M., Cruzado, A. and Alonso, S., 1996. Application of a Lagrangian model to the study of the atmospheric fluxes to the Western Mediterranean. In *The Impact of Desert*

Dust Across the Mediterranean, S. Guerzoni and R. Chester (Eds). Kluwer Academic, Dordrecht, pp. 87–92.

Aldhous, P., 1991. Before and after (Mt Pinatubo). In Nature Reports. *Nature*, **325**, 651.

Anon., 1945. Symposium on 'red rain'. *Victorian Naturalist*, **61**, 165–168.

Arimoto, R., Gray, B.J., Lewis, N.F. *et al.*, 1997. Mass-particle size distributions of atmospheric dust and the dry deposition of dust to the remote ocean. *Journal of Geophysical Research*, **102**, 15874–15876.

Avila, A., Queralt, I., Gallart, F. and Martin-Vide, J., 1996. African dust over Northeastern Spain: Mineralogy and source regions. In *The Impact of Desert Dust across the Mediterranean*, S. Guerzoni and R. Chester (Eds). Kluwer Academic, Dordrecht, pp. 201–205.

Barndorff-Nielsen, O.E. and Sorensen, M., 1991. On the temporal-spatial variation of sediment size distribution. *Acta Mechanica Supplement*, **2**, 23–35.

Behairy, A.K.A., El-Sayed, M.K. and Rao, N.V., 1985. Eolian dust in the coastal area north of Jeddah, Saudi Arabia. *Journal of Arid Environments*, **8**, 89–98.

Betzer, P.R., Carder, K.L., Duce, R.A. *et al.*, 1988. Long-range transport of giant mineral areosol particles. *Nature*, **336**, 568–571.

Biscaye, P.E., 1965. Mineralogy and sedimentation of recent deep-sea clay in the Atlantic Ocean and adjacent seas and oceans. *Geological Society of America Bulletin*, **76**, 803–832.

Blank, M., Leinen, M. and Prospero, J.M., 1985. Major Asian aeolian inputs indicated by the mineralogy of aerosols and sediments in the western North Pacific. *Nature*, **314**, 84–86.

Bowles, F.A., 1975. Paleoclimatic significance of quartz/illite variations in cores from the Eastern Equatorial North Atlantic. *Quaternary Research*, **5**, 225–235.

Bücher, A. and Lucas, G., 1975. Poussières africains sur l'europe. *La Météorologie*, **5**, 53–69.

Bureau of Meteorology, 1982. *Observing the Weather*. Australian Government Publishing Service, Canberra.

Busacca, A., Wagoner, L. and Mehringer, P., 1996. Long-term rates of dust deposition. *Washington State University N.W. Columbia Plateau 'Wind Erosion Air Quality Report'*, pp. 47–53.

Butler, B.E., 1956. Parna – an aeolian clay. *Australian Journal of Science*, **18**, 145–151.

Butler, H.J., Hogarth, W.L. and McTainsh, G.H., 1996. A source based model for describing dust concentrations during wind erosion events: an initial study. *Environmental Software*, **11**, 45–52.

Chatenet, B., Marticorena, B., Gomes, L. and Bergametti, G., 1996. Assessing the microped size distributions of desert soils erodible to wind. *Sedimentology*, **43**, 901–911.

Clemens, S.C. and Prell, W.L., 1990. Late Pleistocene variability of Arabian Sea summer monsoon winds and continental aridity: eolian records from the lithogenic component of deep-sea sediments. *Paleoceanography*, **5**, 109–145.

Coudé-Gaussen, G. and Rognon, P., 1988. The upper pleistocene loess of Southern Tunisia: A statement. *Earth Surface and Processes and Landforms*, **13**, 137–151.

d'Almeida, G.A., 1986. A model for Saharan dust transport. *Journal of Climate and Aplied Meteorology*, **25**, 903–916.

Danin, A., Bar-Or, Y. and Yisraeli, T., 1989. The role of cyanobacterial in stabilization of sand dunes in southern Israel. *Ecologia Mediterranea*, **15**, 55–64.

Danin, A. and Garnor, E., 1997. Trapping of airborne dust by Eig's meadowgrass (*Poa eigii*) in the Judean Desert, Israel. *Journal of Arid Environments*, **35**, 77–86.

Dare-Edwards, A.J. and McTainsh, G.H., 1989. Dust accession onto Australian soils: a review. *Fourth Conference of the Australian and New Zealand Geomorphology Group*, Buchan, Victoria (Abstracts pp. 62–63).

Darwin, C., 1845. An account of fine dust which often falls on vessels in the Atlantic ocean. *Proceedings of the Geological Society* (June 4), 26–30.

Drees, L.R. and Manu, A., 1996. Bird urate contamination of atmospheric dust traps. *Catena*, **27**, 287–294.

Drees, L.R., Manu, A. and Wilding, L.P., 1993. Characteristics of aeolian dusts in Niger, West Africa. *Geoderma*, **59**, 213–233.

Duce, R.A., 1986. The impact of atmospheric nitrogen, phosphorus, and iron species on marine biological productivity. In *The Role of Air-Sea Change in Geochemical Cycling*, P. Buat-Menard (Ed.). Reidel, Dordrecht, pp. 497–528.

Duce, R.A., Liss, P.S., Merril, J.R. *et al.*, 1991. The atmospheric input of trace species to the world ocean. *Global Biogeochemistry*, **5**, 193–259.

Duce, R.A. and Tindale, N.W., 1991. Chemistry and biology of iron and other trace metals. *The American Society of Limnology and Oceanography Inc*, **36**, 1715–1726.

Dulac, F., Moulin, C., Lambert, C.E. *et al.*, 1996. Quantitative remote sensing of African dust transport to the Mediterranean. In *The Impact of Desert Dust across the Mediterranean*, S. Guerzoni and R. Chester (Eds). Kluwer Academic, Dordrecht, pp. 25–49.

Franzen, L.G., Hjelmroos, M., Kallberg, P. *et al.*, 1994. The 'yellow snow' episode of northern Fennoscandia, March 1991 – a case study of the long-distance transport of soil, pollen and stable organic compounds. *Atmospheric Environment*, **28**, 3587–3604.

Friedlander, S.K., 1977. *Smoke, Dust and Haze*. Wiley, New York.

Gabites, J.F., 1954. The drift of radioactive dust from the British nuclear bomb test in October 1953. *New Zealand Journal of Science and Technology*, **36** (Section B2), 159–165.

Ganor, E. and Foner, A., 1996. The mineralogical and chemical properties and the behaviour of aeolian Saharan dust over Israel. In *The Impact of Desert Dust across the Mediterranean*, S. Guerzoni and R. Chester (Eds). Kluwer Academic, Dordrecht, pp. 163–172.

Gao, Y., Arimoto, R., Duce, R.A. *et al.*, 1997. Temporal and spatial distributions of dust and its deposition to the China sea. *Tellus*, **49B**, 172–189.

Gile, L.H. and Grossman, R.B., 1979. *The Desert Project Soil Monograph*. US Soil Conservation Service.

Gillette, D.A. and Dobrowolski, J.P., 1993. Soil crust formation by dust deposition at Shaartuz, Tadzhik, S.S.R. *Atmospheric Environment*, **27A**, 2519–2525.

Gillies, J.A., Nickling, W.G. and McTainsh, G.H., 1996. Dust concentrations and particle-size characteristics of an intense dust haze event: inland delta region, Mali, West Africa. *Atmospheric Environment*, **30**, 1081–1090.

Goossens, D., 1994. Effect of rock fragments on eolian deposition of atmospheric dust. *Catena*, **23**, 167–189.

Goossens, D., 1995. Field experiments of aeolian dust accumulation on rock fragment substrata. *Sedimentology*, **42**, 391–402.

Goossens, D., 1996. Wind tunnel experiments of aeolian dust deposition along ranges of hills. *Earth Surface Processes and Landforms*, **21**, 205–216.

Goudie, A.S., 1978. Dust storms and their geomorphological implications. *Journal of Arid Environments*, **1**, 291–310.

Goudie, A.S., 1983. Dust storms in space and time. *Progress in Physical Geography*, **7**, 502–530.

Goudie, A.S. 1995. *The Changing Earth: Rates of Geomorphological Processes*. Blackwell Science, Oxford.

Greeley, R. and Leach, R.N., 1979. A preliminary assessment of the effects of electrostatics on eolian processes. *NASA Technical Memorandum 79729*, 236–237.

Guerzoni, S. and Chester, R. (Eds), 1996. *The Impact of Desert Dust across the Mediterranean*. Kluwer Academic, Dordrecht.

Hall, D.J., Upton, S.J. and Marsland, G.W., 1994. Designs for a deposition gauge and a flux gauge for monitoring ambient dust. *Atmospheric Environment*, **28**, 2963–2979.

Hall, D.J. and Upton, S.L., 1988. A wind tunnel study of the particle collection efficiency of an inverted frisbee used as a dust deposition gauge. *Atmospheric Environment*, **22**, 1383–1394.

Herrmann, L., Jahn, R. and Stahr, K., 1996. Identification and quantification of dust additions in Peri-Saharan soils. In *The Impact of Desert Dust across the Mediterranean*, S. Guerzoni and R. Chester (Eds). Kluwer Academic, Dordrecht, pp. 173–182.

Herut, B. and Krom, M., 1996. Atmospheric input of nutrients and dust to the SE Mediterranean. In *The Impact of Desert Dust across the Mediterranean*, S. Guerzoni and R. Chester (Eds). Kluwer Academic, Dordrecht, pp. 349–358.

Hesse, P.P., 1993. A Quaternary record of the Australian environment from aeolian dust in Tasman Sea sediments. PhD thesis, Australian National University, Canberra.

Inoue, K. and Naruse, T., 1991. Accumulation of Asian long-range eolian dust in Japan and Korea from the late Pleistocene to the Holocene. *Catena*, **20** (Suppl.), 25–42.

Jaenicke, R., 1979. Monitoring and critical review of the estimated source strength of mineral dust from the Sahara. In *Saharan Dust*, C. Morales (Ed.). Wiley, Chichester.

Khadikar, A.S., Malik, J.N. and Chamyal, L.S., 1998. Calcretes in semi arid alluvial systems: formative pathways and sinks. *Sedimentary Geology*, **116**, 251–260.

Kiefert, L., 1995. Characteristics of wind transported dust in eastern Australia. PhD Thesis, Griffith University, Brisbane, Australia.

Kiefert, L. and McTainsh, G.H., 1996. Oxygen isotope abundance in the quartz fraction of aeolian dust: implications for soil and ocean sediment formation in the Australasian region. *Australian Journal of Soil Research*, **34**, 467–473.

Kiefert, L., McTainsh, G.H. and Nickling W.G., 1996. Sedimentological characteristics of Saharan and Australian dusts. In *The Impact of Desert Dust across the Mediterranean*, S. Guerzoni and R. Chester (Eds). Kluwer Academic, Dordrecht, pp. 183–190.

Knight, A.W., McTainsh, G.H. and Simpson, R.W., 1995. Sediment loads in an Australian dust storm: implications for present and past dust processes. *Catena*, **24**, 195–213.

Kwiecien, M., 1997. Deposition of inorganic particulate aerosols to vegetation – a new method of estimating. *Environmental Monitoring and Assessment*, **46**, 191–207.

Kukal, Z., 1971. *Geology of Recent Sediments*. Academic Press, London.

Leys, J.F. and McTainsh, G., 1998. Dust and nutrient deposition to riverine environments in south-eastern Australia. *International Conference on Aeolian Research*, Oxford, UK.

Littmann, T., 1997. Atmospheric input of dust and nitrogen into the Nizzana sand dune ecosystem, north-western Negev, Israel. *Journal of Arid Environments*, **36**, 433–457.

Livingstone, I. and Warren, A., 1996. *Aeolian Geomorphology: an introduction*. Longman, Singapore.

Lloyd, P., 1991. Iron determinations. *Nature*, **350**, 19.

Longmore, M.E. and McTainsh, G.H., 1997. Dust records in the Pleistocene sediments of Fraser Island: Palaeoclimatic reconstruction of wind erosion over the last 600 ka. *Sixth Australasian Archaeometry Conference*, Sydney, Australia, February 10–February 13 1997.

Löye-Pilot, M.D., Martin, J.M. and Morelli, J., 1986. Influence of Saharan dust on the rainfall acidity and atmospheric input to the Mediterranean. *Nature*, **321**, 427–428.

Löye-Pilot, M.D. and Martin, J.M., 1996. Saharan dust input to the western Mediterranean: An eleven years record in Corsica. In *The Impact of Desert Dust across the Mediterranean*, S. Guerzoni and R. Chester (Eds). Kluwer Academic, Dordrecht, pp. 191–199.

Martin, J.H., Gordon, R.M. and Fitzwater, S.E., 1990. Iron in Antarctic waters. *Nature*, **345**, 156–158.

Maley, J., 1980. Études palynologiques dans le bassin du Tchad et paléoclimatologie de l'Afrique nord tropical de 30,000 ans a l'époque actuelle. Unpublished PhD thesis, University of Montpellier, France.

McTainsh, G.H., 1985. Dust processes in Australia and west Africa: a comparison. *Search*, **16**, 104–106.

McTainsh, G.H., 1980. Harmattan dust deposition in northern Nigeria. *Nature*, **286**, 587–588.

McTainsh, G.H., 1984. The nature and origin of the aeolian mantles of central northern Nigeria. *Geoderma*, **33**, 13–37.

McTainsh, G.H., 1989. Quaternary aeolian dust processes and sediments in the Australian region. *Quaternary Science Review*, **8**, 235–253.

McTainsh, G.H., 1998. Dust Storm Index. *Sustainable Agriculture: Assessing Australia's Recent Performance*, ch. 5, 65–72. Standing Committee on Agriculture and Resource Management (SCARM), February 1998.

McTainsh, G.H., Leys, J.F. and Nickling, W.G., 1999. Wind erodibility of arid lands in the Channel Country of western Queensland, Australia, *Zeitschrift für Geomorphologie*, Supplemental **116**, in press.

McTainsh, G.H., Lynch, A.W. and Hales, R., 1997. Particle-size analysis of aeolian dusts, soils and sediments in very small quantities using a Coulter Multisizer. *Earth Surface Processes and Landforms – Technical and Software Bulletin*, **22**, 1207–1216.

McTainsh, G.H., Nickling, W.G. and Lynch, A.W., 1997. Dust deposition size in Mali, West Africa. *Catena*, **29**, 307–322.

McTainsh, G.H., Lynch, A.W. and Tews, E.K., 1998. Climatic controls upon dust storm occurrence in eastern Australia. *Journal of Arid Environments*, **39**, 457–466.

McTainsh, G.H., Nickling, W.G. and Lynch, A.W., 1996. Dust deposition and particle size in Mali, West Africa. *Catena*, **29**, 307–322.

McTainsh, G.H. and Walker, P.H., 1982. Nature and distribution of Harmattan dust. *Zeitschrift für Geomorphologie*, **26**, 417–435.

Middleton, N., 1997, Desert dust. In *Arid Zone Geomorphology: Process, Form and Change in Drylands*, D.S.G. Thomas (Ed.). John Wiley, London, pp. 413–436.

Middleton, N.J., 1985. Effect of drought on dust production in the Sahel. *Nature*, **316**, 431–434.

Morales, C., 1979. The use of meteorological observations for studies of the mobilization, transport and deposition of Saharan soil dust. *Saharan Dust, Mobilization, Transport, Deposition*, C. Morales (Ed.). John Wiley, Chichester, pp. 119–131.

Muhs, D.R., 1983. Airborne dustfall on the Californian Channel Islands, USA. *Journal of Arid Environments*, **6**, 223–228.

Nickling, W.G. and Gillies, J.A., 1993. Dust emission and transport rates, Mali, West Africa. *Sedimentology*, **40**, 859–868.

Nickling, W.G., McTainsh, G.H. and Leys, J.F., 1999. Dust emissions from the Channel Country of western Queensland, Australia. *Zeitschrift für Geomorphologie*, Supplemental **116**, in press.

Offer, Z.Y. and Goossens, D., 1995. Wind tunnel experiments and field measurements of aeolian dust deposition on conical hills. *Geomorphology*, **14**, 43–56.

Offer, Z.Y., Sarig, S. and Steinberger, Y., 1996. Dynamics of nitrogen and carbon content of aeolian dry deposition in an arid region. *Arid Soil Research and Rehabilitation*, **10**, 193–199.

Parkin, D.W. and Shackleton, N.J., 1973. Trade wind and temperature correlations down a deep sea core off the Saharan coast. *Nature*, **245**, 455–457.

Parmenter, C. and Folger, D.W., 1974. Eolian biogenic detritus in deep sea sediments: A possible index of equatorial Ice Age aridity. *Science*, **185**, 695–697.

Péwé, T.L., Péwé, E.A., Péwé, R.H. *et al.*, 1981. Desert dust: characteristics and rates of deposition in central Arizona. *Geological Society of America*, Special Paper 186, 169–190.

Pitty, A., 1968. Particle size of the Saharan dust which fell in Britain in July 1968. *Nature*, **220**, 364–365.

Prakasa Rao, P.S., Khemani, L.T., Momin, G.A. *et al.*, 1992. Measurements of wet and dry deposition at an urban location in India. *Atmospheric Environment*, **26B**, 73–78.

Prospero, J.M., 1996. Saharan dust transport over the North Atlantic Ocean and Mediterranean: an overview. In *The Impact of Desert Dust across the Mediterranean*, S. Guerzoni and R. Chester (Eds). Kluwer Academic, Dordrecht, pp. 133–151.

Prospero, J.M., Glaccum, R.A. and Nees, R.T., 1983. Atmospheric transport of soil dust from Africa to south America. *Nature*, **289**, 570–572.

Prospero, J.M. and Nees, R.T., 1986. Impact of the North African drought and El Niño on mineral dust in the Barbados trade winds. *Nature*, **320**, 735–738.

Pye, K., 1987. *Aeolian Dust and Dust Deposits*. Academic Press, London.

Pye, K., 1992. Aeolian dust transport and deposition over Crete and adjacent parts of the Mediterranean Sea. *Earth Surface Processes and Landforms*, **17**, 271–288.

Raupach, M.R., McTainsh, G.H. and Leys, J.F., 1994. Estimates of dust mass in recent major Australian dust storms. *Australian Journal of Soil and Water Conservation*, **7**, 20–24.

Reddan, S.P., 1994. Estimates of atmospheric dust flux and wind erosion activity in eastern Australia. Honours Thesis, Griffith University, Brisbane, Australia.

Reheis, M.C., 1997. Dust deposition downwind of Owens (dry) lake, 1991–1994: Preliminary findings. *Journal of Geophysical Research*, **102**, 25 999–26 008.

Reheis, M.C. and Kihl, R., 1995. Dust deposition in southern Nevada and California, 1984–89: Relations to climate, source area, and source lithology. *Journal of Geophysical Research*, **100**, 8893–8918.

Reiff, J., Forbes, G.S., Spieksma, F.T.M. and Reynders, J.J., 1986. African dust reaching Northwestern Europe: A case study to verify trajectory calculations. *Journal of Climate and Applied Meteorology*, **25**, 1543–1567.

Safar, M.I., 1985. *Dust and Dust Storms in Kuwait*. State of Kuwait, Kuwait.

Sarnthein, M., Tetzloff, G., Koppmann, B. *et al.*, 1981. Glacial and interglacial wind regimes over the eastern subtropical Atlantic and North-west Africa. *Nature*, **293**, 193–196.

Schaug, J., Hansen, J.E., Nodop, K. *et al.*, 1987. Co-operative programme on monitoring and evaluation of the long range transmission of air pollutants in Europe (EMEP). *Summary Report From the Chemical Co-Ordinating Centre for the Third Phase of EMEP*.

Schroeder, J.H., 1985. Eolian dust in the coastal desert of the Sudan: aggregates cemented by evaporites. *Journal of African Earth Sciences*, **3**, 370–380.

Schulz, M., Balkanski, Y.G.W., Dulac, D. *et al.*, 1996. Model components necessary to capture a dust plume pattern over the Mediterranean sea. In *The Impact of Desert Dust Across the Mediterranean*, S. Guerzoni and R. Chester (Eds). Kluwer Academic, Dordrecht, pp. 51–58.

Smith, R.M., Twiss, P.C., Krauss, R.K. and Bronn, M.J., 1970. Dust deposition in relation to site, season, and climatic variables. *Proceedings of the Soil Science Society of America*, **34**, 112–117.

Stock, W. and Lewis, O., 1986. Atmospheric input of nitrogen to a coastal fynbos ecosystem of the south-western Cape Province, South Africa. *South African Journal of Botany*, **52**, 273–276.

Stoorvogel, J.J., Van Breemen, N. and Janssen, B.H., 1997. The nutrient input by Harmattan dust to a forest ecosystem in Cote d'Ivoire, Africa. *Biogeochemistry*, **37**, 145–157.

Summerfield, M.A., 1983. Silcrete. In *Chemical Sediments and Geomorphology*, A.S. Goudie and K. Pye (Eds). Academic Press, London, pp. 59–91.

Swap, R., Garstang, M., Greco, S., Talbot, R. and Kallberg, P., 1992. Saharan dust in the Amazon basin. *Tellus*, **44B**, 133–149.

Tiessen, H., Hauffe, H.K. and Mermut, A.R., 1991. Deposition of Harmattan dust and its influence on base saturation of soils in northern Ghana. *Geoderma*, **49**, 285–299.

Tiller, K.G., Smith, L.H. and Merry, R.H., 1987. Accessions of atmospheric dust east of Adelaide, South Australia, and implications for pedogenesis. *Australian Journal of Soil Research*, **25**, 43–54.

Tsoar, H. and Erell, E., 1995. The effect of a desert city on aeolian dust deposition. *Journal of Arid Land Studies*, **5S**, 115–118.

Tsoar, H. and Pye, K., 1987. Dust transport and the question of desert loess formation. *Sedimentology*, **34**, 139–153.

Wagenbach, D., Preunkert, S., Schafer, J. *et al.*, 1996. Northward transport of Saharan dust recorded in a deep alpine ice core. In *The Impact of Desert Dust Across the Mediterranean*, S. Guerzoni and R. Chester (Eds). Kluwer Academic, Dordrecht, pp. 291–300.

Yaalon, D.H., 1987. Saharan dust and desert loess: effect on surrounding soils. *Journal of African Earth Sciences*, **6**, 569–571.

Yaalon, D.H. and Dan, J., 1974. Accumulation and distrbution of loess-derived deposits in the semi-desert and desert fringe of Israel. *Zeitschrift für Geomorphologie, Supplementband*, **20**, 91–105.

Yaalon, D.H. and Ganor, E., 1975. Rate of aeolian dust accretion in the Mediterranean and the desert fringe environments of Israel. *19th International Congress of Sedimentology*, pp. 169–174.

Zhang, L., Dai, X. and Shi, Z., 1991. The sources of loess material and the formation of the Loess Plateau in China. *Catena*, **20**, 1–14.

10 Loess

KENNETH PYE[1] and DEAN SHERWIN[2]
[1] *Department of Geology, University of London, UK*
[2] *Norwich City College, UK*

Loess is a widespread phenomenon in the Quaternary sedimentary record. It occurs in a variety of forms, has a range of colours and has certain grain size characteristics. These may vary with distance from source. It also possesses a suite of particle shapes and surface textures, of mineralogical and geochemical compositions, of sedimentary structures, of microfabrics, geotechnical properties and landform associations. Among the requirements for loess formation are a source of appropriately sized sedimentary material, wind energy to transport that material and a suitable site for deposition and accumulation. It is possible to talk of 'periglacial', 'perimontane' and 'peridesert' loess facies. Loess also provides one of the best terrestrial proxy records of palaeoclimate.

INTRODUCTION

Deposits of Quaternary loess, which can be defined essentially as a terrestrial sediment composed principally of windblown silt, are estimated to cover between 5–10 per cent of the world's present day land surface area, i.e. approximately 7.5 to 15 $\times 10^6$ km^2. Loess has been the subject of considerable scientific interest for more than 150 years, ever since Charles Lyell (1834) first invoked the term to describe loose, silty deposits which he observed along the Rhine in 1833. In recent years the level of interest has increased significantly, reflecting the fact that loess–palaeosol sequences provide some of the most continuous and detailed available records of continental environmental change during the past two million years. However, interpretation of the loess palaeoclimate proxy record is not without its difficulties, and full account needs to be taken of the factors which control variability of the primary depositional characteristics of loess and subsequent post-depositional changes.

THE NATURE AND DISTRIBUTION OF LOESS

There has been much debate about the definition and classification of loess. Some authors have proposed complex definitions based on multiple lithostratigraphic and genetic criteria, while others have preferred much simpler definitions. Pecsi (1990, p. 1), for example, expressed the view that 'loess is not simply dust transported and

---

Aeolian Environments, Sediments and Landforms. Edited by A.S. Goudie, I. Livingstone and S. Stokes.
© 1999 John Wiley & Sons, Ltd.

deposited by the wind. Dust only becomes loess after the passage of a certain amount of time in a given geographical zone, i.e. only through diagenesis in certain ecological environments'. He proposed 10 criteria which must be met before a deposit can be classified as loess. On the other hand, Pye (1987) defined loess simply as a terrestrial windblown silt deposit. This definition was subsequently modified slightly by Pye (1995) who proposed that, for most practical purposes, it is sufficient to define loess simply as 'a terrestrial clastic sediment, composed predominantly of silt-size particles, which is formed esssentially by the accumulation of windblown dust'. In actual fact, although what might be regarded as 'typical loess' is silt-dominated, some loess deposits show a predominance of fine sand or even clay-sized particles. Consequently, it may be more appropriate to define loess simply as a terrestrial sediment composed largely of windblown dust, which has been modified to a greater or lesser degree by post-depositional processes, including weathering and pedogenesis. By this definition, loess can be regarded simply as the sub-aerial equivalent of aeolo-marine deposits and aeolo-lacustrine deposits which are formed by accumulation of aeolian dust in marine and lacustrine environments, respectively (Pye, 1995).

A useful distinction can also be made between 'fresh' or 'unweathered' loess and 'weathered' loess. There is clearly a continuum between the two, since all deposited dust is subjected to a range of post-depositional modification processes which may change its physical, mineralogical and chemical properties to varying degrees. However, in the case of *weathered loess*, the initial depositional characteristics have been significantly changed by weathering and soil development. A further category of *reworked loess* should also be recognised, consisting of loess which has been locally transported and redeposited by colluvial or fluvial processes, but which has retained many of the sedimentological characteristics of the parent windblown deposit. In cases of larger-scale fluvial or colluvial reworking, the primary sedimentological characteristics may be substantially overprinted or obliterated; in these circumstances the resulting deposits are best referred to as *loess-derived colluvium* or *alluvium* (Pye, 1995).

Loess-like deposits are silt-rich sediments which display a number of significant sedimentological similarities to aeolian loess but which have been deposited by fluvial, estuarine or marine deposits, or whose mode of formation is uncertain. The term *loessoid* can also be used to describe sediments which contain mixtures of loess and other materials, such as weathered regolith, dune sand or soil (Pye, 1987).

Loessite is loess which has been lithified by post-depositional diagenetic processes, including cementation and selective mineral dissolution. A number of loessite deposits are known in the geological record, ranging in age from Precambrian to early Quaternary (Sherwin, 1995).

THE SEDIMENTOLOGICAL CHARACTER OF LOESS

Colour

The colour of loess is extremely variable, sometimes within individual loess sections. Light grey, yellow, buff and pale brown are the most common colours, but dark brown, dark grey, orange, red and even black loess is known, reflecting differences in the mineralogy of the source sediments and also the nature and extent of post-depositional modification. Colour can be described using a visual comparator, such as a Munsell Soil Colour Chart, or quantitatively using computerised spectro-photometers. Quantitative measures of colour obtained in this way can be extremely useful as a stratigraphic correlation tool and as a measure of relative weathering intensity.

Grain Size

'Typical loess' is a unimodal, poorly sorted, fine-skewed, meso- or leptokurtic sediment which may contain up to 10 per cent fine sand and up to 20 per cent clay (defined as $< 4 \mu$m in size). However, examples of 'sandy loess' are commonly found which may be unimodal or bimodal, poorly sorted, fine-skewed, containing > 20 per cent sand and often with a primary mode in the fine sand fraction. 'Clayey loess' can be defined as loess in which the clay content exceeds 20 per cent. Such materials are frequently bimodal, poorly sorted and fine-skewed; they may have a primary mode either in the fine silt or clay size range, depending on degree of weathering. The positive (fine) skewness and unimodal character of typical loess reflects selective transport of finer particles by the wind, and also the fact that some of the clay-size particles are actually transported as silt-sized aggregates or as coatings on larger grains. Particle size summary statistics for samples collected from a number of different parts of the world are presented in Table 10.1. Where some samples are bimodal or polymodal, summary statistics based on comparison with a log-normal distribution are inappropriate, and it is more informative to compare values of the median, various modes, and percentages of sand, silt and clay. The most sandy of the loess deposits represented in Table 10.1 are from the Matmata Plateau in Tunisia, and those from Nebraska and northern France are also relatively coarse. The most silty deposits are from the Vicksburg area of southern Mississippi, USA, and the most clayey loess samples are from Netivot in the northern Negev, Israel. Samples from China, Europe and Central Asia are intermediate in this respect. Considering the average mean size of the unimodal sediments only, loess from Taijikistan is the finest analysed, with an average median value of 6.97 phi (7.98 μm) and an average graphic mean value of 7.22 phi (6.70 μm). Grain size frequency histograms for individual samples, obtained using a laser granulometer, show that the Mississippi loess has a relatively peaked, unimodal distribution with a clear mode in the medium silt range (Figure 10.1(a)). Loess from Karamaidan, Tajikistan, typically contains more fine silt and clay and has a less peaked grain size distribution which is still unimodal, although with a 'shoulder' around 5 μm (Figure 10.1(b)). Loess from Netivot is clearly bimodal, with modes centred around 60 μm and 2 μm (Figure 10.1(c)). On average, the most poorly sorted loess is that from Tunisia, with a phi sorting value of 2.35 (very poorly sorted), whilst the least poorly sorted loess is that from Mississippi, with a mean value of 1.78 (poorly sorted). The most positively skewed loess shown in Table 10.1 is that from Tunisia, with an average value of 0.65 (very fine skew) and the least positively skewed loess is that from Tajikistan, with an average value of 0.21 (fine skew).

Table 10.1 Ranges, average values (Av.) and standard deviations (SD) of graphical textural parameters determined for unimodal, unreworked loess samples from several different parts of the world (analysis by laser granulometry, parameters calculated in phi units)

Locality	n	Median (ϕ)			Mean (ϕ)			Sorting (ϕ)			Skewness			Kurtosis		
		Range	Av.	SD	Range	Av.	SD	Range	Av.	SD	Range	Av.	SD	Range	Av.	SD
Vicksburg	10	5.40–6.50	5.68	0.32	5.75–6.78	6.09	0.29	1.62–2.02	1.78	0.12	0.26–0.52	0.46	0.07	0.90–1.73	1.28	0.26
Nebraska	2	4.30	4.30	0.00	5.10	5.10	0.00	1.98–2.00	1.99	0.01	0.63	0.63	0.00	1.54–1.58	1.56	0.02
Uzbekistan	5	4.85–7.20	6.17	0.90	5.58–7.40	6.10	0.70	1.99–2.12	2.08	0.05	0.17–0.55	0.34	0.15	0.89–1.12	0.99	0.09
Tajikistan	3	6.90–7.05	6.97	0.06	7.17–7.32	7.22	0.07	1.96–2.01	1.98	0.02	0.19–0.22	0.21	0.01	0.92–1.01	0.96	0.09
Argentina	2	4.15–4.90	4.53	0.38	5.15–5.65	5.40	0.25	2.38–2.43	2.41	0.02	0.47–0.60	0.54	0.06	0.85–1.06	0.76	0.10
Belgium	1	5.00	5.00	0.00	5.72	5.72	0.00	1.92	1.92	0.00	0.64	0.64	0.00	1.24	1.24	0.00
China	7	4.90–5.95	5.27	0.32	5.45–6.50	5.77	0.32	1.84–2.15	1.95	0.09	0.39–0.52	0.46	0.04	0.97–1.25	1.13	0.09
Tunisia	4	3.90–4.35	4.06	0.17	4.35–5.58	5.12	0.50	1.73–2.72	2.35	0.38	0.53–0.72	0.65	0.07	0.76–1.69	1.13	0.36
France	7	4.30–5.25	4.61	0.35	4.73–5.78	5.17	0.39	1.70–2.15	1.98	0.14	0.40–0.62	0.51	0.08	0.97–2.14	1.62	0.37
All samples	41	3.90–7.20	5.28	0.90	4.35–7.40	5.84	0.74	1.62–2.72	1.99	0.24	0.16–0.72	0.47	0.14	0.76–2.14	1.24	0.32

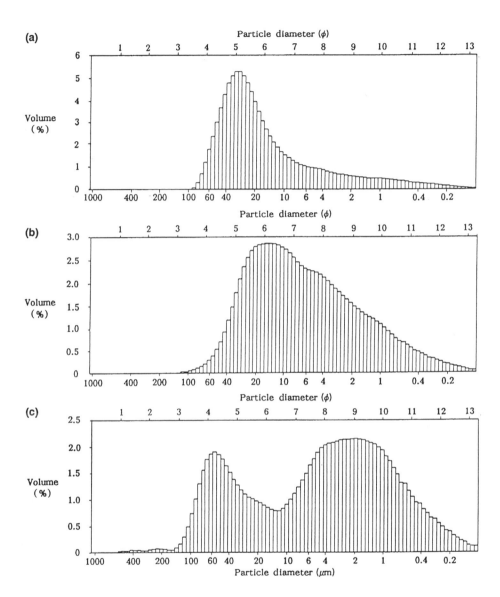

Figure 10.1   Grain size frequency curves for *in situ* loess samples from (a) Vicksburg, Mississippi, (b) Karamaidan, Tajikistan, and (c) Netivot, Israel (determined by laser granulometry)

The proximity of the Tunisian loess to its desert source area accounts for it being the coarsest, least sorted and most positively skewed of the samples analysed. The fine size of the Tajikistan and Chinese Loess Plateau loess reflects the fact that much of the material is derived from distal sources in the deserts of Central Asia and Mongolia, respectively. Moreover, rates of deposition appear to have been relatively low in these areas, allowing some formation of clay by post-depositional weathering processes. The relatively coarser, better sorted character of the southern Mississippi loess mainly reflects the proximal location and the better sorted character of the dust source (the Mississippi River floodplain). Furthermore, this stratigraphical unit of Mississippi loess accumulated relatively quickly during the later part of the last glacial period, with relatively little opportunity for post-depositional weathering and clay formation.

Work in several loess regions has demonstrated a clear downwind reduction in mean (and median) grain size, corresponding with a reduction in sand content and an increase in clay content (e.g. Kes, 1984; Liu et al., 1985; Winspear and Pye, 1996). This is to be expected, since studies of airborne dust show a similar downwind reduction in size, attributed to differential settling of larger particles closer to the source (Tsoar and Pye, 1987).

Grain Shape and Surface Texture

The shape of grains in loess reflects their mineral composition, their mode of formation, and the degree of pre- and post-depositional shape modification induced by weathering and diagenesis. A high proportion of the quartz silt grains tend to be blocky, angular or sub-angular, particularly in fresh (i.e. unweathered or slightly weathered) deposits. Grains derived from soil environments are, in general, more rounded and may be coated with clay minerals and chemical precipitates. Unweathered feldspar and carbonate grains also tend to be blocky, but may be highly etched and well-rounded if derived from soils or if the loess has been modified significantly by post-depositional weathering and pedogenesis.

Downwind changes in grain shape within large loess deposits have also been identified (e.g. Mazzullo et al., 1992), although the mechanisms involved have been the subject of some debate (Pye, 1994). Other things being equal, more angular grains may be expected to be transported further than rounded grains owing to the fact that they have significantly lower settling velocities than spherical particles of equivalent spherical diameter and mass.

The surfaces of the larger silt and sand grains in loess are often covered by finer particles which are held in contact by electrostatic attractive forces, thin moisture films, or salts. Indeed, some silt and clay size 'particles' are actually aggregates of finer particles whose surfaces have been charged by friction. Beneath the adhering material, the surface textural characteristics of quartz and other mineral grains show a wide variety of features depending on the provenance and transport history of the grains. Freshly formed material, produced, for example, by glacial crushing or frost weathering, is typically angular with numerous clean breakage features; on the other hand, grains which have resided for some considerable time in a soil or surface weathering environment are likely to be sub-rounded and to show abundant dissolution–precipitation features. In general, surface textural features on silt grains are of limited value as a guide to provenance and mode of particle formation (Pye, 1984).

Mineral Composition

The most common minerals found in 'typical loess' are quartz (50–70 per cent), feldspar (5–30 per cent), micas (5–10 per cent), carbonate minerals (0–30 per cent), clay minerals (10–15 per cent), and heavy minerals (< 5 per cent). In certain local-ised circumstances, carbonate may be dominant and minerals such as gypsum, palygorskite or iron oxyhdroxides may be significant components.

Table 10.2 shows the major mineral composition of loess samples collected from several different parts of the world, determined by X-ray diffraction. Quartz content ranges from 30 per cent (Argentina) to 89 per cent (Belgium), with an overall average of 61 per cent (SD 14.87 per cent). Feldspar content ranges from 4 per cent (Tunisia) to 57 per cent (Argentina), with an overall average of 19 per cent (SD 10.14 per cent). Calcite ranges from 0 per cent in many samples to 31 per cent (Israel), with an overall average of 9 per cent (SD 8.54 per cent). Dolomite ranges from 0 per cent in many samples to 40 per cent (Mississippi), with an overall average of 5 per cent (SD 9.14 per cent). Although many samples contain virtually no mica, the maximum content is 12 per cent (Uzbekistan), with an overall average of 4 per cent (SD 4.09 per cent). Most samples also contain no gypsum but values of up to 18 per cent are recorded in the Uzbekistan loess and up to 57 per cent in the Tunisian samples. These variations reflect both the differing provenance of the material and the character of soil formation in each regional accumulation site.

The range of values for calcite represents both detrital variation and the effect of post-depositional leaching and re-deposition within the loess profiles. Near-surface decalcified layers contain little or no calcite (and much reduced dolomite) compared with the underlying deposits. The samples with highest calcite contents are those from semi-arid environments which have considerable disseminated detrital carbon-ate remaining, in addition to accumulations of secondary (diagenetic) carbonate.

The clay mineral assemblages present in loess are also partly controlled by source area geology and weathering conditions and the environmental conditions which prevail at the deposition site. Table 10.3 summarises the relative abundances of major clay mineral types in the < 2 μm fraction of samples collected from different parts of the world. Although at a broad global scale some correlation between climate and clay mineral weathering products is evident (e.g. Weaver, 1958; Chamley, 1989), caution needs to be exercised in using clay types as indicators of provenance. The nature of clays in the source rocks, and paleoenvironmental conditions, are clearly important influences. For example, loess derived from desert environments in Central Asia and China is dominated by illite, whereas North-African desert-derived loess in Israel and Tunisia is dominated by smectite and kaolinite, reflecting the fact that large parts of North Africa have experienced a more humid weathering regime earlier in the Cenozoic.

Despite these limitations, several authors have had reported success in using clay mineral ratios to distinguish different loess source regions. For example, Ruhe and

Table 10.2 Ranges, average values (Av.) and standard deviations (SD) of the major mineral abundances in unreworked, late Pleistocene loess samples from different parts of the world (determined by X-ray powder diffraction)

Locality	n	Quartz (%)			Feldspar (%)			Calcite (%)			Dolomite (%)			Mica (%)		
		Range	Av.	SD	Range	Av.	SD	Range	Av.	SD	Range	Av.	SD	Range	Av.	SD
Vicksburg	10	39–87	66	13.72	8–21	14	4.18	0–5	2	2.02	0–40	18	15.78	0–1	Trace	0.31
Nebraska	2	56–65	61	4.50	34–41	38	3.50	0	0	0.00	0	0	0.00	1–3	2	1.00
Uzbekistan	5	33–52	43	6.05	18–29	21	4.10	12–29	20	6.89	1–10	5	4.07	0–12	7	4.12
Tajikistan	3	58–63	61	2.05	15–17	16	5.25	13–17	15	1.70	0	0	0.00	7–11	9	1.63
Khatif	1	66	66	0.00	23	23	0.00	8	8	0.00	4	4	0.00	0	0	0.00
Netivot	4	54–57	55	1.30	13–18	16	2.50	21–31	26	3.61	0–7	3	3.27	0	0	0.00
Argentina	2	30–45	38	7.50	55–57	56	1.00	0–8	4	4.00	0–5	3	2.50	Trace	Trace	–
Belgium	1	89	89	0.00	11	11	0.00	0	0	0.00	0	0	0.00	0	0	0.00
New Zealand	1	74	74	0.00	26	26	0.00	0	0	0.00	0	0	0.00	Trace	Trace	–
China	7	48–54	51	1.84	18–23	21	1.60	11–13	12	0.90	6–8	7	0.72	9–11	10	0.70
Pegwell (UK)	4	65–86	72	8.44	9–16	13	3.27	4–15	11	4.18	0–7	4	2.59	0–4	1	2.00
Tunisia	7	32–86	63	17.02	4–11	8	2.55	2–21	11	6.99	0–2	1	0.99	0	0	0.00
France	11	54–84	70	11.08	12–30	21	6.24	0–10	3	4.07	0–8	2	3.17	0–6	4	0.74
All samples	58	30–89	61	14.87	4–57	19	10.14	0–31	9	8.54	0–40	5	9.14	0–12	4	4.09

Table 10.3 Relative abundance (%) of major clay minerals present in unreworked loess samples from different parts of the world (determined by X-ray powder diffraction)

Sample	% < 2 μm	Smectite (%)	Illite (%)	Kaolinite (%)	Chlorite (%)
V2D	11.10	51	39	5	5
V2J	10.52	57	36	4	4
NEB 92/1	8.80	65	27	8	0
NEB 92/2	8.60	60	31	8	1
UZ18	14.17	25	57	7	11
UZ20	10.76	20	61	8	11
UZ3	14.91	17	61	6	15
UZ2	23.41	27	59	6	8
UZ1	23.70	27	55	4	14
TAJ1	20.66	28	66	4	4
TAJ2	19.93	26	66	4	4
TAJ3	18.44	30	64	4	4
NET2A	36.50	50	27	16	7
NET2C	33.45	53	23	19	5
NET1E	36.87	54	23	18	5
NET3A	28.40	51	27	18	7
4CL	10.34	39	51	5	5
CHI 7.5	9.03	18	61	6	15
CHI 16.5	9.34	16	62	7	15
CHI 21.5	10.36	15	63	7	15
CHI 49.5	10.40	17	62	7	15
CHI 56.5	9.91	15	63	7	15
CHI 94	9.65	15	64	8	13
CHI 230	15.57	21	57	7	15
PEG 8	12.29	40	43	15	2
TUN 9	10.57	15	52	30	3
TUN 11	21.38	14	59	21	5
TUN 12	13.04	7	67	12	12
TUN 13	5.85	12	66	15	6
STROM 2	8.46	35	53	11	3
PRAC 2	7.46	29	56	9	6
SABLE 2	7.34	24	59	15	2

Key to sample codes: V, Vicksburg, Mississippi; NEB, Arnold, Nebraska; UZ, Mingtepe and Orkutsai, Uzbekistan; TAJ, Karamaidan, Tajikistan; NET, Netivot, Israel; 4CL, Sierra de la Ventana, Argentina; CHI, Lanzhou, China; PEG, Pegwell Bay, England; TUN, Matmata, Tunisia; STROM, St Romain, Normandy; PRAC, Port Racine, Normandy; SABLE, Sables d'Or-les-Pins, Britanny

Olsen (1980) used the ratios of illite, expandable clays and kaolinite to differentiate loess derived from different drainage basins in southern Indiana, USA. Palygorskite has also been widely used as an indicator of provenance from desert areas with evaporitic regimes (e.g. Coudé-Gaussen and Rognon, 1988). Care needs to be exercised in such interpretations, however, since clays can form post-depositionally as a result of *in situ* weathering. In central European loess soils, for example, authigenic smectite dominates the 0.2 μm clay fraction while illite dominates the larger fractions (Bronger and Heinkele, 1990). In soils of the Kashmir Valley of Pakistan and the Loess Plateau of China, however, pedogenically-formed illite dominates both the fine and coarse clay fractions. In these areas, even intensely weathered soils contain very little pedogenetic kaolinite; however, this contrasts with the situation in the Mississippi Valley loess, where pedogenetic kaolinite is the dominant clay mineral in older, weathered loess units (Pye and Johnson, 1988).

Geochemical Composition

The bulk geochemical composition of loess is essentially a reflection of the mineralogical composition which, in turn, may be related to the grain size characteristics at a site. Table 10.4 shows the ranges and average values for major element concentrations in loess samples from different areas. Silica is the dominant element in all samples, followed by aluminium and calcium. The loess from Timaru, New Zealand, exemplifies the influence of source lithology on major element composition. The relatively high sodium content in this sample reflects an abundance of Na-rich plagioclase feldspar in local greywackes (Raeside, 1964). The relatively large ranges in values from some areas (e.g. Vicksburg, Mississippi) reflect the effects of decalcification; in such circumstances interpetation of down-profile trends is made easier by consideration of elemental ratios or by normalisation of the data on a carbonate-free basis.

Fossils in Loess

Although loess is often thought to be poor in fossils, especially pollen, it can in fact contain a very wide range of both macrofossil and microfossil remains. Molluscs are extremely widespread and abundant in many loess deposits. For example, Frye et al. (1948) recorded 65 species of molluscs in loess of the Central Great Plains, and used them as a basis for inter-regional correlation. Derbyshire (1983) and Keen (1995) reported a rich and diverse pollen assemblage, plus a mammalian fauna, in the loess of north China, while important finds of human remains have been reported from several areas in Central Asia and elsewhere (e.g. Ranov, 1995). Where the source of dust includes exposed coastal or marine sediments, loess may contain microfossils such as marine diatoms, sponge spicules and foraminifera (Raeside, 1964; Li and Zhou, 1993).

Sedimentary Structures and Topographic Expression

True primary loess is widely considered to be massive and homogeneous, owing both to the mechanism of dust deposition and post-depositional mixing. However, some loess shows well-developed stratification which may have several different origins. Reworked loess is commonly stratified owing to selective movement of different grain sizes during downslope transport by rainsplash, surface wash or solifluction (Mucher and De Ploey, 1977; Mucher et al., 1981). However, downslope reworking does not necessarily form a laminated deposit, especially where mass transport takes place in the form of a slurry.

Laminated silt, or mixed silt and sand deposits, may also occur in primary loess owing to seasonal variations in depositional conditions. For example, Schwan (1986)

Table 10.4 Ranges, average values (Av.) and standard deviations (SD) of the major element oxide contents (weight per cent) of unreworked Pleistocene loess samples from different parts of the world (determined by X-ray fluorescence spectrometry)

Locality	n	Na_2O (wt%)			MgO (wt%)			Al_2O_3 (wt%)			SiO_2 (wt%)			P_2O_5 (wt%)		
		Range	Av.	SD	Range	Av.	SD	Range	Av.	SD	Range	Av.	SD	Range	Av.	SD
Vicksburg	10	0.26–1.09	0.86	0.30	0.94–6.20	3.26	2.03	6.18–16.62	9.91	3.00	52.34–76.99	65.24	7.84	0.03–0.17	0.12	0.05
Nebraska	2	0.91–0.93	0.92	0.01	1.77–1.82	1.79	0.02	11.59–12.23	11.91	0.32	68.47–69.08	68.78	0.30	0.16–0.16	0.16	0.01
Uzbekistan	5	0.76–1.39	1.13	0.21	2.44–3.36	3.04	0.36	11.16–12.23	11.47	0.40	49.57–57.81	54.11	2.92	0.14–0.18	0.17	0.02
Tajikistan	3	0.83–0.83	0.83	0.00	2.49–2.83	2.74	0.05	14.42–15.66	15.03	0.51	59.13–62.93	61.01	1.52	0.13–0.15	0.14	0.01
Khatif	1	0.56	0.56	0.00	2.80	2.80	0.00	8.30	8.30	0.00	55.32	55.32	0.00	0.16	0.16	0.00
Netivot	4	0.45–0.72	0.63	0.10	3.22–3.70	3.39	0.18	9.55–13.07	11.79	1.44	45.17–52.90	48.09	2.93	0.10–0.22	0.16	0.04
Argentina	2	1.66–1.76	1.71	0.05	2.41–2.62	2.51	0.11	12.50–15.25	13.87	1.38	54.09–61.92	58.01	3.92	0.15–0.16	0.15	0.05
Belgium	1	0.77	0.77	0.00	0.95	0.95	0.00	9.66	9.66	0.00	75.57	75.57	0.00	0.15	0.15	0.15
New Zealand	1	2.78	2.78	0.00	1.22	1.22	0.00	13.27	13.27	0.00	71.36	71.36	0.00	0.16	0.16	0.00
China	7	1.66–1.94	1.73	0.09	3.06–3.20	3.12	0.04	11.61–12.31	11.91	0.21	54.22–59.17	56.71	1.60	0.14–0.16	0.15	0.01
Pegwell	4	0.66–1.04	0.81	0.15	1.17–2.11	1.58	0.39	8.26–9.17	8.80	3.28	57.36–69.84	60.81	5.23	0.13–0.19	0.15	0.02
Tunisia	7	0.30–1.51	0.63	0.44	1.61–3.97	2.26	0.73	5.94–9.43	7.91	1.16	38.51–64.95	49.13	8.86	0.05–0.10	0.07	0.02
France	11	0.78–1.49	1.04	0.19	0.97–1.88	1.32	0.32	7.85–11.93	9.59	1.25	59.52–76.04	70.54	6.39	0.05–0.19	0.12	0.04
All samples	58	0.26–2.78	1.03	0.47	0.84–6.20	2.44	1.21	5.94–16.62	10.53	2.38	38.51–76.99	60.59	9.77	0.03–0.22	0.13	0.04

continues overleaf

Table 10.4 (continued)

	K2O (wt%)			CaO (wt%)			TiO2 (wt%)			MnO (wt%)			Fe2O3 (wt%)		
	Range	Av.	SD	Range	Av.	SD	Range	Av.	SD	Range	Av.	SD	Range	Av.	SD
Vicksburg	10	1.74–2.36	2.00	0.18	0.69–17.45	7.66	6.05	0.55–0.82	0.69	0.09	0.02–0.18	0.07	0.04	2.82–5.08	3.67	0.74
Nebraska	2	2.95–3.10	3.02	0.07	2.80–3.15	2.97	0.17	0.51–0.51	0.51	0.01	0.04–0.05	0.05	0.01	3.04–3.04	3.04	0.01
Uzbekistan	5	2.31–2.61	2.45	0.11	11.99–18.20	14.13	2.18	0.66–0.76	0.69	0.03	0.09–0.14	0.11	0.02	4.68–5.70	5.06	0.38
Tajikistan	3	2.09–3.07	2.65	0.41	3.70–8.11	5.90	1.80	0.78–0.80	0.79	0.01	0.13–0.13	0.13	0.01	6.11–6.55	6.34	0.18
Khatif	1	1.40	1.40	0.00	13.03	13.03	0.00	1.06	1.06	0.00	0.08	0.08	0.00	3.77	3.77	0.00
Netivot	4	1.26–1.63	1.47	0.14	9.27–19.52	14.17	3.84	1.27–1.44	1.38	0.07	0.09–0.12	0.11	0.01	5.71–7.63	6.93	0.74
Argentina	2	2.29–2.45	2.37	0.08	2.90–9.39	6.14	3.24	0.75–0.84	0.80	0.04	0.10–0.11	0.10	0.01	5.11–5.84	5.47	0.37
Belgium	1	2.05	2.05	0.00	0.68	0.68	0.00	0.81	0.81	0.00	0.05	0.05	0.00	3.48	3.48	0.00
New Zealand	1	2.18	2.18	0.00	1.59	1.59	0.00	0.60	0.60	0.00	0.04	0.04	0.00	3.82	3.82	0.00
China	7	2.58–2.69	2.64	0.03	7.87–11.48	10.06	1.12	0.64–0.68	0.66	0.01	0.09–0.10	0.09	0.01	4.37–5.12	4.80	0.22
Pegwell (UK)	4	1.78–2.13	1.94	0.15	3.51–13.62	10.64	4.14	0.61–0.68	0.65	0.02	0.06–0.07	0.07	0.01	2.23–3.64	3.16	0.56
Tunisia	7	1.78–2.13	1.63	0.22	4.03–26.26	16.86	7.73	0.42–0.54	0.50	0.04	0.02–0.03	0.03	0.01	1.98–2.71	2.48	0.24
France	11	1.77–2.51	2.08	0.22	0.28–12.03	4.29	5.09	0.49–0.77	0.63	0.09	0.04–0.08	0.06	0.01	2.35–4.07	2.91	0.48
All samples	58	1.26–3.10	2.12	0.44	0.28–26.26	9.21	6.54	0.42–1.44	0.71	0.21	0.02–0.18	0.07	0.03	1.98–7.63	4.01	1.40

described extensive deposits of laminated periglacial laminated silt and sand in western and central Europe which were interpreted to reflect fluctuations in wind conditions.

Loess Microfabric and Geotechnical Properties

Unweathered Quaternary loess typically has an isotropic, open microfabric with a high voids ratio, consisting of single silt grains and silt-sized aggregates connected by clay bridges and, less commonly, calcium carbonate (Derbyshire, 1983; Pye, 1987; Figure 10.2(a)). Some authors have reported preferred grain orientation which reflects the dominant depositional wind conditions (Matalucci et al., 1969), but this does not appear to be common. Weathering of loess causes the fabric to become less open, partly due to weathering of detrital grains and production of authigenic clays, and partly due to physical translocation of clays and organic matter through the pore system (Figure 10.2(b)).

The open microfabric of loess causes it to collapse when it is wetted and subjected to an applied load, i.e. it is a metastable sediment which is subject to hydrocon-solidation. For this reason, suitable precautions must be taken in undertaking engineering works on loess. However, unsaturated loess is able to stand in near-vertical sections up to 6 m high which may remain stable for many years (Figure 10.3).

The microfabric of loess which has been colluvially or fluvially rewowoked is often anisotropic, i.e. it has a preferred grain orientation, and is much less prone to settlement or collapse when wetted or placed under load (Derbyshire et al., 1988).

Loess Landforms

In the simplest case, loess forms a blanket over the pre-existing topography. The thickness of the cover varies with the nature of the relief and with the vegetation density. In areas of thick, uniform vegetation, which inhibit erosion and downslope movement, loess is often thickest on the interfluve areas, but in markedly undulating terrain, and in semi-arid or arid areas, downslope reworking is more rapid and the thickest deposits tend to be found on lower valley slopes. 'Pseudo-anticline' struc-tures have been described from several parts of the central United States (e.g. Priddy et al., 1964), while large-scale linear ridge and furrow structures, which lie approxi-mately parallel with the dominant wind direction, have been described from Bulgaria and elsewhere (Leger, 1990).

Loess is very susceptible to erosion by running water. Consequently, subterranean pipe systems, pseuo-karst features and gullies are widely found (Figure 10.4). Gullies enlarge by a combination of direct fluvial incision and collapse of massive loess slabs due to the development of tensile fracture systems. Landslides are a major hazard in high relief loess areas, and are triggered especially by heavy rains and earthquake shocks (Derbyshire et al., 1991, 1993).

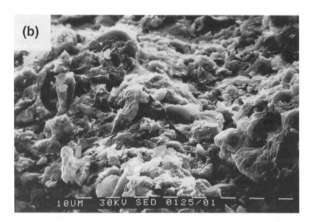

Figure 10.2   Scanning electron micrographs showing (a) relatively open microfabric in unweathered late Pleistocene (oxygen isotope Stage 2) loess and (b) much denser microfabric in older (oxygen isotope Stage 4) weathered loess from Vicksburg, Mississippi (scale bars = 10 μm)

REQUIREMENTS FOR LOESS FORMATION

There are three fundamental requirements for the formation of loess: (1) a source of appropriately sized sedimentary material; (2) wind energy to transport the material; and (3) a suitable site for deposition and accumulation.

The question of the origin of loess material has been the subject of extensive debate for more than a century (e.g. Smalley, 1995). The prevailing view in the 1960s was that glacial grinding is the only really effective process capable of

Figure 10.3 Vertical walls of terraced loess in a roadcut section near Vicksburg, Mississippi. Each terrace is about 6 m high

Figure 10.4 Gullied loess at Devil's Den, Nebraska. The main gully is about 60 m deep

producing quartz-dominated silt in large quantities (Smalley, 1966; Smalley and Vita Finzi, 1968). However, subsequent work has shown that many other processes can also produce silt-sized particles, including fluvial comminution, aeolian abrasion, frost weathering, salt weathering and chemical weathering (e.g. Nahon and Trompette, 1982; Pye and Sperling, 1983; Smith *et al.*, 1991; Wright and Smith, 1993; Gardner, 1994).

Rates of fine particle production are dependent partly on climate conditions but also on geological factors such as rates of tectonic uplift and river incision. Maintenance of an active dust source requires a replenishable sediment supply (McTainsh, 1989). Internal drainage basins which are surrounded by areas of high relief, and which themselves experience an arid climatic regime, at least episodically, provide ideal dust source regions. Specific surface types from which dust is easily entrained include dry river and lake beds, alluvial fans, areas of unvegetated, weathered bedrock, and older aeolian sediments which have become devegetated and reactivated. Climatic changes which produce alternating periods of humid and relatively drier conditions provide optimum conditions for dust production and deflation (Pye, 1989). At the present day, dust storm frequencies are highest in areas which receive approximately 100–200 mm of rainfall per annum (Goudie, 1983). During the Pleistocene, however, areas of glacial outwash would have provided much more important dust sources areas than they do today.

The mechanics of dust entrainment, transport and deposition have been the subject of detailed investigation in the past fifty years (see Chapters 6, 7 and 9, this volume), and are now relatively well understood. So far as loess formation is concerned, most material in the medium and coarse silt range is transported relatively close to the earth's surface in short-term suspension and modified saltation modes. As such, it is prone to be easily trapped by vegetation, topographic obstacles and water bodies (Tsoar and Pye, 1987). Fine silt-size material is more readily dispersed over a larger vertical range and may travel greater distances before being deposited, either by dry deposition in descending air masses, or by wet deposition.

Loess will not accumulate on smooth, dry surfaces such as bare rock since any deposited dust is resuspended if exposed to turbulent windflow or impacting particles. Accumulation of dust to form a loess deposit requires that the deposited dust is trapped by some form of roughness element, most commonly vegetation.

A variety of wind systems are responsible for transporting dust (Pye, 1987), but the most important in terms of loess fomation are low level flows associated with cold air outbreaks from high pressure centres, such as that located over Siberia during winter, and westerly, southwesterly or northwesterly winds associated with mid latitude depressions which periodically track across desert-marginal areas.

The dust accumulation rate is fundamental to the formation of a recognisable loess deposit. Under semi-arid conditions, the threshold accumulation rate is approximately 0.5 mm yr^{-1}, corresponding to an average dust accumulation flux of approximately 625 $g\,m^{-2}\,yr^{-1}$. If the rate of dust accumulation is lower, syn-depositional weathering, bioturbation, colluvial reworking and mixing with non-aeolian sediments combine to produce a loessoid deposit which differs in its sedimentological and geotechnical properties from typical primary loess. At times during the late Pleistocene in north America and western Europe, loess accumulation rates exceeded 2 mm yr^{-1}, equivalent to a minimum depositional flux of approximately 2600 g m^{-2} yr^{-1}. Comparable rates of up to 1.2 mm yr^{-1} occurred in Tajikistan at certain times in the late Pleistocene (Frechen and Dodonov, 1998). Estimates of average long-term loess accumulation rates in North China range from 0.07 mm yr^{-1} at Luochuan to 0.26 mm yr^{-1} at Lanzhou (Burbank and Li, 1985), but dust accumulation in these areas, as elsewhere, has been temporally very variable, and short-term rates of accumulation during colder periods of the Quaternary were considerably higher (An et al., 1991; Shen et al., 1992).

Loess becomes thinner and finer grained with increasing distance from the dust source, reflecting the fact that coarser grains settle out or become trapped preferentially closer to the source. Exponential downwind thinning of loess deposits has been observed in many areas (e.g. Snowden and Priddy, 1968; Frazee et al., 1970).

FACIES MODELS

Although loess can form in a variety of geographical settings, there are three principal end member situations which can be recognized. 'Periglacial' mid continental loess, of the type found widely in the USA, western Europe and Siberia, mainly forms in lowland areas adjacent to ice sheets and associated outwash sediments (Figure 10.5). The 'event sequence' involved in its formation includes glacial erosion of pre-weathered sediments and production of additional fine sediment by glacial grinding and crushing, transport by ice and meltwater, and subsequent wind transport to a suitable marginal area of accumulation. 'Perimontane' loess, sometimes referred to as 'hybrid' loess, forms on the margins of arid basins adjacent to high mountain areas where large amounts of fine sediment are produced by weathering and/or glacial processes. Examples include Tajikistan and Uzbekistan in Central Asia (Figure 10.6). 'Peridesert' loess occurs on the margins of deserts in continental areas dominated by relatively low relief. Dust is deflated from dry lake and river beds, and from alluvial fans and plains, and is transported by wind to the better vegetated desert margins where it is deposited (Figure 10.7). Each of these different loess depositional settings is associated with a distinctive set of sedimentological and stratigraphic characteristics.

PALAEOENVIRONMENTAL RECORDS IN LOESS DEPOSITS

Loess provides one of the best terrestrial proxy records of palaeoclimate in terms of the length of record and potential temporal resolution. In parts of North China and Central Asia, loess has been accumulating semi-continuously for at least the last 2.4 million years. These thick sequences display alternating sequences of loess and soil horizons, clearly reflecting variations in dust deposition rates and post-depositional weathering conditions over time (Figure 10.8).

Although many thick loess deposits in the world have now been investigated in terms of their palaeoclimatic record, particular attention has focused on the thick sequences in China and Central Asia. In particular, numerous authors in recent

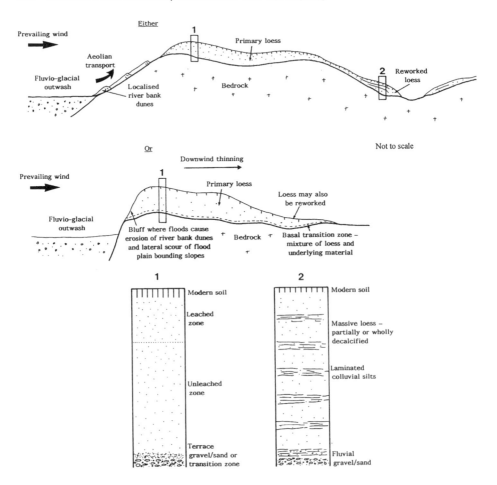

Figure 10.5 Schematic facies model of a periglacial lowland mid-continental loess setting (above) with representative stratigraphic sections (below)

years have sought to identify evidence of changes in monsoon intensity and possible relationships with orbital forcing (e.g. Kukla *et al.*, 1988; Zhou *et al.*, 1991; An *et al.*, 1991, 1993; Zhang *et al.*, 1994; Ding *et al.*, 1994; Shackleton *et al.*, 1995; Porter and An, 1995). The widely accepted current model proposes that periods of enhanced winter monsoon in the Quaternary, i.e. colder, drier intervals, were associated with more extensive dust source regions in northern China, more frequent and more vigorous outbreaks of northwesterly dust-laden air from Mongolia, and higher rates of dust deposition under cold, dry steppe conditions on the Chinese Loess Plateau. During intervals dominated by the summer monsoon, i.e. global interglacials and some interstadials, dust sources in northwest China contracted, dust transporting wind systems became less effective, and dust deposition rates on the Loess Plateau region decreased significantly, encouraging soil development under conditions of warmer, wetter climate and thicker vegetation cover.

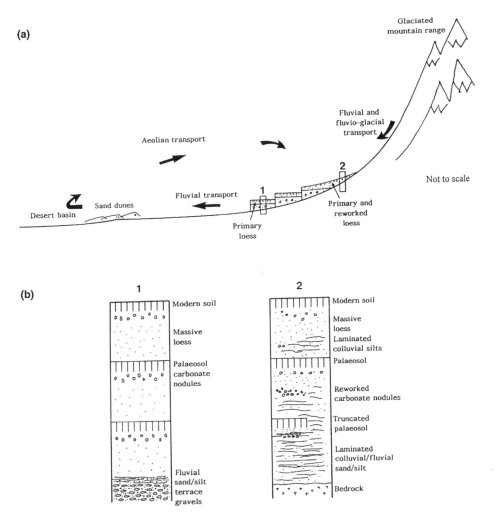

Figure 10.6 Schematic facies model of a perimontane (hybrid) loess setting (above) with representative stratigraphic sections (below)

In addition to changes in loess thickness and degree of weathering (i.e. mass accumulation rate), a number of other sedimentological properties of loess can provide important palaeoclimate proxy information. Grain size, especially of the quartz fraction, has been widely used as an indicator of average wind speed, larger grains being associated with stronger winds. Grain size can be determined directly, by techniques such as laser granulometry or X-ray sedigraph analysis, or it can be inferred from geochemical parameters such as SiO_2/TiO_2, SiO_2/K_2O or SiO_2/Al_2O_3 ratios. Coarser grained, quartz rich loess has a relatively higher content of SiO_2 than finer grained loess with a higher content of clay minerals. For example, Bigelow *et al.* (1990) interpreted a short-lived increase in grain size in Alaskan loess to

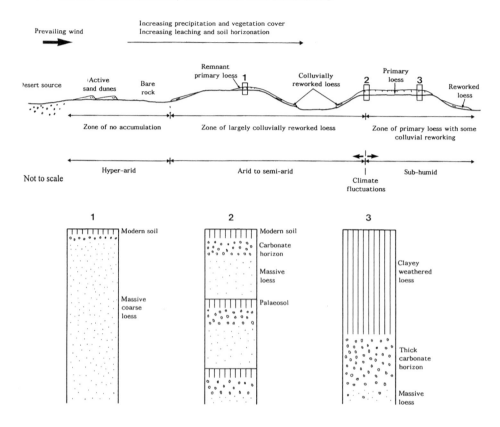

Figure 10.7 Schematic facies model of peridesert loess setting (above) with representative stratigraphic sections (below)

represent a brief increase in wind intensity which they correlated with the European Younger Dryas event. Xiao *et al.* (1995) and Porter and An (1995), amongst others, have used the median size of loess, or, more specifically the quartz fraction, as a proxy for winter monsoon intensity. Analysis only of the quartz fraction gives a more reliable indicator of dust transport conditions than the bulk sediment because it excludes the effect of clay and carbonate formed post-depositionally during weathering and soil formation. Other measures of grain size which have been used as proxies include the ratio of clay to coarse silt (Kd ratio), which has been interpreted as a measure of mechanical versus chemical weathering, either in the dust source area or at the deposition site.

Several geochemical indices, based on per cent abundance of major element oxides, have been proposed in an attempt to quantify relative levels of loess weathering and as proxies for environmental conditions. For example, An et al. (1991) proposed an index of summer monsoon intensity, SMI, as follows:

$$SMI = (Al_2O_3/Al_2O_3 + CaO + Na_2O + K_2O) \times 100$$

Figure 10.8 Top part of a thick interbedded loess/palaeosol sequence at Orkutsai, Uzbekistan (the part of the section shown is approximately 30 m high)

An index of winter monsoon intensity, WMI, was defined by:

$$WMI = SiO_2/TiO_2$$

While such indices are potentially useful, care needs to be taken to exclude auto-correlation effects and local interferences.

Another widely used property of loess–soil sequences is magnetic susceptibility (Kukla *et al.*, 1988). This property of the sediment is largely a function of the amount of ferrimagnetic minerals (mainly magnetite and maghaemite) present in the material, although size and shape of the particles also have an influence. In general terms, soil units within loess have higher magnetic susceptibilities than the intervening loess units, although the reasons for this remain a matter of debate. Some authors have favoured a dominant primary depositional control (e.g. Kukla and An, 1989; An *et al.*, 1991), while others have favoured a post-depositional weathering explanation, with larger numbers of authigenic ferrimagnetic grains attributed to the effects of weathering and soil forming processes (Maher and Thompson, 1992, 1995; Maher *et al.*, 1994). Magnetic susceptibility records have formed a central plank in attempts to correlate the terrestrial loess record with the deep ocean core record, and magnetic susceptibility has been used by several authors as a direct proxy for summer monsoon intensity (Xiao *et al.*, 1995; Shackleton *et al.*, 1995; Figure 10.9). However, at a global scale there are a number of apparent contradictions. For example, Beget *et al.* (1990) found that the pattern of magnetic susceptibility variation in Alaskan loess is opposite to that found in Chinese loess. In Alaska the highest susceptibility values are found in the unweathered loess units, possibly due to the fact that soil development is relatively weak and strong local winds carried more coarse grained magnetic minerals from local sources during colder periods. Other

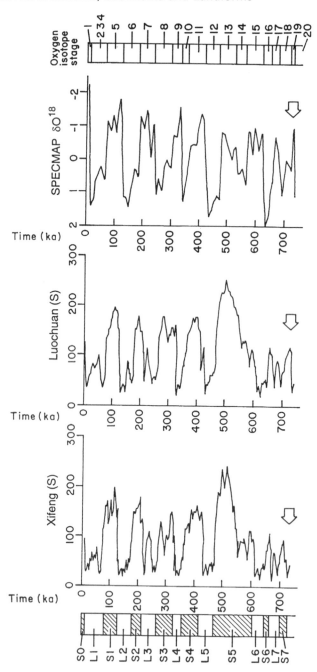

Figure 10.9 Simplified composite summary diagram showing comparison between magnetic susceptibility profiles at Xifeng and Luochuan sections, Chinese Loess Plateau, the stratigraphy of loess (L) and soil (S) layers at Xifeng, the SPECMAP oxygen isotope record from a marine core in the North Pacific, and the oxygen isotope 'time' scale

difficulties arise from the fact that variations in carbonate content of the loess can interfere with the susceptibility signal. In China, the upper palaeosols generally contain less carbonate than the loess, tending to enhance the magnetic susceptibility of the soils; however, in Nebraska, the reverse situation is found, with soils containing more carbonate than the loess, owing to post-depositional mobilization and reprecipitation. All magnetic susceptibility data should therefore be corrected for carbonate content before interpretation. There are clearly a number of controls on the magnetic susceptibility of loess and soils and this characteristic cannot be taken simply as a proxy for degree of weathering (Feng and Johnson, 1995; Meng et al., 1997).

There are clearly potential dangers in taking a simplistic approach to the interpretation of loess stratigraphy. Much early work sought to make inter- and intra-regional correlations based mainly on palaeomagnetic or magnetic susceptibility data which have subsequntly been found to be enigmatic. It has also been widespread practice to count loess and soil units down from the top of a section, and to assume that the sequence is more or less complete. More detailed studies, using closely spaced sections, have subsequently shown that such assumptions are not always correct, especially in high relief areas where loess is subject to erosion and redeposition (e.g. Derbyshire et al., 1991, 1995). In some areas, soil units can be divided into two or more sub-units while elsewhere only a single soil can be identified. Thinner loess sequences are more likely to contain a condensed record in which two or more soil forming intervals may be superimposed. Recent attempts to correlate between loess records in widely separated areas such as the Chinese Loess Plateau, Tajikistan, Uzbekistan, western Europe and the deep sea record, in order to identify possible common forcing factors (e.g. Shackleton et al., 1995; Bronger et al., 1995; Kukla and Cilek, 1996), have been hampered by the fact that, in detail, the sedimentary records within and between regions show important differences in terms of stratigraphy and age structure. The sampling intervals used have commonly been too coarse or incomplete to allow detailed comparisons to made. Better quantification of local and regional scale lithostratigraphic variability and continuity is required in order to provide more reliable estimates of loess accumulation rates and a firmer foundation for inter-regional correlations. Work should also be undertaken using higher resolution sampling intervals, ideally 1 or 2 cm, supported by large numbers of closely spaced luminescence age estimates (e.g. Frechen and Dodonov, 1998), in order to define short term variability and to identify possible inter-regional teleconnections. Additional work also needs to be undertaken on the geochemical fingerprinting of dust (e.g. Clarke, 1995), both to establish provenance and to facilitate stratigraphic correlations with dust records in ice cores and ocean cores.

REFERENCES

An, Z., Kukla, G.J., Porter, S.C. and Xiao, J.L., 1991. Magnetic susceptibility evidence of monsoon variations on the Loess Plateau of central China during the last 130 Ka. *Quaternary Research*, **36**, 29–36.

Beget, J.E., Stone, D.B. and Hawkins, D.B., 1990. Palaeoclimatic forcing of magnetic susceptibility variations in Alaskan loess during the Late Quaternary. *Geology*, **18**, 40–43.

Bigelow, N., Beget, J., and Powers, R., 1990. Late Pleistocene increase in wind intensity recorded in aeolian sediments from central Alaska. *Quaternary Research*, **34**, 160–165.

Bronger, A. and Heinkele, Th., 1990. Mineralogical and clay mineralogical aspects of loess research. *Quaternary International*, **7/8**, 37–51.

Bronger, A., Winter, R., Derevjanko, O. and Aldag, S., 1995. Loess-palaeosol sequences in Tadjikistan as a paleoclimatic record of the Quaternary in central Asia. *Quaternary Proceedings*, **4**, 69–81.

Burbank, D.W. and Li, J., 1985. Age and palaeoclimatic significance of the loess of Lanzhou, north China. *Nature*, **316**, 429–431.

Chamley, H., 1989. *Clay Sedimentology*. Springer-Verlag, Berlin.

Clarke, M.L., 1995. Sedimentological characteristics and rare earth element fingerprinting of Tibetan Silts and their relationship with the sediments of the Western Chinese Loess Plateau. *Quaternary Proceedings*, **4**, 41–52.

Coudé-Gaussen, G. and Rognon, P., 1988. The Upper Pleistocene loess of Tunisia: a statement. *Earth Surface Processes and Landforms*, **13**, 137–151.

Derbyshire, E., 1983. Origin and characteristics of some Chinese loess at two locations in China. In *Eolian Sediments and Processes*, M.E. Brookfield and T.S. Ahlbrandt (Eds). Elsevier, Amsterdam, pp. 69–80.

Derbyshire, E., Billard, A., Van Vliet Lanoe, B. *et al.*, 1988. Loess and paleoenvironment: some results of the European Joint Program of Research. *Journal of Quaternary Science*, **3**, 147–169.

Derbyshire, E., Wang J.T., Jin, Z.X. *et al.*, 1991. Landslides in the Gansu loess of China. *Catena Supplement*, **20**, 119–145.

Derbyshire, E., Dijkstra, T.A., Billard, A. *et al.*, 1993. Thresholds in a sensitive landscape: the loess region of central China. In *Landscape Sensitivity and Change*, D.S.G. Thomas and R.J. Allison (Eds). John Wiley, Chichester, pp. 97–127.

Derbyshire, E., Kemp, R. and Meng, X., 1995. Variations in loess and palaeosol properties as indicators of palaeoclimatic gradients across the loess plateau of North China. *Quaternary Science Reviews*, **14**, 681–697.

Ding, Z.L., Liu, T.S., Rutter, N.W. *et al.*, 1995. Ice volume forcing of east Asian winter monsoon variations in the past 800 000 years. *Quaternary Research*, **44**, 149–159.

Dodonov, A.E. and Baiguzina, L.L., 1995. Loess stratigraphy of Central Asia: palaeoclimatic and palaeoenvironmental aspects. *Quaternary Science Reviews*, **14**, 707–720.

Feng, Z.D. and Johnson, W.C., 1995. Factors affecting the magnetic susceptibility of a loess-soil sequence, Barton County, Kansas, USA. *Catena*, **24**, 25–37.

Frazee, C.J., Fehrenbacher, J.B. and Krumbein, W.C., 1970. Loess and distribution from a source. *Soil Science Society of America Proceedings*, **34**, 296–301.

Frechen, M. and Dodonov, A.E., 1998. Loess chronology of the Middle and Upper Pleistocene in Tadjikistan. *Geologische Rundschau*, **87**, 2–20.

Frye, J.C., Swineford, A. and Leonard, A.B., 1948. Correlation of the Pleistocene deposits of the central Great Plains with the glacial section. *Journal of Geology*, **56**, 501–525.

Gardner, R.A.M., 1994. Silt production from weathering of metamorphic rocks in the southern Himalaya. In *Rock Weathering and Landform Evolution*, D.A. Robinson and R.B.G. Williams (Eds). John Wiley, Chichester, pp. 487–503.

Goudie, A.S., 1983. Dust storms in space and time. *Progress in Physical Geography*, **7**, 502–530.

Keen, D.H., 1995. Molluscan assemblages from the loess of North Central China. *Quaternary Science Reviews*, **14**, 699–706.

Kes, A.S., 1984. Zonation and facilality of loessic deposits. In *Lithology and Stratigraphy of Loess and Palaeosols*, M. Pesci (Ed.). Geographical Research Institute, Hungarian Academy of Sciences, Budapest, pp. 104–111.

Kukla, G.J. and An, Z.S., 1989. Loess stratigraphy in central China. *Palaoegeography, Palaeoclimatology, Palaeoecology*, **72**, 203–225.

Kukla, G.J. and Cilek, V., 1996. Plio-Pleistocene megacycles: record of climate and tectonics. *Palaeogeography, Palaeoclimatology, Palaeoecology*, **120**, 171–194.

Kukla, G.J., Heller, F., Liu, T.S., and An, Z.S., 1988. Pleistocene climate in China dated by magnetic susceptibility. *Geology*, **16**, 811–814.

Leger, M., 1990. Loess landforms. *Quaternary International*, **7/8**, 53–61.

Li., P-Y and Zhou, L.P., 1993. Occurrence and paleoenvironmental implications of the Late Pleistocene loess along the eastern coasts of the Bohai Sea, China. In *The Dynamics and Environmental Context of Aeolian Sedimentary Systems*, K. Pye (Ed.). Geological Society Special Publication No. 72, Geological Society Publishing House, Bath, pp. 293–309.

Liu, T.S. (Ed.), 1985. *Loess and the Environment*. China Ocean Press, Beijing.

Lyell, C., 1834. Observations on the loamy deposit called 'loess' of the Basin of the Rhine. *Edinburgh New Philosophical Journal*, **17**, 110–113, 118–120.

McTainsh, G., 1989. Quaternary aeolian dust processes and sediments in the Australian region. *Quaternary Science Reviews*, **8**, 235–253.

Maher, B.A. and Thompson, R., 1992. Palaeoclimatic significance of the mineral magnetic record of the Chinese loess and palaeosols. *Quaternary Research*, **37**, 155–170.

Maher, B.A. and Thompson, R., 1995. Palaeorainfall reconstruction from pedogenic magnetic susceptibility variations in the Chinese loess and paleosol. *Quaternary Research*, **44**, 383–391.

Maher, B.A., Thompson, R. and Zhou, L.P., 1994. Spatial and temporal reconstruction of changes in the Asian palaeomonsoon: a new mineral magnetic approach. *Earth and Planetary Science Letters*, **125**, 461–471.

Matalucci, R.V., Shelton, J.W. and Abdel-Hady, M., 1969. Grain orientation in Vicksburg loess. *Journal of Sedimentary Petrology*, **39**, 969–979.

Mazzullo, J., Alexander, A., Tieh, T. and Menglin, D., 1992. The effects of wind transport on the shapes of quartz silt grains. *Journal of Sedimentary Petrology*, **62**, 961–971.

Meng, X., Derbyshire, E. and Kemp, R.A., 1997. Origin of the magnetic susceptibility signal in Chinese loess. *Quaternary Science Reviews*, **16**, 833–839.

Mucher, H.J. and De Ploey, J., 1977. Experimental and micromorphological investigation of erosion and redeposition of loess by water. *Earth Surface Processes and Landforms*, **2**, 117–124.

Mucher, H.J., De Ploey, J. and Savat, J., 1981. Response of loess materials to simulated translocations by water: micromorphological investigations. *Earth Surface Processes and Landforms*, **6**, 331–336.

Nahon, D. and Trompette, R., 1982. Origin of siltstones: glacial grinding versus weathering. *Sedimentology*, **29**, 25–32.

Pecsi, M., 1990. Loess is not just accumulation of airborne dust. *Quaternary International*, **7/8**, 1–21.

Priddy, R.R., Christmas, J.Y. and Ward, J.G., 1964. Pseudoanticlines in Vicksburg loess. *Journal of the Mississippi Academy of Sciences*, **10**, 178–179.

Porter, S.C. and An, Z., 1995. Correlation between climatic events in the North Atlantic and China during the last glaciation. *Nature*, **375**, 305–308.

Pye, K., 1984. SEM investigations of quartz silt micro-textures in relation to the source of loess. In *Lithology and Stratigraphy of Loess and Palaeosols*, M. Pesci (Ed.). Geographical Research Institute Hungarian Academy of Sciences, Budapest, pp. 139–151.

Pye, K., 1987. *Aeolian Dust and Dust Deposits*. Academic Press, London.

Pye, K., 1989. Processes of fine particle formation, dust source regions, and climatic changes. In *Paleoclimatology and Paleometeorology, Modern and Past Processes of Global Atmospheric Transport*, M. Sarnthein and M. Leinen (Eds). NATO ASI Series C 282, Kluwer, Dordrecht, pp. 3–30.

Pye, K., 1994. Shape sorting during wind transport of quartz silt grains – discussion. *Journal of Sedimentary Research A*, **64**, 704–705.

Pye, K., 1995. The nature, origin and accumulation of loess. *Quaternary Science Reviews*, **14**, 653–667.

Pye, K. and Johnson, R., 1988. Stratigraphy, geochemistry and thermoluminescence ages of Lower Mississippi Valley loess. *Earth Surface Processes and Landforms*, **13**, 103–124.

Pye, K. and Sperling, C.H.B., 1983. Experimental investigation of silt formation by static breakage processes: the effect of temperature, moisture and salt on quartz dune sand and granitic regolith. *Sedimentology*, **30**, 49–62.

Raeside, J.D., 1964. Loess deposits of the South Island, New Zealand. *New Zealand Journal of Geology and Geophysics*, **7**, 811–838.

Ranov, V., 1995. The 'Loessic Palaeolithic' in South Tadjikistan, Central Asia: its industries, chronology and correlation. *Quaternary Science Reviews*, **14**, 731–745.

Ruhe, R.V. and Olsen, C.G., 1980. Clay mineral indicators of glacial and nonglacial sources of Wisconsinan loesses in southern Indiana, USA. *Geoderma*, **24**, 283–297.

Schwan, J., 1986. The origin of horizontal alternating bedding in Weichselain aeolian sands in northwestern Europe. *Sedimentary Geology*, **49**, 73–108.

Shackleton, N.J., An, Z., Dodonov, A.E. *et al.*, 1995. Accumulation rate of loess in Tadjikistan and China: relationship with global ice volume cycles. *Quaternary Proceedings*, **4**, 1–6.

Sherwin, D., 1995. Loessite: its occurrence, recognition and interpretation. PhD Thesis, University of Reading.

Smalley, I.J., 1966. The properties of glacial loess and the formation of loess deposits. *Journal of Sedimentary Petrology*, **36**, 669–670.

Smalley, I.J., 1995. Making the material: the formation of silt-sized primary mineral particles for loess deposits. *Quaternary Science Reviews*, **14**, 645–651.

Smalley, I.J. and Vita-Finzi, C., 1968. The formation of fine particles in sandy deserts and the origin of 'desert' loess. *Journal of Sedimentary Petrology*, **38**, 766–774.

Smith, B.J., Wright, J.S. and Whalley, W.B., 1991. Simulated aeolian abrasion of Pannonian sands and its implications for the origins of Hungarian loess. *Earth Surface Processes and Landforms*, **16**, 745–752.

Snowden, J.O. and Priddy, R.R., 1968. Geology of Mississippi Valley loess. *Mississippi Geological Survey Bulletin*, **111**, 13–203.

Tsoar, H. and Pye, K., 1987. Dust transport and the question of desert loess formation. *Sedimentology*, **34**, 139–153.

Weaver, C.E., 1958. Geologic interpretation of argillaceous sediments. Part 1. Origin and significance of clay minerals in sedimentary rocks. *Bulletin of the American Association of Petroleum Geologists*, **42**, 254–271.

Winspear, N.R. and Pye, K., 1996. Textural, geochemical and mineralogical evidence for the sources of aeolian sand in central and southwestern Nebraska, USA. *Sedimentary Geology*, **101**, 85–98.

Wright, J. and Smith, B.J., 1993. Fluvial comminution and the production of loess-sized quartz silt: a simulation study. *Geografiska Annaler*, **75A**, 25–34.

Xiao, J., Porter, S.C., An, Z. *et al.*, 1995. Grain size of quartz as an indicator of Winter Monsoon strength on the Loess Plateau of Central China during the last 130 000 Yr. *Quaternary Research*, **43**, 22–29.

Zhang, X, An, Z., Chen, T. *et al.*, 1994. Late Quaternary records of the atmospheric input of eolian dust to the centre of the Chinese Loess Plateau. *Quaternary Research*, **41**, 35–43.

11 The Aeolian Rock Record
(Yes, Virginia, it Exists, But it Really is Rather Special to Create One)

GARY KOCUREK

Department of Geological Sciences, University of Texas, USA

The creation of an aeolian rock record can be viewed as occurring in three phases: (1) sand sea construction; (2) accumulation of a body of strata; and (3) preservation of the accumulation. Each of these phases has its own set of governing rules, the fundamentals of which give rise to three separate hypotheses. Hypothesis I states that sand sea construction is a function of the three separate issues of (1) sediment supply; (2) sediment availability; and (3) transport capacity of the wind, which together define the sediment state of the system. Hypothesis II states that the accumulation of sand sea strata requires specific conditions that satisfy the principle of continuity, the dynamics of which rest with the three basic types of aeolian systems: (1) dry; (2) wet; and (3) stabilising. Hypothesis III states that the preservation of sand sea accumulations occurs with (1) subsidence and burial, and/or (2) a rise of the water table through the accumulation. Because the governing rules for each phase differ, a selection process occurs in which the ultimate creation of a record has to be viewed as the coincidence of independent variables, most of which are strongly influenced by their boundary conditions.

INTRODUCTION

If viewing extensive tracts of ancient aeolian strata, such as the Permian or Jurassic record on the Colorado Plateau of the USA, one might assume that the creation of an aeolian rock record is very common and that within this record lies a detailed chronology of events. The Quaternary record, however, argues the contrary in that it is clear that whole sand seas can leave no record and that events of varying magnitudes and durations can leave no impression or that these are subsequently removed. One is then quickly back to time-honoured questions of just what is the rock record, how representative is it, and what scale and sorts of events are recorded?

While it is true that the route from surface expression to rock record for aeolian systems follows the same general rules of geology as other systems, it is also true that each system is its own barometer in terms of sensitivity to environmental change, that there are unique internal responses to external controls, and that the rules of rock-record architecture differ from system to system. For the aeolian rock record, the route from geomorphic expression to rock record might be viewed as

Aeolian Environments, Sediments and Landforms. Edited by A.S. Goudie, I. Livingstone and S. Stokes.
© 1999 John Wiley & Sons, Ltd.

occurring in three phases, each with its own set of controls and responses. The first phase must be the construction of a sand sea. Whereas a desert is a climatic condition and the consequence of local or global events, a sand sea requires very large quantities of sand. The second phase is the accumulation of a body of strata. The mere existence of a sand sea does not mean that the system dynamics are such that a body of strata will form. The third phase is the incorporation of this strata into the rock record. The existence of a sedimentary body is no guarantee of a later stratigraphic record.

The purpose of this chapter is to address the three basic phases in the creation of an aeolian rock record, beginning with the formation of sand seas, progressing through their building of a body of strata, and ending with the incorporation of this body into the stratigraphic record. The route taken here toward this goal is by way of three hypotheses. Each hypothesis originates in observations of modern aeolian systems and their Quaternary and more ancient representations. In turn, each hypothesis carries its own set of implications. These hypotheses are advanced as simplified attempts at understanding a reality of complex systems, which do not begin from some 'clean slate', but rather progress from antecedent conditions, then evolve through complicated routes toward a final configuration.

SAND SEA CONSTRUCTION

Hypothesis

Hypothesis I states that sand sea construction is a function of three separate issues: (1) sediment supply; (2) sediment availability; and (3) transport capacity of the wind. Together, these issues define the *sediment state* of an aeolian system, as set forth in detail in Kocurek and Lancaster (1999).

The *sediment supply* of an aeolian system is the volume of sediment of a suitable grain size generated per time that contemporaneously or at some later point in time serves as the source material for the system. While aeolian sediment can be first-cycle or derived through deflation of rock, the vast quantities of sediment in sand seas nearly always are derived secondarily from fluvial/alluvial, coastal or lacustrine systems. Examples, therefore, of the generation of an aeolian sediment supply include the accumulation of fluvial/deltaic/lacustrine terrigenous sands (e.g. Mojave River and fan delta sediments that sourced the Kelso Dune Field, California; Sharp, 1966) and coastal or shelf sands (e.g. Namib Sand Sea; Corbett, 1993), as well as the precipitation of a lacustrine gypsum (e.g. Lake Lucero sourcing of the White Sands Dune Field, New Mexico; McKee, 1966) or a marine carbonate (e.g. Arabian Gulf area; Glennie, 1998).

Whether or not the sediment supply is ever utilised for sand sea construction is independent of the creation of the supply itself, but rather is, as a first step, a function of the availability of the sediment to the wind. *Sediment availability* is the susceptibility of surface grains to entrainment by the wind. Lacustrine or shelf sands are not available for aeolian transport until they are exposed, and vegetation, coarse-grained lag or mud surfaces, the water table or capillary moisture, surface binding or cementation, and grain properties such as sorting and shape also affect availability. The exhaustion of the sediment supply *per se* can be viewed as zero sediment availability. Sediment availability is difficult to quantify, but its manifestation is the actual transport rate for a given wind, q_a, which can be represented as a volumetric rate.

The *transport capacity* of the wind is its sediment carrying capacity and is independent of sediment availability. Wind has a potential sediment transport capacity, q_p, which increases with wind power and can be expressed as a volumetric rate. Transport capacity or *saturated flow* is reached in a matter of seconds over a surface of perfectly available sand (Anderson and Haff, 1991), but where sediment availability is limited, the wind must be at least potentially erosional and the flow *unsaturated* until capacity is reached (see fig. 2 in Kocurek and Havholm, 1993).

Plotting sediment supply, availability and transport capacity as separate curves against time defines nine possible sediment states of an aeolian system (Figure 11.1). These states are defined by: (1) whether the sediment supply is stored, or serves as contemporaneous influx or lagged influx to the aeolian system; and (2) by the controls on storage and influx.

Sediment that is *stored* must be so because it is either *availability-limited* (S_{AL}) or *transport-limited* (S_{TL}) while the total volume of stored sediment could contain components of both *availability-* and *transport-limited* storage (S_{TAL}). Availability-limited stored sediment occurs where $q_a < q_p$ because surface factors limit the ability of the wind to entrain sediment (Figures 11.1(a)–(c)). Transport-limited stored sediment occurs where $q_a = q_p$, and implies that the sediment supply is generated at a rate greater than the wind can entrain it, and that availability is not a limiting factor (Figure 11.1(a)). Stored sediment that contains both availability- and transport-limited components describes the case where availability limits sediment entrainment by the wind, but even without this factor, winds are insufficient to transport sediment from the source area at the rate at which it is generated (Figure 11.1(a)).

Contemporaneous influx to the aeolian system must at least be *transport-limited* (CI_{TL}) because influx cannot exceed the transport capacity of the wind, and with fully saturated flow $q_a = q_p$ (Figure 11.1(a)). Where $q_a < q_p$ because of availability, q_a is the *availability-limited* contemporaneous influx (CI_{AL}) to the aeolian system (Figure 11.1(a)). *Lagged influx* to an aeolian system occurs with deflation of sediment that has been previously stored for any reason, with its controls paralleling those of contemporaneous influx, such that lagged influx must be *transport-limited* (LI_{TL}) or *availability-limited* (LI_{AL}) (Figure 11.1(b)). Sediment influx to the aeolian system could also consist of both *contemporaneous and lagged influx*, with both *transported-limited influx* (CLI_{TL}) and *availability-limited influx* (CLI_{AL}) (Figure 11.1(c)).

Implications

The primary implication of hypothesis I is that aeolian sand seas occur through the coincidence of events that may be of wholly different origins and separated in both space and time. While the sediment state diagram (i.e. Figure 11.1) depicts aeolian

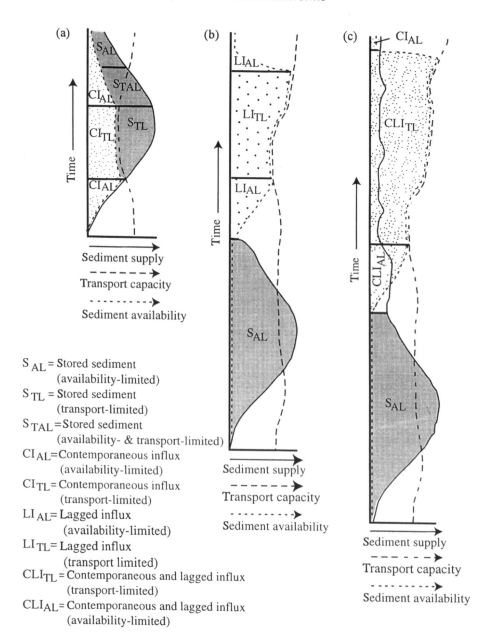

S AL = Stored sediment
    (availability-limited)
S TL = Stored sediment
    (transport-limited)
S TAL =Stored sediment
    (availability- & transport-limited)
CI AL=Contemporaneous influx
    (availability-limited)
CI TL= Contemporaneous influx
    (transport-limited)
LI AL= Lagged influx
    (availability-limited)
LI TL= Lagged influx
    (transport limited)
CLI TL = Contemporaneous and lagged influx
    (transport-limited)
CLI AL= Contemporaneous and lagged influx
    (availability-limited)

Figure 11.1 Plot of sediment supply, sediment availability, and transport capacity of the wind over time. Nine possible sediment states of the aeolian system are defined by the relative positions of the three curves. (a) The generated sediment supply is partly stored and partly used as contemporaneous influx to the aeolian system. (b) The sediment supply is entirely stored as it is generated, then deflated later to serve as lagged influx. (c) Both contemporaneous and lagged influx after a period when the entire generated sediment supply was stored

systems under any condition, sand sea construction occurs when a large supply of sediment is available to wind of a sufficient transport capacity.

Considering just terrigenous sands, events that lead to the generation of a sediment supply for aeolian systems are tectonic, climatic and eustatic in nature (Figure 11.2). Tectonic uplift generates a source area for sediment, but also increases gradient, thereby increasing stream power for erosion and transport at the expense of infiltration (Figure 11.2(a)). The tectonic template also defines the regional drainage pattern and may determine the location of the sediment supply. Climate affects stream power, the extent of vegetative cover, and rates of weathering and erosion (Figure 11.2(b)). For potential desert areas, most thinking follows Langbein and Schumm (1958) in envisioning maximum erosion to occur at the transition from subhumid to semiarid climates. The reduction in vegetation promotes surface runoff, but also the magnitude of low-frequency floods and the maximum/mean precipitation ratio increase with the onset of climatic drying. Eustatic sea level, as with sea level determined by regional tectonism, can affect the volume of sediment supply and its placement (Figure 11.2(c)). Transgressions and still-stands promote lesser volumes of sediment supply because of a raised base level, but these accumulations of fluvial, deltaic and coastal sands occupy more proximal positions with respect to coastal or continental aeolian systems. A regression promotes greater erosion because of lower base level, but these sands may be transported to more distal positions.

Sediment availability is most directly related to climate and sea level, but tectonism plays an indirect role. In many of deserts today, sediment availability is an inverse function of the extent of vegetative cover, which, in turn, is largely a function of climate (Figure 11.2(b)). A drying climate also increases sediment availability by the lowering or desiccation of lakes, as well as any regional fall in the water table in which capillary moisture is a surface binder. Arid or semi-arid climates do not, however, always coincide with high sediment availability. For example, the development of accretionary pavements and soil encrustations produces surfaces of low sediment availability in the Mojave Desert (McDonald et al., 1995). Exposure of surfaces during regressions and any corresponding fall of the coastal and inland water table increase sediment availability (Figure 11.2(c)). Beyond its effects on relative sea level, tectonism and the nature of any uplifted source rock play a role in the size and type of grains produced, which, in turn, affect the susceptibility of a surface toward lag or pavement development.

The transport capacity of the wind is inherently climatic, but climate has multiple causes, ranging from orographic effects, to the configuration of land and sea because of global tectonism or eustatic sea level, to extra-terrestrial factors such as Milankovitch cycles. At least for the subtropical belt of deserts, wind energy increases during icehouse times and decreases during greenhouse times (Figure 11.2(b)). The larger question is during what climatic conditions through time or by position on Earth are winds significantly higher or lower than the time-averaged global mean?

While the discussion above on the factors that control sediment state is by no means complete, it does illuminate the complexity in modelling aeolian systems, as with other complex systems in nature, beyond simple governing rules. Factors such

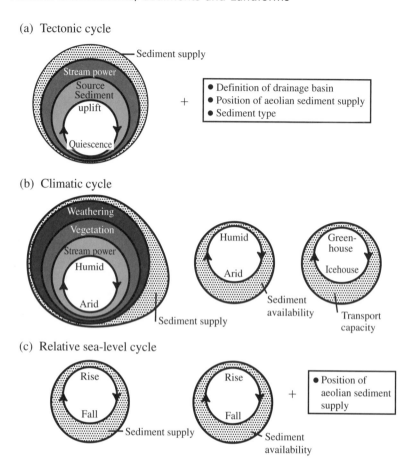

(a) Tectonic cycle

(b) Climatic cycle

(c) Relative sea-level cycle

Figure 11.2   Basic tectonic (a), climatic (b), and sea level (c) cycles with their corresponding effects that govern sediment supply, sediment availability, and transport capacity of the wind

as tectonism, climate, sea level, along with other controlling variables, broadly define the *boundary conditions* of the evolving systems, and their interplay with the dynamics of the aeolian system per se largely promise that each system is unique to some degree. Inspection of Figure 11.2, however, suggests that sand sea construction in many cases is the product of antecedent conditions. Major sand sea construction most readily occurs during the windy, arid, icehouse phase of climatic cycles, but only providing that this phase follows a more humid phase during which a large sediment supply was generated. Under favourable conditions, sand sea construction should be rapid, limited only by the transport-capacity of the wind, implying as well that an exhaustion of the sediment supply may occur quickly, with the sand sea yielding to a destructional phase. Sand sea construction, however, may imprint the landscape with a template that is resistant to change, as in the case of linear dunes, which, because of their low number of dune terminations per crest

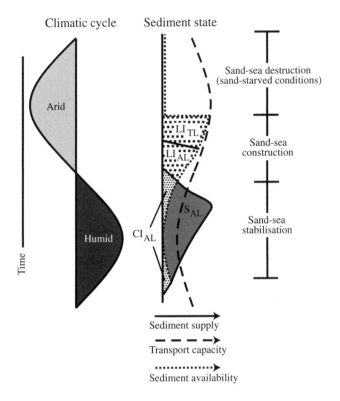

Climatic cycle   Sediment state

Arid

Humid

Time

Sand-sea destruction
(sand-starved conditions)

Sand-sea
construction

Sand-sea
stabilisation

LI_{TL}

LI_{AL}

S_{AL}

CI_{AL}

Sediment supply

Transport capacity

Sediment availability

Figure 11.3 Idealised sediment-state diagram for the Sahara region as discussed in the text

length, are very difficult to reform (Werner and Kocurek, 1997). This dune geomorphic template itself then acts as an antecedent condition in which subsequent humid/arid cycles are reflected in dune stabilisation and reactivation, respectively, especially when humid cycles are minor and do not yield large volumes of additional sediment supply.

Example

The Saharan region is an example of an aeolian system that has undergone significant change over time for which a sediment-state approach helps in unravelling the dynamics of the system, but also illuminates the complexity of the sequence of events that gave rise to the Sahara (Figure 11.3). At the broad scale, during glacial maximum times of windy, arid conditions the Sahara expands and sand sea construction occurs, whereas during climatic optimums of more humid conditions, the Sahara contracts, dunes are stabilised and there is a proliferation of fluvial-lacustrine systems (e.g. Sarnthein, 1978; Petit-Maire, 1989; Hooghiemstra *et al.*, 1992; Yan and Petit-Maire, 1994). Upon these broad climatic cycles, which largely follow Milankovitch solar insolation cycles (e.g. CLIMAP, 1976; Kutzbach and

Street-Perrott, 1985), are superimposed shorter cycles (Gasse and Van Campo, 1994), some of which correlate to major climatic shifts such as the Younger Dryas (e.g. Gasse et al., 1990; Street-Perrott and Perrott, 1990; Gasse and Fontes, 1992; Blum et al., 1998).

The Sahara sand supply probably originates from alluvial-fluvial-lacustrine systems that ultimately derive their sediment from the central African uplands of a Neogene age and, to a lesser degree, from erosion of the more stable craton. The inactivity of these systems and the stability of the landscape during arid phases such as the present clearly show that maximum sediment supply is generated during humid phases of the climatic cycle (Figure 11.3). The coincidence of maximum tectonism during a relatively humid period characterised by significant fluvial systems (represented by the widespread 'Continental Terminal') is during the Neogene, whereas uplifted areas persisted and must have been subject to continued erosion during subsequent Quaternary humid periods. Following from Figure 11.3, the sediment supply for the Sahara is generated during these humid periods and during the transition from relatively humid to more arid conditions. In addition, the tectonic template imposed upon the Sahara region with the creation of the Atlas Mountains along the northern rim of Africa during the Neogene must have radically altered the pre-existing fluvial drainage pattern in which systems issued from the interior uplands to empty along the northern shelf margin (e.g. Kogbe and Burollet, 1990; Lang et al., 1990), to one in which the Neogene sediment depocenter shifted to the inboard foreland basin and yet interior craton. In this regard, the Neogene sequence of tectonic uplift and definition of the drainage pattern, coinciding with a humid time of enhanced fluvial activity, may be speculated as the major origin of the Sahara sediment supply, with subsequent additions during Quaternary humid periods.

The onset of arid conditions for the Sahara is during the Pliocene with subsequent development of full glacial-interglacial climatic cycles during the Quaternary (Ruddiman et al., 1989; Tiedemann et al., 1989). Decreasing vegetation with aridity fostered sediment availability, while the enhanced wind power of glacial times fostered enhanced sediment transport capacity of the wind, the combination of which is most conducive for sand sea construction (Figure 11.3). Aeolian sediment influx is then largely availability- and then transport-limited lagged influx. The Sahara sediment cycle, therefore, as first conceptualised by Wilson (1973), consists of the generation of the sediment supply during humid periods, followed by sand sea construction during arid periods when the generation of a sediment supply is limited, but sediment availability and transport capacity are high. Dunes are largely stabilised during humid periods, then reactivated during arid times. Many linear dunes of the Sahara and elsewhere (see Chapter 3, this volume) have largely been shown to be Pleistocene in age with the amalgamation of generations of dunes onto the linear template, as discussed in general in the previous section. Continuation of arid conditions beyond the point where the sediment supply has been exhausted necessarily leads to sand-starved conditions (Figure 11.3) and the export of sediment out of the system, a condition that characterises much of the Sahara today (Mainguet and Chemin, 1983). Given the regional, trade-wind driven, sand-transport path of the Sahara, the Atlantic Ocean is the final depocentre of Saharan sands.

ACCUMULATION OF SAND SEA STRATA

Hypothesis

Hypothesis II states that the accumulation of sand sea strata requires specific conditions that satisfy the principle of continuity, the dynamics of which rest with three basic types of aeolian systems: (1) dry; (2) wet; and (3) stabilising systems. Accumulation is used specifically here to refer to the passage of sediment from above to below the accumulation surface such that a body of strata forms and the accumulation surface rises over time (Figure 11.4(a)). The alternatives to accumulation and a rising accumulation surface are: (1) bypass in which the dunes simply migrate over the accumulation surface that remains constant over time; and (2) erosion in which the accumulation surface falls over time such as with negatively climbing dunes that cannibalise the substrate. A shift from a surface of accumulation to one of erosion or bypass results in a bounding surface (i.e. 'super surface' of Kocurek, 1988; Kocurek and Havholm, 1993; and sequence surface of Kocurek, 1996), such that the basic aeolian architecture consists of the vertical and lateral configurations of accumulations and their bounding surfaces.

As defined by Kocurek and Havholm (1993), *dry aeolian systems* are those in which the water table or the capillary fringe are at a depth below the accumulation surface such that they have no effect on the substrate and no other stabilising factors exist, leaving the wind as the sole control on the behaviour of the accumulation surface over time. *Wet aeolian systems* are those in which the water table or its capillary fringe is at or near the accumulation surface and the water table controls the behaviour of the accumulation surface over time. *Stabilising aeolian systems* are those in which surface stabilising factors such as vegetation control the behaviour of the accumulation surface over time.

In all cases, the principle of continuity applied to sediment transport systems is the *sediment conservation equation* of Middleton and Southard (1984),

$$\frac{\partial h}{\partial t} = -\left(\frac{\partial q}{\partial x} + \frac{\partial c}{\partial t}\right)$$

where h is the height of the accumulation surface, t is time, q is the transport rate in the x direction, and c is the concentration of sediment in transport (Figure 11.4(a)). Solutions to the sediment conservation equation by sign alone define a limited set of possible scenarios for a rising (i.e. accumulation), falling (i.e. erosion), and constant (i.e. bypass) accumulation surface (Figure 11.4(b)). It is assumed here that transport occurs as dunes such that the transport rate (i.e. the spatially averaged bulk volume of sediment per unit width of flow per time, units of $l^3/lt = l^2/t$) is a function of dune shape, height and migration speed (Simons *et al.*, 1965). The concentration of sediment in transport (i.e. the bulk volume of sediment in transport above a unit area of the accumulation surface, units of $l^3/l^2 = l$) is less obvious, but given that transport occurs as dunes, then the average dune height is the concentration of sediment in transport.

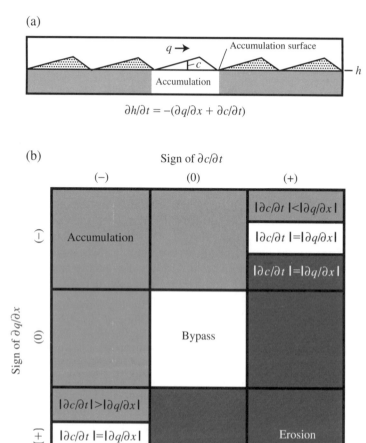

$$\partial h / \partial t = -(\partial q / \partial x + \partial c / \partial t)$$

Figure 11.4 Application of the sediment conservation equation to aeolian systems. (a) Definition diagram showing the accumulation surface at height (h) and covered by dunes, the migration rate of which determine the sediment transport rate (q), whereas the average dune height determines the concentration of sediment in transport (c). (b) Possible solutions to the sediment conservation equation by sign alone, defining the fields of accumulation, bypass, and erosion

Implications

The implications of Hypothesis II are that, given a sand sea, (1) accumulations will not form unless the transport rate decreases downwind and/or the concentration of sediment in transport decreases over time, and (2) the conditions needed to satisfy these aspects of the principle of continuity must be found in the dynamics of wet, dry and stabilising aeolian systems. Of these, only dry and wet systems are addressed below.

For dry systems, therefore, the question is under what conditions, given the aerodynamic control on the accumulation surface, do a spatial decrease in the transport rate and/or a temporal decrease in the concentration of sediment in transport occur in order to form accumulations? It has been long recognised (e.g. Rubin and Hunter, 1982) that a downwind decrease in the transport rate occurs with a downwind decrease in wind power, with accumulation occurring wherever the wind continues to decelerate beyond the point where $q_a = q_p$ (Figure 11.5(a1)). Sediment is essentially transferred from dunes to the accumulation, with the dunes decreasing in size downwind. Such a downwind decrease in wind power and the transport rate occurs with flow into topographic basins in which the flow expands vertically, or along any regional flow paths where the pressure gradient decreases. Alternately, accumulation in dry systems because of a temporal decrease in concentration is speculative because dune height as a function of wind power is yet poorly understood. One possibility, however, is that dune height and wind power positively correlate such that accumulations form as dune height decreases over time because of a decrease in wind power over time (Figure 11.5(a2)). A temporal decrease in wind power, therefore, would mean that the volume of sand that can no longer remain in transport is transferred to the accumulation. Both aspects of continuity could contribute to accumulation, such as where a spatial deceleration of wind power occurs as well as a decrease in wind power over time.

For wet systems, in which the water table controls the accumulation surface over time, the water table must rise for accumulations to form (Kocurek and Havholm, 1993). On the one hand, the migration of dunes from a dry area to one with a rising water table causes accumulation because of a decrease in the transport rate downwind (Figure 11.5(b1)). In this case, concentration does not decrease over time because, at any time, the progression of dunes from the dry area to the wet area remains the same. The transport rate decreases because as the dunes decrease in size downwind as their sediment is transferred to the accumulation, they migrate faster and increase their spacing because no change in wind power occurs, such that downwind progressively less and less sediment is in transport. On the other hand, a uniform rise of the water table under the entire sand sea causes a decrease in concentration over time because the dunes grow smaller over time, whereas the spatial transport rate at any given time stays constant (Figure 11.5(b2)). The case in which both a spatial decrease in the transport rate and a temporal decrease in concentration occurs can be envisioned where the water table rises over time under the entire sand sea, but the rate of rise increases in the transport direction. The cause of the rising water table could be relative because of subsidence, or absolute because of climatic change or a rise in sea level. This last aspect, which implies a link between the continental water table and sea level, presents an intriguing problem that is also manifested in the preservation of aeolian accumulations and will be addressed in that section.

Examples

In addressing accumulation, it is imperative first to recognise what is not an accumulation. As noted earlier, luminescence dating has shown the commonality of large

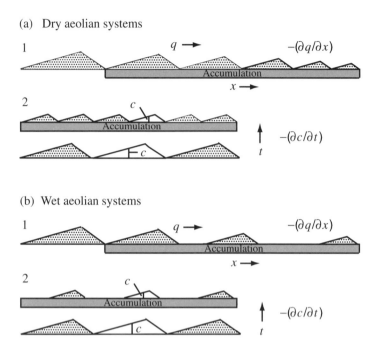

Figure 11.5 Accumulation in dry and wet aeolian systems based upon the sediment conservation equation. (a) Dry system with accumulation because of (1) a downwind decrease in the transport rate because of flow deceleration, and (2) a temporal decrease in concentration because of a decrease in wind power over time. (b) Wet system with accumulation because of (1) a spatial decrease in the transport rate as dunes migrate from a dry to a wet area, and (2) a temporal decrease in concentration because of a uniform rise of the water table under the entire aeolian system

dunes that are amalgamated features that record events dating into the Pleistocene. These dunes are not accumulations, but rather bedforms. Indeed, amalgamated bedforms should characterise accumulation surfaces that change little with time such that the bedforms that occupy these surfaces are not buried, but rather remain on the surface, subject to repeated episodes of stabilisation and reactivation.

For the dry aeolian systems that characterise the Sahara, the potential for accumulation lies along regional flow paths where wind deceleration occurs, and in the topographic basins where most of the sand seas occur (i.e. Figure 11.5(a1)), as first recognised by (Wilson, 1973). It is also possible that periods of accumulation occurred with regional decreases in wind power during climatic transitions (i.e. Figure 11.5(a2)). The former two cases would have been most likely during the sand sea constructional phases of the Saharan cycles prior to a dwindling of the sediment supply (Figure 11.3), while the latter may have characterised the transition from Pleistocene to Holocene times. The situation that appears to characterise much of the Sahara today is that accumulations are being cannibalised and dunes are being reworked. As discussed in the next section, this situation occurs as a function of preservation.

For the ancient record of the Colorado Plateau, the paucity of interdune-flat deposits argues that most of the sand seas were dry systems (Kocurek and Havholm, 1993; Havholm and Kocurek, 1994). The cause for their accumulation is speculative, but in addition to any regional or local decreases in wind speed, it is possible that the foreland and cratonic basins that house the accumulations may have also been topographic basins that allowed for flow deceleration. Cases of abnormally thick sets of cross-strata, such as that for the Jurassic Navajo Sandstone in the Zion area of Utah, may represent pre-existing accumulation space, such as an unfilled, subsided basin, which was filled relatively rapidly once dunes entered the basin. The Jurassic Entrada Sandstone is yet the one recognised example of an interpreted aeolian wet system. For much of the Entrada, the water table was demonstrably at or near the accumulation surface and rose in a punctuated fashion during Entrada times, fostering the accumulation of dunes and wet interdune-flat deposits (i.e. Figure 11.5(b1); Crabaugh and Kocurek, 1993). In addition, sections shown by Crabaugh and Kocurek (1998) along southeastern Utah show that Entrada dunes prograded onto progressively wetter surfaces (i.e. Figure 11.5(b2)).

PRESERVATION OF SAND SEA ACCUMULATIONS

Hypothesis

Hypothesis III states that the preservation of sand sea accumulations occurs with: (1) subsidence *and* burial; and/or (2) a rise of the water table through the accumulation (Figure 11.6). It is the necessity of burial of the accumulation beneath yet additional sediments that departs from conventional thinking about accommodation space as used for non-aeolian accumulations. With other types of continental deposits, the nature of the accumulations themselves (i.e. coarse grain size, cohesive muds) and conditions of the environment (i.e. vegetation, high water table) typically do not allow deflation of the accumulation by the wind to be a major consideration. Without burial by subsequent accumulations, the continuation of aeolian conditions beyond the point when the sand supply has been exhausted necessarily means that the incoming wind is under-saturated and can erode the accumulation until the flow reaches its transport capacity. Even winds entering a topographic basin and decelerating have the potential to deflate the substrate so long as the flow is unsaturated. In this case, sediment outflux from the basin will exceed the influx, and at the expense of the accumulation. A rise of the water table through the accumulation fosters preservation because no deflation occurs from below the water table, and deflation is greatly reduced once grains are within the capillary fringe. The rise of the water table can be *absolute* as with an eustatic sea level rise along coastal aeolian accumulations or a rise in the continental water table because of the onset of a more humid climate, or *relative* as with the subsidence of the accumulation through a water table that remains at the same elevation.

Beyond subsidence with burial and a rise of the water table, it is difficult to envision long-term preservation of aeolian accumulations beyond unusual circumstances. Surface cementation and vegetation may stabilise accumulations, but the

Rise in relative sea level

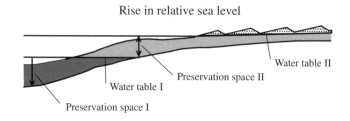

Rise in relative continental water table

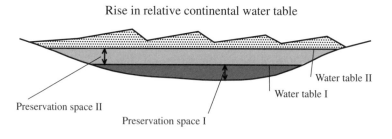

Subsidence with burial

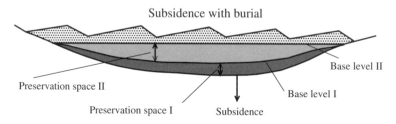

Figure 11.6 The basic modes of preservation of aeolian accumulations in which the space for the preserved accumulations increases with a coastal or continental rise in the water table, and by subsidence. Preservation occurs so long as the raised water table height is maintained, and the subsided accumulations are buried

former is typically associated with the water table and the latter effective only so long as the climate supports the vegetation. Examples of flooding of sand seas by marine or fresh waters (e.g. Permian Weissliegend; Glennie and Buller, 1983) or by basalts (e.g. Clemmensen, 1988) represent an instantaneous rise of the water table and an instantaneous burial, respectively. Of course, all 'preserved' aeolian accumulations, as with those of other environments, are subject to renewed erosion with a change in conditions such as tectonic uplift.

Implications

The implications of hypothesis III are that geological preservation of aeolian accumulations (i.e. creation of the rock record) is restricted to: (1) tectonic basins that are subject to a long history of sediment influx and subsidence; (2) relatively 'coastal' regions that are marine transgressed or in which the adjacent continental water table rises because of the transgression or subsidence; and (3) continental interior areas subject to a rise of the water table because of tectonism or climate. The conditions that define *preservation space* for aeolian accumulations do not necessarily coincide with those of *accumulation space* (terminology of Kocurek and Havholm, 1993), as is the case for some other types of accumulations. For example, sea level defines both the height to which marine carbonate accumulations may build as well as base level for their preservation. Aeolian strata may accumulate on a stable craton because of deceleration of the regional winds, but unless the accumulations are subsequently marine transgressed or a 'permanent' rise in the water table occurs, they have little chance of preservation.

Examples

If the preservation of aeolian accumulations is relatively selective, then any widespread, aeolian rock record raises the question as to why were these accumulations preserved? The world's most extensive preserved accumulations of aeolian strata are the Late Paleozoic–Jurassic strata on the Colorado Plateau of the Western Interior of the US. These strata, spanning about 150 my of time and occurring over about 1 million km^2, have a cumulative thickness of over 3500 m. A comparison of this region to that of the Quaternary Sahara shows a contrast in preservation potential that is remarkable.

Focusing on the greater part of the Jurassic section of the Western Interior (Figure 11.7), the record consists of an eastward back-stepping of vertically stacked aeolian units. All the aeolian units are replaced westward by strata representing marginal marine-sabkha environments. Major regional unconformities (e.g. J1, J2) occur when large volumes of strata were removed (see review in Blakey *et al.*, 1988), and where the units have been studied in detail (e.g. Page Sandstone: Havholm *et al.*, 1993; Entrada Sandstone: Crabaugh and Kocurek, 1993, 1998), unconformities also occur within the units. In terms of preservation, however, at least two major factors were at work.

First, from the Jurassic through the Cretaceous, the Western Interior of the USA was an active, subsiding foreland basin with a large influx of sediment (Riggs and Blakey, 1993). In spite of significant times of accumulation stripping, the overall balance of the basin was toward burial of accumulations by the influx of additional sediment.

Second, a combination of subsidence and eustatic sea level lead to a punctuated, but progressive rise of sea level over the region (Figure 11.7). In effect then, aeolian units, which accumulated during low stands of sea level or co-existed with adjacent marine-sabkha units during periods of rising sea level, were ultimately transgressed to a large extent. Curtis and equivalent marginal marine strata rest in-place over most of the Entrada Sandstone, and the Page Sandstone entirely underlies the Carmel Formation. Sabkha strata occur lateral to aeolian strata of the Temple Cap Sandstone, and Temple Cap remnants after J2 erosion were transgressed by the Carmel mudstones. Sabkha portions of the Temple Cap Formation transgressed over the J1-eroded upper surface of the Navajo Sandstone in the south, and these strata are coeval with the more widespread marine Twin Creek and Gypsum Springs Formations to the north, which transgressed over large portions of the Navajo

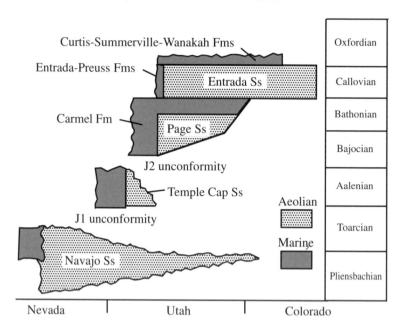

Figure 11.7 Idealised Wheeler-type diagram for a portion of the Jurassic of the Western Interior showing preserved aeolian accumulations and equivalent and overlying marine-sabkha strata. The line of section is approximately across southern Utah. Data from Blakey *et al.*, 1988)

Sandstone and correlative aeolian units. At the smaller, intraformational scale, the Page Sandstone shows repeated episodes of aeolian progradation at low stands, followed by sabkha-marine transgressions of the Carmel 'seaway' over the aeolian accumulations (Havholm *et al.*, 1993; Havholm and Kocurek, 1994). What is being argued here, therefore, is that a significant degree of preservation of the Jurassic section occurred because the aeolian units were placed below the regional water table by a combination of subsidence and an eustatic rise in sea level.

What is more poorly understood is the response, if any, of the inland water table to the marine transgression. For example, most Carmel transgressive tongues over Page aeolian accumulations can be traced to polygonal-fractured surfaces that bound the aeolian strata and extend far inland beyond the transgression itself (Havholm *et al.*, 1993). As noted earlier, the Entrada, thought to have accumulated as a wet system during a rise in relative sea level, shows polygonal surfaces capping progradational aeolian units (Crabaugh and Kocurek, 1998). Yet more widespread are polygonal fractures that occur over much of the J2 surface that bounds the Navajo Sandstone. In all these cases, the polygonal-fractured surfaces are interpreted as formed by deflation to the water table, with the polygons representing the fracturing of an evaporite-encrusted surface (Kocurek and Hunter, 1986).

In the simplest interpretation (Model I in Figure 11.8), the relative rise of the water table inland is because of subsidence, with the strata subsided below the water table preserved, and the strata above deflated. Alternatively, the inland water table

Model I. Water table independent of sea level

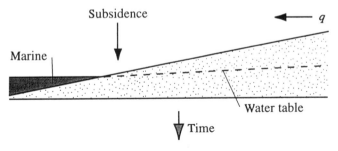

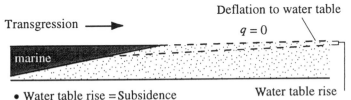

- Water table rise = Subsidence
- Relationship between q and transgression ?

Model II. Water table dependent on sea level

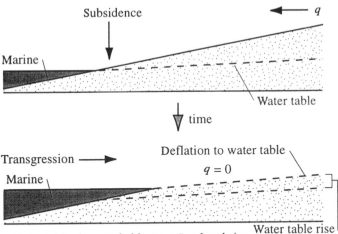

- Water table rise = Subsidence + Sea-level rise
- Relationship between q and transgression ?

Figure 11.8 Alternate modes of formation for polygonal-fractured surfaces formed by deflation to the water table. In Model I, after a period of aeolian accumulation, deflation proceeds to the water table, the relative height of which is determined by subsidence alone. In Model II, the height of the water table is a function of both subsidence and the inland effects of a rise in sea level along the adjacent coast

could rise in response to the rise in sea level at the coast (Model II in Figure 11.8). In this case, the strata preserved below the water table inland would reflect both subsidence and the absolute rise of the water table itself. Whereas it is accepted that such a rise in the water table occurs immediately adjacent to the shore and inland perhaps some kilometres, it is undetermined how the water table, some tens to hundreds of kilometres inland, might respond to a rise along the coast for this case where the water table underlies a low-gradient, desert land surface that receives little hydrologic recharge. Because the polygonal fractures require a free surface to form and because they formed during deflationary periods, there appears to yet an additional, poorly understood relationship – the transgression and apparent inland rise of the water table coincide with a diminished sediment influx into the basin (Figure 11.8).

Whereas the Jurassic of the Western Interior shows at least two major factors contributing to preservation of the accumulations, the bulk of the Sahara shows none. Tectonically, the Sahara Desert occupies the African craton and a variety of relict and active basins (see review in Sahagian, 1993; Hartley and Allen, 1994; fig. 3 in Kocurek, 1998). Aeolian accumulations, which likely formed during Pleistocene glacial constructive times and during glacial–interglacial transitions, are present today largely as erosional remnants. A good example is the Quaternary record of the Chott Rharsa area, as mapped by Blum *et al.* (1998) and Swezey *et al.* (1999). In spite of occupying an active foreland basin at an elevation that is below sea level over much of the area, the Quaternary record from at least five separate aeolian constructional episodes occurs as isolated, erosional outcrops, many of which are sculptured into yardangs through deflation. While periodic times of preservation occurred with a rise in the chott water table and deposition of subaqueous strata over portions of the aeolian strata, the net and current sediment state for the basin is one of net erosion. This erosion is occurring because: (1) the water table has fallen to below the accumulations; (2) the current winds that enter the basin are unsaturated; and (3) the subsiding basin is unfilled or has a low sediment influx.

While the Atlas foreland basin is more sediment-filled westward into Algeria, and some basins house a significant Quaternary record, such as the Chad Basin where a high water table is maintained by drainage from tropical uplands to the south, the bulk of the Sahara appears to show a scant Quaternary record and one largely present as erosional remnants that exist by virtue of the lag time involved in their removal. As argued in Kocurek (1998, fig. 5), the most significant preservation of Quaternary aeolian accumulations for the Sahara lie along the western margin of the African continent. These consist of those accumulations on the shelf that were transgressed (but not entirely reworked) by the 120 m rise in sea level since the last glacial maximum, and those incorporated into the water table and presently underlying the coastal sands and sabkhas. Although different in scale, basin type, and nature of the marine realm, if a comparison is to be made between the Jurassic of the Western Interior and the Quaternary of the Sahara, it is that the preserved Jurassic record is analogous to the Atlantic shelf and coast of Africa where Saharan aeolian sands are now submerged and buried, whereas the continental Sahara eastward is analogous to the North American craton eastward of the Jurassic foreland basin where no Jurassic aeolian strata are preserved.

CONCLUSIONS

The sub-title of this paper, a parody of the 1897 *New York Sun* editorial on Santa Claus, was adopted because it emphasises the fact that the creation of an aeolian rock record is a very selective process. The explanation offered here for this selective process is that rock-record creation occurs in three separate phases, each of which has its own set of governing rules. These rules, in turn, can be summarised as three separate hypotheses. The variables that impact each phase of rock-record creation are largely independent of each other, but are also strongly influenced by antecedent conditions and dynamic interactions with other variables. These latter aspects may be viewed as boundary conditions imposed upon the evolution of a natural complex system. The route from sand sea construction to accumulation of a body of strata to its preservation essentially requires the favourable coincidence of independent variables. On the one hand, the hypotheses summarised here, because they are based upon fundamental principles or basic geological processes, do provide an understanding of in terms of simple governing rules for aeolian systems from geomorphic expression to rock record. On the other hand, the wide variety of factors that affect the evolution of these complex systems means that, when viewed in detail, each system has a significant degree of uniqueness.

REFERENCES

Anderson R. and Haff P., 1991. Wind modification and bed response during saltation of sand in air. *Acta Mechanica Supplement*, **1**, 21–52.

Blakey R., Peterson F. and Kocurek G., 1988. Synthesis of late Paleozoic and Mesozoic eolian deposits of the Western Interior of the United States. *Sedimentary Geology*, **56**, 3–125.

Blum M., Kocurek G., Swezey C. *et al.*, 1998. Quaternary wadi, lacustrine, aeolian depositional cycles and sequences, Chott Rharsa basin, southern Tunisia. In *Quaternary Deserts and Climatic Change*, A. Alsharhan, K. Glennie, G. Whittle and C. Kendall C. (Eds). Balkema, Rotterdam, pp. 539–552.

Clemmensen L., 1988. Aeolian morphology preserved by lava cover, the Precambrian Mussartut Member, Eriksfjord Formation, south Greenland. *Bulletin of the Geological Society Denmark*, **37**, 105–116.

CLIMAP, 1976. The surface of the ice-age Earth. *Science*, **191**, 1131–1137.

Corbett I., 1993. The modern and ancient pattern of sandflow through the southern Namib deflation basin. In *Aeolian Sediments Ancient and Modern*, K. Pye and N. Lancaster (Eds). Special Publication International Association Sedimentologists 16, pp. 45–60.

Crabaugh M. and Kocurek G., 1993. Entrada Sandstone – example of a wet aeolian system. In *The Dynamics and Environmental Context of Aeolian Sedimentary Systems*, K. Pye (Ed.). Special Publication Geological Society London 72, pp. 103–126.

Crabaugh M. and Kocurek G., 1998. Continental sequence stratigraphy of a wet eolian system – a key to relative sea level change. In *Relative Role of Eustasy, Climate, and Tectonism in Continental Rocks*, K. Stanley and P. McCabe (Eds). SEPM Special Publication, **59**, 213–228.

Gasse F. and Fontes J., 1992. Climatic changes in northwest Africa during the last deglaciation. In *The Last Deglaciation: Absolute and Radiocarbon Chronologies*, E. Bard and W. Broecker (Eds). Springer-Verlag, Berlin, pp. 295–325.

Gasse F., Tehet R., Duran A. *et al.*, 1990. The arid-humid transition in the Sahara and the Sahel during the last deglaciation. *Nature*, **346**, 141–146.

Gasse F. and Van Campo E., 1994. Abrupt post-glacial climate events in West Africa and North Africa monsoon domains. *Earth and Planetary Science Letters*, **126**, 435–456.

Glennie K., 1998. The desert of southeast Arabia: a product of Quaternary climatic change. In *Quaternary Deserts and Climatic Change*, A. Alsharhan, K. Glennie, G. Whittle and C. Kendall (Eds). Balkema, Rotterdam, pp. 279–291.

Glennie K. and Buller A., 1983. The Permian Weissliegendes of NW Europe: the partial deformation of eolian dune sands caused by the Zechstein transgression. *Sedimentary Geology*, **35**, 43–81.

Hartley R. and Allen P., 1994. Interior cratonic basins of Africa: relationship to continental breakup and role of mantle convection. *Basin Research*, **6**, 95–113.

Havholm K., Blakey R., Capps M. *et al.*, 1993. Eolian genetic stratigraphy: an example from the Middle Jurassic Page Sandstone, Colorado Plateau. In *Aeolian Sediments Ancient and Modern*, K. Pye and N. Lancaster (Eds). Special Publication International Association Sedimentologists 16, pp. 87–111.

Havholm K. and Kocurek G., 1994. Factors controlling eolian sequence stratigraphy: clues from super bounding surface features in the Middle Jurassic Page Sandstone. *Sedimentology*, **41**, 913–934.

Hooghiemstra H., Stalling H. Agwu C. and DuPont L., 1992. Vegetational and climatic changes at the northern fringe of the Sahara 250 000-5000 years BP: evidence from four marine pollen records located between Portugal and the Canary Islands. *Review of Palaeobotany and Palynology*, **74**, 1–53.

Kocurek G., 1988. First-order and super bounding surfaces in eolian sequences – bounding surfaces revisited. *Sedimentary Geology*, **56**, 193–206.

Kocurek G., 1996. Desert aeolian systems. In *Sedimentary Environments: Processes, Facies and Stratigraphy*, H. Reading (Ed.). Blackwell Science, Oxford, pp. 125–153.

Kocurek G., 1998. Aeolian system response to external forcing factors – a sequence stratigraphic view of the Saharan region. In *Quaternary Deserts and Climatic Change*, A. Alsharhan, K. Glennie, G. Whittle and C. Kendall (Eds). Balkema, Rotterdam, pp. 327–337.

Kocurek G. and Havholm K., 1993. Eolian sequence stratigraphy – a conceptual framework. In *Siliciclastic Sequence Stratigraphy*, P. Weimer and H. Posamentier (Eds). American Association Petroleum Geologists Memoir 58, pp. 393–409.

Kocurek G. and Hunter R., 1986. Origin of polygonal fractures in sand, uppermost Navajo and Page Sandstones, Page, Arizona. *Journal Sedimentary Petrology*, **56**, 895–904.

Kocurek G. and Lancaster N., 1999. Aeolian system sediment state: theory and Mojave Desert Kelso dune field example. *Sedimentology*, **46**, 505–515.

Kogbe C. and Burollet P., 1990. A review of continental sediments in Africa. *Journal of African Earth Sciences*, **10**, 1–25.

Kutzbach J. and Street-Perrott A., 1985. Milankovitch forcing of fluctuation in the level of tropical lakes from 18 to 0 kyr BP. *Nature*, **317**, 130–134.

Lang J., Kogbe C., Alidou S. *et al.*, 1990. The continental terminal in West Africa. *Journal African Earth Sciences*, **10**, 79–99.

Langbein W. and Schumm S., 1958. Yield of sediment in relation to mean annual precipitation. *Transactions of the American Geophysical Union*, **39**, 1076–1084.

Mainguet M. and Chemin M., 1983. Sand seas of the Sahara and Sahel: an explanation of their thickness and sand dune type by the sand budget principle. In *Aeolian Sediments and Processes*, M. Brookfield and T. Ahlbrandt (Eds). Elsevier, Amsterdam, pp. 353–363.

McDonald E., McFadden L. and Wells S., 1995. The relative influences of climate change, desert dust, and lithologic control on soil-geomorphic processes on alluvial fans, Mojave Desert, California: summary of results. In *Ancient Surfaces of the East Mojave Desert*, R. Reynolds and J. Reynolds (Eds). San Bernardino County Museum Association Quarterly 42, pp. 35–42.

McKee, E., 1966. Structure of dunes at White Sands National Monument, New Mexico (and a comparison with structures of dunes from other selected areas. *Sedimentology*, **7**, 1–69.

Middleton G. and Southard J., 1984. Mechanics of sediment movement. SEPM Short Course 3.

Petit-Maire N., 1989. Interglacial environments in presently hyperarid Sahara: paleoclimatic implications. In *Paleoclimatology and Paleometeorology: Modern and Past Patterns of Global Atmospheric Transport*, M. Leinen and M. Sarnthein (Eds). Kluwer, Dordrecht, pp. 637–661.

Riggs N. and Blakey R., 1993. Early and Middle Jurassic paleogeography and volcanology of Arizona and adjacent areas. In *Mesozoic Paleography of the Western United States II*, G. Dunn and K. McDougall (Eds). Pacific Section SEPM 71, pp. 347–375.

Rubin D. and Hunter R., 1982. Bedform climbing in theory and nature. *Sedimentology*, **29**, 121–138.

Ruddiman W., Sarnthein M., Backman J. *et al.*, 1989. Late Miocene to Pleistocene evolution of climate in Africa and the low-latitude Atlantic: overview of leg 108 results. In *Proceedings of the Ocean Drilling Program Scientific Results*, **208**, 463–484.

Sahagian D., 1993. Structural evolution of African basins: stratigraphic synthesis. *Basin Research*, **5**, 41–54.

Sarnthein M., 1978. Sand deserts during glacial maximum and climatic optimum. *Nature*, **272**, 43–46.

Sharp R., 1966. Kelso Dunes, Mojave Desert, California. *Bulletin Geological Society of America*, **77**, 1045–1074.

Simons D., Richardson E. and Nordin C., 1965. Bedload equation for ripples and dunes. *US Geological Survey Professional Paper*, 462-H, H1-H9.

Street-Perrott A. and Perrott A., 1990. Abrupt climate fluctuations in the tropics: the influence of Atlantic Ocean circulation. *Nature*, **343**, 607–612.

Swezey C., Lancaster N., Kocurek G. *et al.*, 1999. Response of aeolian systems to Holocene climatic and hydrologic changes on the northern margin of the Sahara: a high resolution record from the Chott Rharsa basin, Tunisia. *The Holocene*, **9**, 141–147.

Tiedemann R., Sarnthein M. and Stein R., 1989. Climatic changes in the western Sahara: aeolo-marine sediment record of the last 8 million years (sites 657–661). In *Proceedings of the Ocean Drilling Program, Scientific Results*, **108**, 241–277.

Werner B. and Kocurek G., 1997. Bed-form dynamics: does the tail wag the dog? *Geology*, **25**, 771–774.

Wilson I., 1973. Ergs. *Sedimentary Geology*, **10**, 77–106.

Yan Z. and Petit-Maire N., 1994. The last 140 ka in the Afro-Asian arid/semi-arid transition zone. *Palaeogeography Palaeoclimatology Palaeoecology*, **110**, 217–233.

12 Dune Palaeoenvironments

VATCHE P. TCHAKERIAN
Departments of Geography and Geology and Geophysics, Texas A&M University, USA

Various biogeomorphic and sedimentologic criteria have been used to analyse the nature and frequency of dune construction and destruction in the world's drylands during late Quaternary time. Some of the most widely used criteria are reviewed, including the use of vegetation as an indicator for stability, dune-vegetation succession cycles, dune mobility indices, dune size, statistical methods using granulometric data, sedimentary structures in dunes, grain roundness and redness, the presence of palaeo-sols and carbonate horizons in dunes, the properties and characteristics of silt and clay-sized particles in dune sediments, desert aeolian dust, and the relative frequencies of quartz grain surface microfeatures using the scanning electron microscope. Many of the above criteria are then used to compare and contrast the major dune-building episodes in some of the world's deserts, including the Sahara–Sahel region in northern Africa, the Mega-Kalahari sand sea in southern Africa, the Australian arid zone, and the drylands of North America. Using the Mojave Desert in North America as a case study, a more detailed analysis of late Quaternary aeolian activity is presented, and a tentative model showing the complex pathways between aeolian, lacustrine, fluvial and hillslope sediment production, storage, transport and deposition, outlined.

INTRODUCTION

During the late Quaternary, numerous dune-building episodes have been documented from the world's arid zones (Thomas, 1997). Some sand seas (ergs) experienced dramatic expansions and aeolian accumulation, such as the Sahel in northern Africa, the Mega-Kalahari in southern Africa, and the Great Plains in the USA (Figure 12.1). Various biogeomorphic and sedimentologic criteria have been used to distinguish ancient aeolian sands from their modern counterparts. Establishing a reliable chronology for dune emplacement has been rather problematic, since the preservation of organic materials suitable for conventional radiocarbon dating are scarce in dune sediments. Current advances in luminescence dating methods offer a new challenge for the proper establishment of global and regional chronologies of dune construction. This chapter presents a review of the various biogeomorphical and sedimentological criteria utilised for reconstructing dune palaeoenvironments and a summary of the present understanding of major dune-building episodes during late Quaternary time, with a particular focus on Africa and North America. The chapter concludes with a brief discussion on future trends for aeolian paleoenvironmental research.

Aeolian Environments, Sediments and Landforms. Edited by A.S. Goudie, I. Livingstone and S. Stokes.
© 1999 John Wiley & Sons, Ltd.

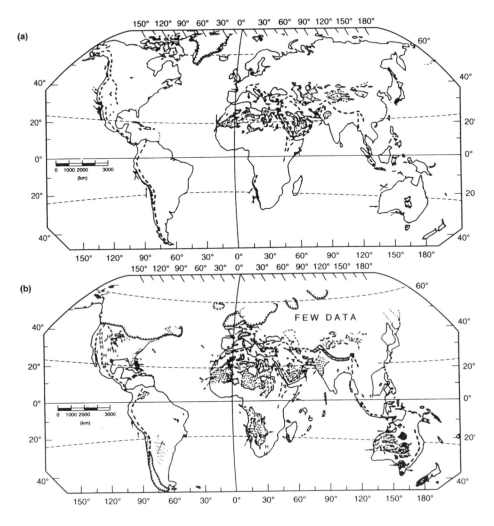

Figure 12.1 Global distribution of sand seas for (a) the present and (b) the Last Glacial Maximum, 18000 years BP (after Sarnthein, 1978). H denotes humid conditions

RECONSTRUCTING DUNE PALAEOENVIRONMENTS

A number of biogeomorphic and sedimentological criteria have been used to interpret dune palaeoenvironments, including dune morphology, dune mobility, vegetation, mineralogy, granulometric analysis, dune reddening and scanning electron microscopy (SEM) of quartz grain microfeatures. Although beyond the scope of this review, the pre-Quaternary aeolian record has been recently reviewed by Kocurek (1996). The following section examines a number of criteria used for palaeoenvironmental interpretations.

Biogeomorphical Evidence

Some of the most widely used biogeomorphical measures to interpret aeolian palaeoenvironments include:

Dune Morphology

The presence of degraded or completely vegetated sand dunes in areas that are presently not subject to aeolian depositional activity and which at present receive mean annual rainfall values > 250 mm usually indicate that the area experienced enhanced aeolian activity in the past (Lancaster, 1995). This latitudinal zonation of rainfall has been particularly useful for mapping the advance of sand dunes south from the Sahara into the Sahel (Sarnthein, 1978). Presently, most of the sand dunes in the Sahel are completely vegetated and stabilised (Grove and Warren, 1968). Stabilised dunes can also be used to reconstruct past atmospheric circulation patterns and former arid zone extensions (Thomas, 1997). The orientation of the relict dunes is compared with modern active sand moving winds and/or sand dunes, either directly from remotely sensed data and air photography, or from wind data, using the Drift Potenial (DP) and the Resultant Drift Potential (RDP) technique of Fryberger (1979), or indirectly from mapping the orientation of ventifacts near major dune fields (Laity, 1992).

Dune Mobility

A number of climatic and geomorphic parameters have been used to establish whether dunes are active or not. The most widely used climatic dune mobility index is that of Lancaster (1988), based on ideas first proposed by Ash and Wasson (1983) and Wasson (1984). The dune mobility index (M) of Lancaster (1988) states that M is proportional to the per cent time that the wind is blowing above the threshold velocity (W) for sand transport (5 m s^{-1}), and inversely proportional to precipitation over potential-evapotranspiration (P/PE), hence:

$$M = W/(P/PE)$$

Lancaster (1988) proposed four classes of dune activity: (a) inactive dunes (M < 50); (b) dune crests only active (50 < M < 100); (c) dune crests active, lower windward and slipfaces (dune plinths) and interdune depressions vegetated (100 < M < 200); and (d) fully active dunes (M > 200). The dune mobility index has been applied by Muhs and Maat (1993) in the Great Plains of the USA, and Bullard et al. (1997) in the Mega-Kalahari of southern Africa. A recent evaluation by Tsoar and Illenberger (1998), suggest the use of DP and RDP/DP rather than W, for a more accurate calculation of the wind regime. Livingstone and Warren (1996) also point out some of the inherent problems using climatic indices considering the paucity of weather stations in desert regions and the spatial and temporal interannual to decadal variability of precipitation.

Figure 12.2   The Pismo Dunes, central California. Active transverse dunes moving over stabilized, dormant parabolic dunes

Another approach to dune mobility is based on geomorphic criteria. Dunes can be classified as active, dormant or episodically active, or relic(t) (e.g. Livingstone and Thomas, 1993; Lancaster, 1995; Livingstone and Warren, 1996).

Active   All parts of the dune are subject to aeolian activity, with sand ripples on windward slopes. Primary sedimentary structures are prevalent and vegetation cover mostly absent. All major dune types (e.g. barchan, transverse, parabolic, reversing, linear, star) can be formed depending on wind regime, sediment availability and the nature of the terrain (Figure 12.2).

Dormant or Episodically Active   Most aeolian activity limited to dune crestal areas. Dune plinths and interdune areas covered with vegetation, ranging in type from grasses to shrubs. Primary sedimentary structures altered by bioturbation. Biogenic crusts or cyptogamic soils (mosses, lichens, algae, cyanobacteria), partially cover dune plinths and interdune depressions. The crusts can reduce infiltration rates and enhance surface runoff. The diminished aeolian activity can be the result of reduction in sediment supply, wind regime, vegetation cover (natural and/or anthropogenic), and to a lesser degree, to climatic perturbations (e.g. increase in precipitation leading to increased moisture content and a decrease in net aeolian transport). A change in environmental conditions can lead to the re-activation of the dormant dunes, particularly in coastal dune complexes, where such activities as excessive vehicular traffic (e.g. ATVs) and increased human trampling are more prevalent.

Relic(t) These features exhibit no current evidence of aeolian activity and represent deposits formed during past climatic regimes (Figure 12.3). Their sufaces are completely stabilised by such features as woodland vegetation, palaeosols, calcic horizons, and alluvial and colluvial mantles (Lancaster and Tchakerian, 1996).

Dune Size

The size of sand dunes itself might hold certain palaeoenvironmental significance (Livingstone and Warren, 1996; Warren and Allison, 1998). Wilson (1972) proposed three discrete aeolian bedform hierarchies (ripple, dune, draa) based on grain size (which has been subsequently shown not to be discrete – see Livingstone and Warren, 1996). Grain courseness was ultimately responsible for the largest bedform (draa), as well as their spacing. Course sands were believed to move only during exceptional wind events and thus the courser the grain, the further apart and bigger the bedform. Draas or mega-dunes formed only during very high winds, such as the intensified winds in the tropical deserts during the Last Glacial Maximum (LGM), at about 20 ka. However, dune size could be a function of wind regimes that persist for longer periods (Warren and Allison, 1998). Dunes that constantly adjust to hourly or diurnal changes in wind regime are smaller in size (1–2 m). Dunes that survive this cycle might continue to buid up and reach a dynamic equilibrium with respect to an annual wind regime (perhaps even a decadal wind regime), attaining heights of 3–10 m. Dunes that survive the annual/decadal wind perturbations and the resultant morphologic changes, will then be able to continue to grow, until a threshold condition whereby variables such as speed-up rations, sand supply, climatic and biotic changes, would control the final size. Warren and Allison (1998) suggest that long periods of uninterrupted sediment movement and dune growth, necessary for the formation of mega dunes, can only be accomplished during Milankovitch type cycles and sub-cycles (Figure 12.4).

Dune Vegetation

The presence of dune surface vegetation as an indicator for diminished aeolian activity and hence dune stabilisation. Studies have shown that sand transport can take place even in areas where the vegetation cover is between 30 and 35 per cent (Ash and Wasson, 1983; Thomas and Shaw, 1991). In the case of sparsely vegetated linear dunes, plant cover may be an integral part of the aeolian system, with a decrease in dune extension and movement as plant cover increases, a type of positive feedback (Wiggs *et al.*, 1995). In linear dunes, different levels of aeolian activity are found and the distinction between active and inactive is rather porous (Livingstone and Thomas, 1993). Inactive areas might favour vegetation establishment while active areas remain free from plant cover. However, the presence of woodland vegetation on former sand dune surfaces is a reliable indicator of the former extent of ancient dunefields.

A typical dune vegetation sequence might be as follows (keeping in mind the complex nature of vegetation and dune form):

Figure 12.3 (a) The northern section of the relict Dale Lake sand ramp formed against the west slopes of the Sheephole Mountains, Mojave Desert, California. (b) Close up view of the middle section of the sand ramp showing aeolian units separated by palaeosols. Note arrow pointing to a geologic hammer

Figure 12.4 The northern mega-dunes of the Wahiba Sands, Sultanate of Oman. These large dunes are believed to require long periods of uninterrupted sediment movement and dune growth, most likely during Milankovitch-type cycles, as recently proposed by Warren and Allison (1998)

Hyper-arid conditions. Biogenic crusts, cryptogams and vegetation cover at a minimum, because of active aeolian entrainment and sand movement. Very low water table precludes vegetation establishment, even in interdune depressions.

Arid conditions. Biogenic activity mostly curtailed to interdune depressions and dune plinths. In interdune depressions and dune plinths, high sand porosities can facilitate moisture infiltration and subsequent vegetation establishment. Aeolian activity confined to upper windward slopes, crestal regions and upper slipface slopes. Formation of evaporitic crusts in interdune areas.

Arid to semi-arid conditions. An increase in vegetation cover and the development of biogenic crusts. Also, as precipitation increases, there will be a concomitant increase in the water table leading to the depositions of pedogenic calcrete. Well established vegetation on dunes and interdune depressions.

Semi-arid to sub-humid conditions. Most dunes stabilised by vegetation. High water table leads to the deposition of marsh and lacustrine sediments in interdune depressions. Well-developed pedogenic calcrete and other weathering deposits, including the development of soils on dune and interdune surfaces. Increased precipitation could also lead to increased runoff on dune surfaces leading to dune degradation, infilling of interdune depressions, and the establishment of a dense vegetation cover or woodland.

Dune Chronologies

Dune-building periods can be constrained either by relative or absolute dating. Relative dating involves mostly stratigraphic position, particularly the use of overlying or underlying lacustrine sediments as marker horizons. Other relative dating tools include weathering and post-depositional characteristics of sand grains and palaeosol development (e.g. Tchakerian, 1991). Owing to the paucity of organic material for radiocarbon techniques and the yet unresolved issue of the applicability of rock varnish dating methods, the direct dating of terrestrial aeolian sediments has been doubtful and problematic (Thomas, 1997). Recently, luminescence dating seems to provide an ideal tool for establishing numerical chronologies in aeolian sedimentary environments (see Chapter 14, this volume, for a more detailed review of dune chronologies).

Sedimentological Evidence

Several sedimentological measures have been used in aeolian research to distinguish modern, active aeolian sands from their older counterparts. Additionally, primary textural, mineralogical, and sedimentary features of active sands are commonly subject to significant post-depositional modifications. In the following section, a review of some of the significant sedimentological parameters and principal early post-depositional changes affecting aeolian deposits is presented.

Granulometric Analysis

Keeping in mind issues of provenance, transport and inheritance, changes in the values of mean grain size, sorting (standard deviation), skewness and kurtosis have been used to compare paleosands from modern sands (e.g. Ahlbrandt, 1979). Mean grain size tends to slightly decrease with increasing age because of the weathering of feldspars (Orme and Tchakerian, 1986). Most paleosands tend to be moderately sorted, because of the addition of aeolian dust and mineral disaggregation, and slightly positively skewed, because of the addition of the fine component. Bivariate plots of mean grain size and sorting, skewness, kurtosis and percent silt and clay (% < 65 μm), have been used to separate modern aeolian sands from paleosands with varying degrees of success (Ahlbrandt, 1979; Orme and Tchakerian, 1986; Thomas, 1987). In general, bivariate scattergraphs are ineffective in distinguishing between various aeolian paleosands owing largely to differences in sediment source and reworking of sediments (Thomas, 1987; Tchakerian, 1991; Livingstone and Warren, 1996). Nevertheless, Lancaster (1993) was able to distinguish the three main aeolian sands of the Kelso Dunes using bivariate plots of mean grain size and sorting. From the variables, only per cent silt and clay can be used with any confidence (along with other geomorphic methods), since it has been shown to distinguish certain paleosands from modern sands, particularly in coastal dune complexes (Orme and Tchakerian, 1986). On the other hand, multivariate statistical methods, such as discriminant or principal components analysis, can be more effective in separating the different aeolian palaeoenvironments (Tchakerian, 1991).

Figure 12.5   Sand dunes advancing over older aeolianites, Wahiba Sands, Sultanate of Oman

Sedimentary Structures

Most primary sedimentary structures (e.g. foreset beds, cross-bedding, interdune bedding) are destroyed or significantly altered by such post-depositional processes as bioturbation and the growth and decay of vegetation. A detailed study of active and past aeolian sandstone sedimentary structures, including analysis of accumulation patterns (cross-strata, wind laminae), surfaces (bounding, super) and modes of preservation (subsidence, rise of water table), can lead to a better understanding of sand sea processes and dynamics (see Chapters 3 and 11, this volume). Analysis of the internal structure and composition of aeolianites, calcium carbonate cemented dunes found mostly in tropical coastal environments (Figure 12.5), could also provide palaeoenvironmental information about sea level fluctuations, climate, weathering, and wind regime (Gardner, 1983).

Grain Roundness

Field and microscopic analysis of sand dune sediments suggest that a high proportion of active aeolian sand grains tend to be sub-rounded to sub-angular, and not rounded (Goudie and Watson, 1981). Grain roundness appears to be also influenced by dune type, with constantly migrating barchan or transverse dune sands exhibiting more rounded clasts than those found on linear or star dunes (Thomas, 1987). Among some of the stabilised paleosands in the eastern Mojave Desert, higher proportions of grain roundness is most likely the result of silica precipitation which tends to give the grains a more 'rounded' appearance (Tchakerian, 1991).

Palaeosols and Carbonate Horizons

In areas that have experienced episodic aeolian depositional phases, paleosols and/ or carbonate horizons are present. For palaeosols, only the Bt and the Bk horizons remain, some exhibiting calcified plant rhizoliths (Lancaster and Tchakerian, 1996). Periods of soil development represent pulses or intervals of geomorphic stability under more humid climates than today. Some aeolian sands exhibit various calcium carbonate accumulation and a relative dating system based on stages of calcium carbonate formation has been developed for the southwestern deserts of the United States (Machette, 1985). For example, palaeosols and carbonate horizons developed on sand ramps in the Mojave Desert, can serve as stratigraphic markers for regional correlations and help constrain periods of aeolian accumulation (Figure 12.3).

Silt and Clay Particles

Most stabilised aeolian dune sands (paleosands) contain a higher proportion of 'fines' or silt and clay size particles (% <65 μm). Active dunes rarely contain fines greater than 2 or 3 per cent (Tchakerian, 1989). Vegetated linear dunes in Rice Valley, in the southeastern Mojave Desert, contain between 6 and 9 per cent silt and clay (Tchakerian, 1992). Other stabilised sand dunes containing 5–7 per cent fines have also been reported (e.g. Tsoar and Møller, 1986). High silt and clay content is related to a number of post-depositional weathering processes, including the infiltration of aeolian dust (Pye and Tsoar, 1990). This is largely accomplished through grain translocation (rain, dew, groundwater) and mineral disaggregation over time, the latter particularly effective on feldspars, with both orthoclase and albite releasing clay minerals and clay size particles (Orme and Tchakerian, 1986; Pye and Tsoar, 1990). Additional sources for fines include airborne evaporites, such as calcium carbonate, sodium sulphate and sodium carbonate, especially in mountaineous desert basins, where dune fields are located along, near or downwind from playas, the latter a significant source area for deflation and groundwater solutes (Tchakerian, 1992).

Grain Redness

It has been widely reported that certain palaeodune sediments are redder than their more active counterparts, a function largely of dune age (e.g Walker, 1979). However, dune reddening seems to be a function of a number of complex factors including age (stability), transport distance downwind, dune type and mobility, climate (especially moisture availability and the translocation of moisture down into the dune sediments and temperature), mineralogy (particularly the presence of pre-reddened sediments), weathering regime (Eh, pH), vegetation, and aeolian dust input (Pye and Tsoar, 1990; Lancaster, 1995). Reddening is believed to be achieved mostly as a result of the deposition of iron oxides on the surfaces of the grains largely from ferromagnesian minerals and from iron-bearing clay minerals in airborne dust (Nanson *et al.*, 1992). In the Mojave Desert, grains from stabilised sand ramp deposits dated by luminescence methods to over 35 ka (Rendell *et al.*, 1994), do not show any significant reddening with age. In the Wahiba Sands of Oman, mineralogical and remote sensing studies indicate the presence of reddened sands in the northern section of the dunes (Warren, 1988). The high percentage of reddened sands is believed to result from a number of complex factors including a high proportion of mafic minerals (from ophiolites) in the mountain complexes to the north, red chert outcrops in the mega-dune corridors, and in-situ weathering of the dune sands and deposition of iron oxides on the grain surfaces (Pease, 1998; Pease *et al.*, 1998).

Quartz Surface Microtextures

Using a scanning electron microscope (SEM), the nature and frequency of quartz grain surface microfeatures has been used to distinguish between various depositional environments (Krinsley and Doornkamp, 1973; Bull, 1981; Smart and Tovey, 1981). Quartz grain microfeatures can be used to discern distinct weathering and other morphological characteristics in aeolian paleosands (Tchakerian, 1991). Some of the diagnostic microfeatures observed on aeolian quartz grains under the scanning electron microscope for palaeoenvironmental analysis include (Figure 12.6):

Upturned plates   These microfeatures appear more or less as parallel ridges ranging in length from 0.5–10 μm in diameter, and spaced from about 0.1–2 μm apart, and form as a result of breakage along cleavage planes in the quartz lattice, the product of mechanical abrasion from saltation, and are characteristic of aeolian quartz grains from active desert dunes (Krinsley and McCoy, 1978; Krinsley and Trusty, 1985). Their frequency tends to decrease with increasing age, mostly the result of post-depositional modifications (Tchakerian, 1991).

Breakage blocks   Formed largely as a result of mechanical impact between colliding grains (Elzenga *et al.*, 1987), and to a lesser degree, from post-depositional chemical weathering processes, such as salt weathering (Krinsley and Trusty, 1985).

Conchoidal fractures   Most likely caused by grain-to-grain collision in an aeolian (or fluvial) medium and commonly stepped in a series of arcuate or elongate ridges (Krinsley and Doornkamp, 1973). According to Elzenga *et al.* (1987), it is possible that some fractures are inherited from the original source of the material prior to aeolian transport, or from earlier periods of aeolian transport and post-depositional weathering.

Mechanical depressions   Elongate or equidimensional features and disc-shaped concavities initially caused by the development of conchoidal fractures or breakage patterns (Krinsley and Trusty, 1985). They are considered typically diagnostic of

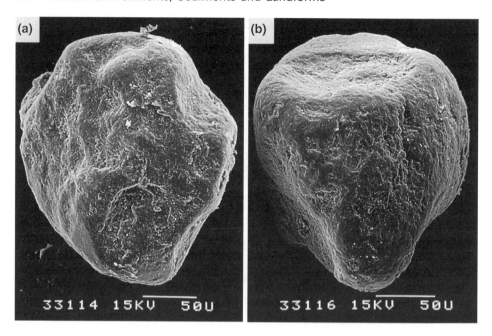

Figure 12.6 (Left) SEM image of a quartz grain from lowermost aeolian unit (dated by luminescence methods between 35 to 25 ka, Rendell *et al.*, 1994) at Dale Lake sand ramp shown in Figure 12.3. Note adhered particles, upturned plates and solution pits and etches. (Right) SEM image of a quartz grain from a late Pleistocene parabolic dune, Pismo Dunes, central California. Note silica dissolution and surface etching. Scale bar in μm

aeolian environments (Margolis and Krinsley, 1971). Some depressions serve as foci for silica, iron, and clay deposition (Krinsley and Doornkamp, 1973; Tchakerian, 1989).

Rolled topography   Also refered to as smooth surfaces or frosting. Produced by the smoothing of mechanically derived upturned plates by the solution and precipitation of silica (Krinsley and Doornkamp, 1973). This process is especially prominent in arid environments subject to extreme temperature and evaporation variations, and where moisture, charged with carbon dioxide forms a weak carbonic acid which infiltrates aeolian sand, dissolves certain minerals and leaves a residue of dissolved silica in the weathering profile (Pye and Tsoar, 1990). This is then redeposited as an irregular layer of either opal or silicic acid (Krinsley and McCoy, 1978), smoothing out the upturned plates and thus masking any pre-existing mechanically derived microfeatures. This process is further enhanced by the presence of evaporites and carbonates which are relatively abundant in desert environments (Krinsley and Doornkamp, 1973; Walker, 1979). Airborne evaporitic materials enhance the pH conditions and the salinity of the available moisture, thereby facilitating silica dissolution and reprecipitation (Tchakerian, 1989; Pye and Tsoar, 1990). Quartz dissolution and reprecipitation in high-energy chemical environments can sometimes lead to silica coatings up to 50 μm thick (Pye, 1983; Tchakerian, 1991).

Silica precipitation is effected by vadose water which dissolves quartz grains and silicate minerals, and then promotes reprecipitation during evaporative periods. Solution also liberates iron from various ferromagnesian minerals, which are susceptible to hydrolysis. Subsequent precipitation provides the characteristic iron-oxide coatings found on grains, and thereby the yellow, brown, orange and red colour so typical of palaeodunes. As indicated earlier, colour alone is no adequate indicator of age for much depends on initial sand mineralogy, provenance, aeolian dust inputs, temperature and available moisture. As aridity increases, iron oxidation becomes less notable and carbonate precipitation occurs on grains as a prelude to various stages of calcrete formation (Orme and Tchakerian, 1986).

Adhered particles   Consist of silt and clay particles attached to quartz grain surfaces. These are the combined result of electrostatic forces, mechanical and chemical weathering processes and post-depositional modifications, such as movement of surface moisture and groundwater, and infiltrating aeolian dust (Orme and Tchakerian, 1986; Pye and Tsoar, 1990).

Solution etchings and pittings   Formed primarily by dissolution of quartz in highly aggresive chemical environments (Krinsley and Doornkamp, 1973). Since most quartz grains contain microfractures and other minute crystallographic weaknesses, etching, pitting and dissolution is particularly concentrated along those weaknesses, especially under intense chemical weathering regimes and/or increasing time (Pye and Tsoar, 1990). In coastal dune environments in central California, solution etching and pitting was especially pronounced on quartz grains from stabilised late Pleistocene dunes, most likely the result of chemical weathering and increased time (Orme and Tchakerian, 1986). Similar quartz grain microfeatures were observed by Pye (1982) among podsolised, Pleistocene, coastal dunes in southern Queensland, Australia.

Multivariate statistical analysis, using the relative frequencies of mechanical, chemical and gross morphologic quartz surface microfeatures, can be an effective analytical and relative dating tool for distinguishing between spatially and temporally distributed aeolian deposits (Tchakerian, 1991). Surface textures, coatings, adhered particles and other surficial microtextures can be accurately analysed by energy-dispersive X-ray spectra (e.g. Pye and Tsoar, 1990; Tchakerian, 1991).

Aeolian Dust

Palaeoenvironmental information from desert aeolian dust deposits can also be discerned from soil development on aeolian accretionary mantles on alluvial fan deposits, marine sediments in ocean cores, and the mineral dust component of atmospheric aerosols.

Aeolian accretionary mantles   The influence of aeolian dust on the origin and evolution of stone pavements on alluvial fans has been the focus of much attention

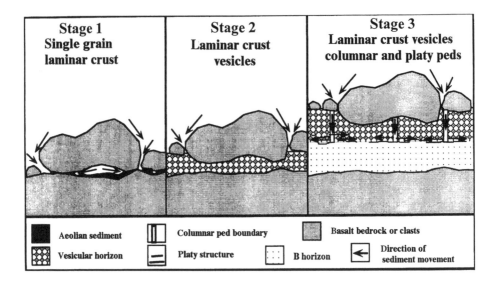

Figure 12.7 Idealised stages of clay and carbonate accumulation and vesicular soil formation (Av horizon) from aeolian sources, causing horizon thickening and vertical accretion, and the subsequent 'riding' upwards of pavement clasts (modified from Anderson *et al.*, 1994)

since initially proposed by McFadden *et al.* (1987). It has been shown that stone pavements on basaltic alluvial fans in the eastern Mojave Desert, can form by the upward lifting of surface clasts from the accumulation of aeolian fines below pavement clasts (Figure 12.7). This upward lifting process differs from the more traditional explanations such as deflation, water sorting, freeze and thaw, wetting and drying (Cooke *et al.*, 1993). Thus, desert pavement formation and the development of Av horizons are contemporaneous, and accrete vertically over time providing a stable platform for uplift of pavement clasts (Anderson *et al.*, 1994). Pavement and Av horizons develop at much faster rates on alluvial fans that have more pronounced bar and swale topography and other surface roughness elements, because irregular topography provides 'pockets' in which aeolian dust can preferentially accumulate (McDonald *et al.*, 1995). The source of the aeolian dust for the development of Av horizons and desert pavements are the sediments from desiccating lake basins (McFadden *et al.*, 1987; Wells *et al.*, 1987; Anderson *et al.*, 1994; McDonald *et al.*, 1995). The granulometric, sedimentologic and mineralogical analysis of the aeolian fines accumulated on desert pavements can provide additional information about regional aeolian fluxes, rates of deposition, sediment sources and other relevant palaeoenvironmental data.

Marine sediments   Off certain major continental deserts, aeolian dust comprises a significant proportion of the total accumulated sediments in ocean cores (Pye, 1987). The aeolian component in marine sediments is mostly represented by quartz-rich bottom sediments, which can be compared to modern dust sediments. The thickness of the quartz-silt layers, mineralogy, grain size, SEM analysis, and the distribution of weight per cent quartz in ocean sediments, can provide palaeoenvironmental information on source regions, fluxes, spatial distribution, and general wind directions of the dust transporting wind systems. Based on mineralogy and grain size distributions, dust input to the oceans, particularly from the global tropical continental arid zone, was increased during the late Quaternary, especially during the LGM (e.g. Sarnthein, 1978; Kolla *et al.*, 1979). Quartz-rich layers in marine sediments from the eastern Atlantic are believed to have been deposited during the LGM, as a result of increased aeolian activity and dust deflation from desert sediments in the western Sahara Desert, owing to the intensification of the trade winds, changes in global atmospheric pressures (steeper pressure gradients), and the southward shift of global winds and pressure cells, because of northern hemisphere glaciations (Sarnthein and Koopman, 1980; Sarnthein *et al.*, 1981; Pokras and Mix, 1985). Evidence from marine cores (along with terrestrial records) has been instrumental in deciphering the waxing and waning of some of the tropical deserts, such as the expansion of the Sahara Desert during the last glacial maximum. Dust fluxes were dramatically lowered during the Holocene. In the Arabian Sea, the Holocene dust production fell 60 per cent from its peak Pleistocene flux of $160 \times 10^6 \, t \, yr^{-1}$ (Sirocko *et al.*, 1991).

Mineral aerosols   Aeolian dust from desert regions has been identified as one of the major constituents of atmospheric aerosols (e.g. Prospero *et al.*, 1983; Pease *et al.*, 1998). Mineral aerosols do have a significant impact on global and local climate, agriculture and human health (Pye, 1987). Geochemical and meteorological analysis of mineral aerosols can be used to determine the spatial and temporal significance of dust production, particularly with issues related to provenance. For example, dust geochemistry of samples collected from the Arabian Sea region and surrounding desert areas, strongly suggest the Wahiba Sands region in Oman as a major dust source for the Arabian Sea, largely because of the presence of distinct mafic minerals (e.g. chromite) from ophiolitic mountain complexes to the north, in aerosol samples (Pease *et al.*, 1998).

AEOLIAN ENVIRONMENTS DURING THE LATE QUATERNARY

The world's drylands experienced significant environmental changes during the late Quaternary (Thomas, 1997). Some of the major global sand seas (ergs) show evidence of repeated phases of dune emplacement during late Quaternary time (Tchakerian, 1994). In the following section, a selective overview of late Quaternary aeolian activity is presented. More detailed global palaeoenvironmental reconstructions can be found in Tchakerian (1994), Lancaster (1995), Livingstone and Warren (1996) and Thomas (1997).

The present paradigm equates enhanced dune-building episodes during the late Quaternary in the continental tropical and subtropical deserts of the world, to regional aridities developed as a result of high latitude glaciations (Williams, 1975; Sarnthein, 1978). This hemispheric scale tropical aridity reached its peak during the LGM, as evidenced by the presence of a wide belt of fixed, fossil or degraded vegetated dunes, extending from the Sahara south to latitude 10° to 12°N in the Sahel (Grove and Warren, 1968).

Africa

The Sahara–Sahel region in northern Africa experienced late Quaternary dune building activity, with peak dune construction between 18 and 12 ka, a time interval characterised by drier and windier than present climatic conditions, with the driest period occurring between 14 and 12.5 ka (Alimen, 1982; Rognon and Williams, 1987). In the Akchar Erg in Mauritania, Kocurek et al. (1991) found evidence for the emplacement of large linear dunes during the period 20 to 13 ka. The dunes were subsequently stabilised by vegetation and paleosols between 11 and 4.5 ka, with renewed aeolian activity after 4 ka, continuing more or less unabated to the present. Although aeolian activity in the Sahara–Sahel region was widespread during the LGM, the northern sections of both the Grand Ergs and Erg Chech, remained inactive (Rognon, 1987). Aeolian activity was significantly reduced between 11 and 6 ka, as the region experienced a period of humid conditions characterised by increased vegetation cover, high lake levels, incised fluvial channels, and a rise in Neolithic cultures (Williams, 1982). Two major lacustrine and marsh sedimentary sequences are now widely recognised in the Malian Sahara, the first between 9.5 and 6.4 ka, and the second between 5.4 and 4 ka (Petit-Maire and Riser, 1983). After 4 ka, aridity returned to most of the Sahara–Sahel regions, as the subtropical high pressure cells assumed their position, with the present, hyper-arid central Saharan core region well established by 2 ka (Rognon and Williams, 1977). The last 2 ka is characterised by renewed episodic aeolian activity, including the reactivation during the last 30 to 40 years of some Sahelian dormant/relict dunes by increased anthropogenic activites, since stabilised dunes provide a richer plant cover (for grazing and firewood), and are easier to cultivate (O'Hara, 1997). Saharan environmental changes are believed to follow arid/humid cycles and subcycles as explained by Milankovitch forcing and climate models (Kocurek, 1998). In tropical continental deserts, major dune construction in sand seas would proceed during 100 000 year glacial maxima cycles (northern glaciations = tropical aridity) as predicted by the Milankovitch forcing of climate, superimposed by lesser sub-cycles perhaps corresponding to other forcing periods (even perhaps down to Dansgaard-Oeschger and/ or Heinrich events?). A time-lagged, sediment process-response system (Figure 12.8), has been tentatively proposed for understanding whether major global dune construction episodes can be constrained by Milankovich type, externally forced, climatic events (Kocurek, 1998). The testing (given recent advances and successes with various luminescence chronologies) of this model would be one of the most important challenges for future aeolian research.

Some of the most detailed aeolian palaeoenvironmental studies have focused on the Mega-Kalahari sand sea in central southern Africa, including dating of the aeolian sands by luminescence methods (Thomas et al., 1997; Stokes et al., 1997).

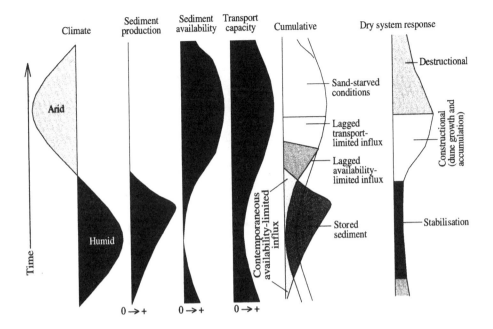

Figure 12.8 A process-response model based on sediment production, sediment availability (supply) and transport capacity of wind, for Saharan sand seas, in response to climatic cycles from humid to arid, and assumed to repeat in time (from Kocurek, 1998). See text for further discussion

The Mega-Kalahari covers an area of 2.5×10^6 km^2 and consists largely of vegetated, linear dunes interspersed (Figure 12.9) with dry lakes (Thomas and Shaw, 1991). Three major linear dune sand seas are found within the Mega-Kalahari: northern, eastern and southern, and, based on recent luminescence dating, each has been active at different times during the late Quaternary (Thomas *et al.*, 1987; Stokes *et al.*, 1997, 1998). In the southwestern Mega-Kalahari, two late-Pleistocene dune-building episodes have been identified by optically stimulated luminescence (OSL) dating, the first between 27 and 23 ka, and the second between 17 and 10 ka (Stokes *et al.*, 1997). From the Namibian part of the southwestern Mega-Kalahari, Blümel *et al.* (1998), using OSL dating, report only early Holocene ages of linear dune emplacement, between 9 and 8 ka. Thomas *et al.* (1997) also identified two periods of Holocene dune construction, the first at around 6 ka and the second between 2–1 ka. In the northeastern Mega-Kalahari, four arid events with linear dune construction episodes (arid = dune building), are identified based on OSL dating of quartz-grains: 115–95, 46–41, 26–20 and 16–9 ka (Stokes *et al.*, 1997, 1998). The arid phases are of short duration, lasting between 5 and 20 ka, while the humid phases are of longer duration, lasting between 20–40 ka. The formation of the linear dunes is thought to be linked to changes in southeast Atlantic sea surface temperatures, which causes changes in the NE–SW summer rainfall gradient, with periods of low sea surface temperatures (SSTs) corresponding to periods of

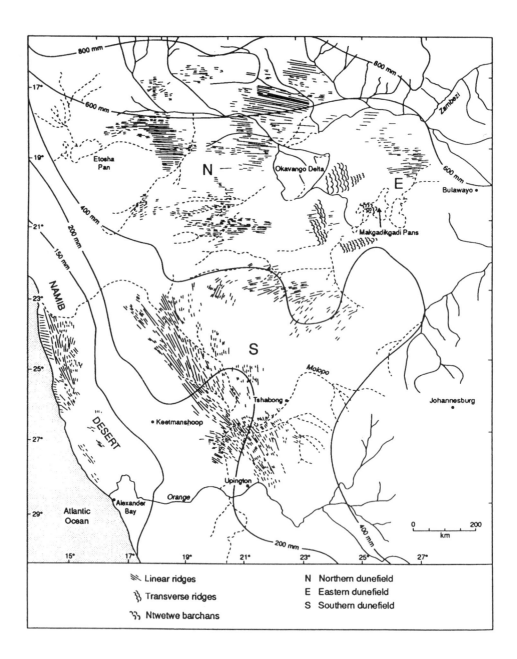

Figure 12.9 Dune field locations and alignments, Mega-Kalahari, southern Africa (from Livingstone and Thomas, 1993)

enhanced aridity in the Mega-Kalahari (Stokes *et al.*, 1997). Dune emplacement during the Pleistocene involved both new dune construction as well as reworking of older aeolian sands, while Holocene aeolian activity was confined to the reworking of Pleistocene dunes.

Australia

In the Australian arid zone, linear dunes are once again the dominant aeolian depositional landform, with most dunes being either dormant or relict (Wasson, 1984). The present wind regime is not conducive for active sand movement, as threshold velocities are generally low (Wasson, 1984). Most of the linear dunes were probably constructed between *ca* 26 and 13 ka, with peak dune-building activity between 20 and 16 ka, during the LGM (Bowler, 1976; Rognon and Williams, 1977; Wasson, 1984). Dune construction during the LGM was probably achieved by stronger winds owing largely to increased anticyclonic activity, and less from diminution in precipitation and/or lower vegetation densities (Wasson, 1984). Based on thermoluminescence and radiocarbon dating, Gardner *et al.* (1987) proposed that linear dune construction in western Australia could have started as far back as 35 ka. Although some of the longitudinal (linear) dunes in the Simpson Desert have been reworked during the Holocene, their underlying cores (based on TL dating) suggest deposition about 80 ka or earlier (Nanson *et al.*, 1992). During the LGM, linear dune emplacement phases between the Australian arid zone and the Mega-Kalahari appear to be synchronous.

Asia

In the Thar Desert of India and Pakistan, the period between 20 and 11 ka witnessed marked aeolian activity and dune formation, with peak dune construction between 15 and 13 ka (Wasson *et al.*, 1983; Chawla *et al.*, 1991; Thomas *et al.*, 1998). There is also some evidence for an older dune constructional period between 40 and 36 ka (Chawla *et al.*, 1991). Soon after the LGM, the intensified SW Monsoon was responsible for sediment mobility and entrainment, leading to the development of the parabolic dunes, even though precipitation and vegetation cover remained relatively high (Wasson *et al.*, 1983). The aeolian accumulation patterns are highly variable for the Holocene, with episodic pulses around 7, 4 and 1.8 ka, and a number of short-lived dune reactivation events within the past 1000 years (Thomas *et al.*, 1998).

From the Negev Desert in Israel, two late Pleistocene dune constructional phases from 20.9 to 16 ka and 11.7 to 10.3 ka, have been identified based indirectly on radiocarbon ages from lacustrine sediments (Magaritz and Enzel, 1990). Based on TL dates, linear dunes in the northern Negev have been stable for the last 10 to 6 ka (Rendell *et al.*, 1993). A well established arid period with dune building phases from 17 to 9 ka, has also been suggested for the southern Arabian Peninsula (McClure, 1976). Some of the linear mega-dunes in the northern Wahiba Sands could have formed during the LGM (Warren and Allison, 1998), while their lowermost sections yielded preliminary TL ages of about 110 ka (Juyal *et al.*, 1998).

North America

In North America, the geomorphic evidence for major dune-building episodes during the late Quaternary comes primarily from the Great Plains, the Great Basin, and the Mojave Desert (Tchakerian, 1997).

Great Plains

Extensive tracts of stabilised (dormant and relict) sand dunes and sand sheets are found throughout the Great Plains region of the central USA and south-central Canada (Muhs *et al.*, 1997a). The largest stabilised sand sea in the western hemisphere is the Nebraska Sand Hills, with an area of about 50 000 km^2. A group of early investigators asserted that the sand dunes in the Nebraska Sand Hills were most likely emplaced/reactivated from previous sands during the LGM (e.g. Warren, 1976; Wells, 1983). Another group proposed that the sand dunes were deposited during two dune-building episodes in the Holocene: the first from 7.5 to 5 ka, and the second, from 3 to 1.5 ka (e.g. Ahlbrandt *et al.*, 1983). Recently, Stokes and Swinehart (1997) based on ^{14}C and AMS radiocarbon dates and optical dating techniques, document two middle Holocene periods of aeolian deposition: the first around 6 to 5.7 ka, and the second, between 4 and 2 ka, as well as at least three depositional pulses during the latest (< 800 a) Holocene, the latter reactivations most likely occurring in response to periods of extended drought, such as 20 to 30 year drought cycles. Similar quasi-periodic, drought induced, dune reactivations have also been documented from the aeolian sediments of northern Colorado (Madole, 1994, 1995). Based on stratigraphic and AMS radiocarbon ages of unaltered bison bones and organic sediments, Muhs *et al.* (1997b) also propose two periods of dune reactivation in the Nebraska Sand Hills during the past 3 ka, and three more in the past 800 years. They too attribute Holocene reactivations to drought cycles more intense than the Dust Bowl events of the 1930s. Furthermore, the aeolian sands in the Nebraska Sand Hills are mineralogically mature and well-mixed, with high quartz content and depleted K-feldspars (Muhs *et al.*, 1997b). The sand sea has witnessed multiple aeolian constructional and destructional cycles during both glacial and interglacial periods, the last major aeolian cycle initiated during the LGM. Supporting evidence for enhanced mid-to-late Holocene aeolian activity also comes from Wyoming, where Gaylord (1990) found four pulses of dune deposition between 7.5 and 4.5 ka; from the High Plains in Texas, where Holliday (1989) reports dune construction between 6 and 4.5 ka; from western Kansas, where Olson *et al.* (1997) report dune deposits overlying radiocarbon dated palaeosols on loess after 6 ka; and from northern Colorado, where Muhs and Maat (1993) recognise increased aeolian activity between 3.5 to 1.5 ka. Valero-Garcés *et al.* (1997), employing multiple independent proxies (pollen stratigraphy, sedimentary facies, diatom assemblages, trace-element geochemistry, stable isotopes and carbonate chemistry) from Moon Lake cores in North Dakota, provide one of the most complete Holocene palaeoenvironmental studies, including supporting evidence for a widespread mid-Holocene arid interval, between 7.1 and 4 ka. Laird *et al.* (1996) from diatom analysis in Moon Lake sediments, provide further evidence for seven drought episodes during the past 2300 years. The available geomorphic evidence from the Great Plains suggest that the area has experienced major aeolian depositional phases during the Quaternary, with the Holocene experiencing episodic aeolian deposition, particularly during middle to latest time. Middle-Holocene aridity (the Altithermal) has been extensively documented for the Great Plains and other parts of the southwestern USA (e.g. Holliday, 1989; Spaulding, 1991). Episodic aeolian activity largely in response to desiccation events seems to have been widespread. The Holocene dune constructional episodes most likely took advantage (pirating) of the Pleistocene sand dunes and sand sheets deposited during the LGM or earlier (Muhs et al., 1997b). The present stabilised aeolian sands can be reactivated if the plant cover is significantly reduced either naturally, such as under greenhouse warming scenarios (whereby temperatures would increase with a concomitant reduction in precipitation as predicted by a number of GCM's) or anthropogenically, such as a reduction in vegetation cover owing to increased agricultural activities (the efficacy of centre-pivot irrigation). According to Muhs and Maat (1993), wind power is high for the Great Plains and present wind speeds exceed the threshold wind velocity (U_{*t}) 30–60 per cent of the time. The sand dunes and sand sheets are thus 'poised' for renewed aeolian activity if and when the present vegetation cover is removed or significantly disturbed.

Great Basin

Periods of late Quaternary dune activity in the Great Basin are less well constrained than those in the Great Plains or the Mojave Desert. In eastern Arizona, Stokes and Breed (1993) recognise three periods of dune construction between 400 and 4700 years ago. Middle-Holocene dune-building episodes have also been identified by Wells et al. (1990) in the southern Colorado Plateau, as well as a number of aeolian depositional pulses during the last 1500 years. Additional episodic late Holocene aeolian activity has been recognised from southeastern Oregon (Mehringer and Wigand, 1986), southern Nevada (Quade, 1986) and southern Arizona (Brakenridge and Schuster, 1986). A recent study by Quade et al. (1998), based on spring mound activity and the formation of black mats, confirm the prevalence of drier than present conditions between 6.3 and 2.3 ka in southern Nevada.

Mojave Desert

The Mojave Desert, located in southern California at the southern end of the Basin and Range Province, contains a plethora of aeolian depositional features, and has been the focus of much aeolian geomorphic research during the past 15 years (e.g. Smith, 1982; Wells et al., 1987, 1989; Tchakerian, 1989, 1991, Lancaster, 1993, 1997; Zimbelman et al., 1995; Lancaster and Tchakerian, 1996), and the beneficiary of luminescence chronologies (e.g. Rendell et al., 1994; Wintle et al., 1994; Clarke et al., 1995, 1996; Rendell and Sheffer, 1996). With the exception of the Gran Desierto del Altar in the Sonoran Desert and the Algodones Dunes in the Colorado Desert in extreme southern California, the Mojave Desert is the only arid zone with significant aeolian landforms, because aeolian depositional landforms constitute only a minor component of the total surficial deposits in the North American arid zone (Tchakerian, 1997). In the Mojave Desert, aeolian deposition is largely confined to topographically well-defined sand-transport corridors (Figure 12.10) that follow the region's geologic and tectonic setting (Zimbelman *et al.*, 1995).

One of the largest and best studied dunefields in the Mojave Desert is the Kelso Dunes. This well-developed aeolian corridor begins at the fan-delta complex of the ephemeral Mojave River as it exits Afton Canyon (Figure 12.10). Sand is then transported through the Devil's Playground for the next 60 km, finally reaching the Kelso Dunes, the major depositional sink for this transport corridor (Lancaster, 1993). Although active, dormant and relict dunes are found within the Kelso Dunes, vegetation-stabilised dormant dunes are the most extensive, as well as relict climbing and falling dunes and sand ramps (Lancaster and Tchakerian, 1996; Lancaster, 1997). Luminescence ages for dune sediments at the Kelso Dunes, indicate numerous aeolian depositional episodes, especially between 26 and 15 ka, and from 12 to 7 ka (Wintle *et al.*, 1994; Clarke *et al.*, 1995). Extensive reworking of the sands during the Holocene, between 50 to 4000 years ago, has also been documented (Edwards, 1993; Lancaster, 1993).

For aeolian palaeoenvironmental reconstruction, sand ramps offer a unique geomorphic environment (Tchakerian, 1991). Sand ramps consist mainly of aeolian sands, intermixed with fluvial and talus deposits, and occur in relation to well-defined regional and local-scale sand transport corridors, that extend from source areas in the western and central Mojave Desert, towards sinks to the east and the south (Lancaster and Tchakerian, 1996). Most, but not all, of the sand accumulations occur on the western (windward) flanks of mountain ranges that lie astride these sand transport corridors (Zimbelman *et al.*, 1995). Most sand ramps are dominated by aeolian sands that were deposited as sand sheets on the piedmont slopes of desert mountain ranges. In the Bristol Lake sand transport corridor (Figure 12.10), preliminary sediment geochemistry obtained with instrumental neutron activation analysis indicates that multiple, distinct sediment sources supply sand at different points along the sand transport corridors (Figure 12.11). The sand ramp sediments from the west side of the transport corridor have a higher mafic content than those from the east side. These variations in geochemistry appear to be influenced primarily by differences in mineralogy and weathering. It is highly probable that the two areas have: (a) different sources (the west source being more mafic); (b) there is sediment mixing and dilution down the corridor; and (c) the differences are the result of progressive weathering down the corridor, or a combination of all three.

Multiple periods of aeolian accumulation can be recognised, separated by paleosols formed in periods of geomorphic stability (Figure 12.3). Most sand ramps in the Mojave Desert are relict features and are not accumulating today, with boulder-to-gravel size talus mantles, and incised stream channels. Chronological dating based on quartz TL and feldspar infrared stimulated luminescence dating (IRSL), indicate the existence of two main periods of late Pleistocene dune construction, between > 35 to 25 ka and 15 to 10 ka (Rendell *et al.*, 1994).

The major source of sediments for the aeolian sands in the Mojave Desert are the Pleistocene paleolakes of the region and the well-established Mojave River Wash

(a)

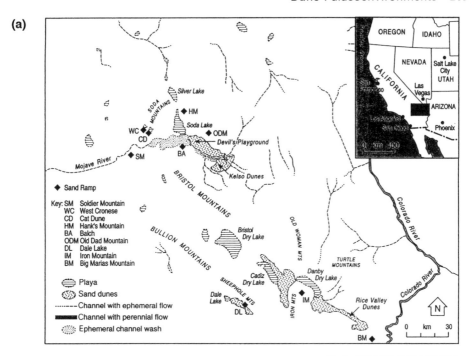

(b)

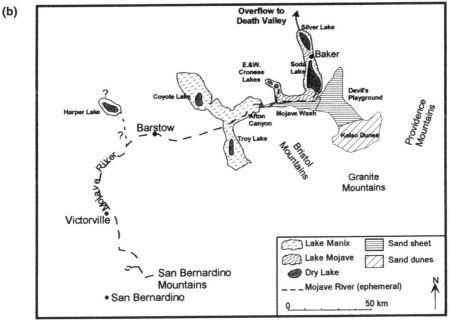

Figure 12.10   (a) Location of major sand dunes, sand sheets and sand transport corridors in the eastern Mojave Desert (from Rendell and Sheffer, 1996). (b) A close up view of the Mojave River/Kelso Dunes corridor (modified from Clarke *et al.*, 1995 and Meek, 1997)

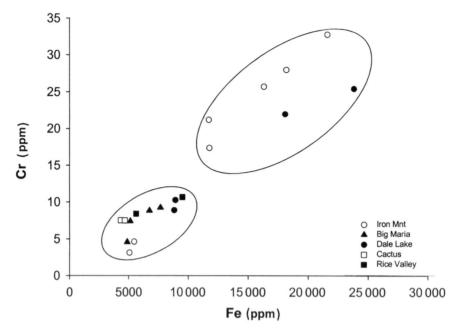

Figure 12.11 Chromium (Cr)/Iron (Fe) plot indicating the relative concentration of mafic minerals in sediment samples from sand ramps, eastern Mojave Desert. See Figure 12.10 for location of sample sites. The samples from Cactus are from linear dunes in Arizona, across from the Colorado River from the Big Maria site. See text for further discussion

(Wells *et al.*, 1989; Tchakerian, 1991; Lancaster and Tchakerian, 1996; Lancaster, 1997). Lake-level fluctuations can provide extensive sandy sediments for subsequent deflation. Sediments are also supplied or replenished by storms to the Mojave River Wash. Additional sources for sand include fan-deltaic and beach deposits formed in and around lakes, and alluvial fans and wadis. Geomorphic, stratigraphic, chrono-logic, palaeohydrologic and palaeoclimatic studies (Wells *et al.*, 1987, 1989; Enzel *et al.*, 1992; Rendell *et al.*, 1994; Clarke *et al.*, 1995, 1996; Lancaster and Tchakerian, 1996; Lancaster, 1997; Kocurek and Lancaster, 1998) indicate that a number of complex, but inter-related processes associated with lacustrine, hillslope, fluvial, and aeolian sediment production, storage, transport and deposition, ultimately control rates, fluxes and the timing of the aeolian depositional episodes. A time-lagged, temporally and spatially disjointed, sediment system has been tentatively suggested for the Kelso Dunes in the Mojave Desert (Kocurek and Lancaster, 1998), based on a concept outlined by Kocurek (1998) and shown in Figure 12.8. The following discussion is a brief summary of the above two papers, with some additional elaborations. During sub-humid periods and/or sub-humid/semi-arid climatic tran-sitions, sediment production is high, while sediment availability for aeolian transport is low because of increased precipitation and vegetation density (higher U_{*t}). Sedi-ment is stored in and around lake basins, lower piedmont slopes, distal alluvial fan deposits and in arroyos and wadis, with limited amounts of sediment entrainment.

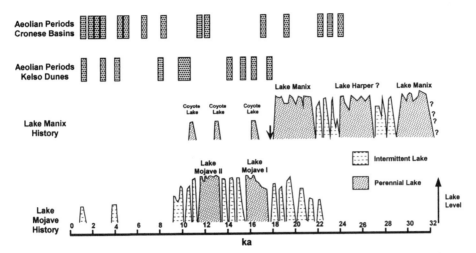

Figure 12.12 Generalised history of dune-building episodes and palaeolake fluctuations from the eastern Mojave Desert (after Tchakerian, 1989; Wells *et al.*, 1989 and Kocurek and Lancaster, 1998). Lake level chronologies based on Brown (1989), Wells *et al.* (1989), Meek (1997 and personal communication). Aeolian constructional periods based on luminescence ages including data from Rendell *et al.* (1994), Clarke *et al.* (1995) and Rendell and Sheffer (1996). Lakes Manix, Harper, Coyote and Mojave were formed from waters of the Mojave River. Arrow indicates the draining of Lake Manix. Intermittent lake level fluctuations are not as well constrained as those from high lake stands and should be considered preliminary

During humid conditions (e.g. high lake stands), dense plant cover reduces sediment transport rates, leading to periods of landscape stability and pedogenesis. During arid periods, sediment production is dramatically reduced, as fluvial and lacustrine sytems experience periods of desiccation. Wind transport capacity is generally high (e.g. the Holocene Altithermal). Concomittantly, the vegetation cover is reduced and the stored sediment during the humid and sub-humid/semi-arid phases is mobilised/ re-activated. In the Mojave Desert, it is likely that during arid or hyper-arid and windier periods (e.g. the Altithermal), sediment movement and deposition is low, because the previously stored sediment has already been mobilised (such as during climatic transitions and intermittent lake levels), and sediment production (weathering and transport) at a minimum. Dune-building activity is thus at a minimum during the drier and windier subcycles.

Between *ca* 34 and 9 ka, a series of full and intermittent lakes occupied the drainage basin of the Mojave River (Figure 12.12), the largest arid fluvial system in the Mojave Desert (Wells *et al.*, 1989). The terminus of the Mojave River in desert basins has changed during the late Quaternary including Lake Mojave (present day Soda Lake and Silver Lake basins), Lake Manix (present day Lake Coyote, Troy Lake and Afton Canyon), East and West Cronese Playas, and perhaps even Harper Lake (Meek, 1997). Sediments from the above fluctuating and desiccating lake basins are believed to be the primary source for the sand dunes and sand ramps of the eastern Mojave Desert. Based on recent TL, IRSL and OSL chronologies (Rendell *et al.*, 1994; Wintle *et al.*, 1994; Clarke *et al.*, 1995), aeolian deposition seems to have occurred during both high and intermittent lake levels, as well as during periods of greater than present aridity (Figure 12.12). Deposition has been mostly episodic or in discrete pulses controlled largely by the availability of sediment. It is likely that after the final desiccation of Lake Mojave II, a major regional dune-building activity took place between 9 and 7 ka, since most sand ramps were stabilised by the early Holocene (Rendell et al., 1994). During the more drier and windier middle Holocene Altithermal, aeolian activity was at a minimum, owing to the fact that sediment production was low or exhausted, even though sediment transport capacity was high. A number of brief latest-Holocene intermittent lakes in the Cronese and Silver Lake basins have been identified, including lacustrine events at around 700, 550, 380 (Little Ice Age?) and 220 years ago (Brown, 1989). Between 1850 and 1994, 14 extreme floods in the Mojave River produced documented lake stands in Silver Lake and/or East Cronese Playa (Wells et al., 1989, 1994). The events of the last 150 years indicate that sediment production and availability has been operating with frequencies approximating the floods of the Mojave River. It is possible that aeolian deposition can take place even during high lake stands, given enough variations in lake-levels and storm driven sediment inputs, and/or sediment mobilisation from stored sources and ephemeral washes. Dune building episodes in the Mojave Desert appear to be highly episodic in nature (and not continuous), and primarily driven by sediment production, availability and transport capacity systems (e.g. Kocurek and Lancaster, 1998), rather than the result of intensified winds or aridity and/or decrease in plant cover. Luminescence dating methods seem to also corraborate the episodic and disjointed nature of aeolian accumulation (e.g. Clarke et al., 1995). Additional luminescence dating of sand dunes and sand ramps should provide a more comprehensive chronology of dune-building activities in the Mojave Desert.

FUTURE DIRECTIONS

The application of geomorphological, biological and sedimentological criteria and luminescence dating methods, has been instrumental in reconstructing the late Quaternary aeolian depositional environments, particularly in deciphering the complexities of dune construction and destruction. Some of the future challenges in aeolian palaeoenvironmental research include:

1. More detailed understanding of vegetation succession on dunes, from the establishment of cryptogamic soils to the development of woodland vegetation. The relations between dune form, dune stability, and vegetation, remains an important issue in aeolian research.
2. The formation and preservation of palaeosols in sand dunes, particularly in sand ramps. Palaeosol stratigraphy and geochemistry from sand ramps (e.g. Mojave Desert) can provide additional information for environmental reconstruction and regional correlation.
3. Dune size and its relation to time (age), sediment supply and wind regime. The formation of multiple generation of dunes in sand seas (such as in the Wahiba

Sands in Oman) can only be unravelled if the above relation is more thoroughly understood.

4. Further application of geochemical and meteorological analysis to desert dust sediments, particularly in mineral aerosol studies, to constrain source regions and quantify aeolian dust fluxes.

5. Continue 'fine-tuning' luminiscence dating methods since they seem to be one of the few reliable absolute dating tools available for establishing aeolian chronologies.

6. Better explicate the relations between climate variability, arid/humid cycles and sediment systems, particularly the complex interactions between Milankovitch-type climatic forcing, millenial-to-decadal scale climatic oscillations, and their influence on sediment production, storage, transport and deposition. Also, the relations between sea surface temperatures (SST) and arid/humid cycles, as well as the complex feedbacks between sediment systems, aridity and dune-building episodes, need to be elucidated in greater detail.

REFERENCES

Ahlbrandt, T.S., Swinehart, J.B. and Maroney, D.G., 1983. The dynamic Holocene dune fields of the Great Plains and Rocky Mountain Basins, USA. In *Eolian Sediments and Processes*, M.E. Brookfield and T.S. Ahlbrandt (Eds). Elsevier, New York, pp. 379–406.

Alimen, H.M., 1982. Le Sahara – grande zone desertiques Nord Africaine. In *The Geological Story of the World's Deserts*, T.L. Smiley (Ed.). University of Uppsala Press, Uppsala, pp. 35–51.

Anderson, K.C., Wells, S.G., Graham, R.C. and McFadden, L.D., 1994. Processes of vertical accretion in the stone-free zone below desert pavements. *Geological Society of America Abstracts with Programs*, **26**, A-87.

Ash, J.E. and Wasson, R.J., 1983. Vegetation and sand mobility in the Australian desert dunefield. *Zeitschrift für Geomorphologie*, Supplementband **45**, 7–25.

Blümel, W.D., Eitel, B. and Lang, A., 1998. Dunes in southeastern Namibia: evidence for Holocene environmental changes in the southwestern Kalahari based on thermolumines-cence data. *Palaeogeography, Palaeoclimatology, Palaeoecology*, **138**, 139–149.

Bowler, J.M., 1976. Aridity in Australia: age, origins and expression in aeolian landforms and sediments. *Earth Science Reviews*, **12**, 279–312.

Brakenridge, G.R. and Schuster, J., 1986. Late Quaternary geology and geomorphology in relation to archaeological site locations, southern Arizona. *Journal of Arid Environments*, **10**, 225–239.

Brown, W.J., 1989. The late Quaternary stratigraphy, paleohydrology, and geomorphology of Pluvial Lake Mojave, Silver Lake and Soda Lake basins, southern California, Unpublished MS theses, University of New Mexico.

Bull, P.A., 1981. Environmental reconstruction by electron microscopy. *Progress in Physical Geography*, **5**, 368–397.

Bullard, J.E., Thomas, D.S.G., Livingstone, I. and Wiggs, G.F.S., 1997. Dunefield activity and interactions with climatic variability in the southwest Kalahari Desert. *Earth Surface Processes and Landforms*, **22**, 165–174.

Chawla, S., Dhir, R.P. and Singhvi, A.K. (1991). Thermoluminescene chronology of sand profiles in the Thar Desert and their implications. *Quaternary Science Reviews*, **11**, 25–32.

Clarke, M.L., Richardson, C.A. and Rendell, H.M., 1995. Luminescence dating of Mojave Desert sands. *Quaternary Science Reviews*, **14**, 783–789.

Clarke, M.L., Wintle, A.G. and Lancaster, N., 1996. Infra-red stimulated luminescence dating of sands from the Cronese Basins, Mojave Desert. *Geomorphology*, **17**, 199–205.

Cooke, R.U., Warren, A. and Goudie, A.S., 1993. *Desert Geomorphology*. UCL Press, London.

Edwards, S.R., 1993. Luminescence dating of sand from the Kelso Dunes, California. In *The Dynamics and Environmental Context of Aeolian Sedimentary Systems*, K. Pye (Ed). Geological Society Special Publication No. 72, pp. 59–68.

Elzenga, W., Schwan, J., Baumfalk, Y.A. *et al.*, 1987. Grain surface characteristics of periglacial aeolian and fluvial sands. *Geologie en Mijnbouw*, **65**, 273–286.

Enzel, Y., Brown, W.J., Anderson, R.Y. *et al.*, 1992. Short-duration Holocene lakes in the Mojave River drainage basin, southern California. *Quaternary Research*, **38**, 60–73.

Fryberger, S.G., 1979. Dune form and wind regime. *US Geological Survey, Professional Paper*, **1052**, 137–169.

Gardner, G.J., Mortlock, A.J., Price, D.M. *et al.*, 1987. Thermoluminescence and radiocarbon dating of Australian desert dunes. *Australian Journal of Earth Sciences*, **34**, 343–357.

Gardner, R.A.M., 1983. Aeolianites. In *Chemical Sediments in Geomorphology*, A.S. Goudie and K. Pye (Eds). Academic Press, London, pp. 265–300.

Gaylord, D.R., 1990. Holocene paleoclimatic fluctuations revealed from dune and interdune strata in Wyoming. *Journal of Arid Environments*, **18**, 123–138.

Goudie, A.S. and Watson, A., 1981. The shape of desert sand dune grains. *Journal of Arid Environments*, **4**, 185–190.

Grove, A.T. and Warren, A., 1968. Quaternary landforms and climate on the south side of the Sahara. *Geographical Journal*, **134**, 194–208.

Holliday, V.T., 1989. Middle Holocene drought on the southern High Plains. *Quaternary Research*, **31**, 74–82.

Juyal, N. Glennie, G.W. and Singhvi, A.K., 1998. Chronology and palaeoenvironmental significance of Quaternary desert sediments in S.E. Arabia. In *Quaternary Deserts and Climatic Change*, A.S. Alsharhan, K.W. Glennie, G.L. Whittle and C.G. St. C. Kendall (Eds.). Balkema, Rotterdam, pp. 338–353.

Kocurek, G., 1996. Desert aeolian systems. In *Sedimentary Environments: Processes, Facies and Stratigraphy*, H.G. Reading (Ed.). Blackwell, London, pp. 125–153.

Kocurek, G., 1998. Aeolian system response to external forcing factors – a sequence stratigraphic view of the Saharan region. In *Quaternary Deserts and Climatic Change*, A.S. Alsharhan, K.W. Glennie, G.L. Whittle and C.G. St. C. Kendall (Eds). Balkema, Rotterdam, pp. 327–337.

Kocurek, G. and Lancaster, N., 1998. Aeolian system sediment configuration: theory and Mojave Desert Kelso Dune field example. *Sedimentology*, in press.

Kocurek, G., Havholm, K., Deynoux, M. and Blakey, R., 1991. Amalgamated accumulations resulting from climatic and eustatic changes, Akchar Erg, Mauritania. *Sedimentology*, **38**, 751–772.

Kolla, V., Biscaye, P.E. and Hanley, A.F., 1979. Distribution of quartz in late Quaternary Atlantic sediments in relation to climate. *Quaternary Research*, **11**, 261–277.

Krinsley, D.H., and Doornkamp, J.C., 1973. *Atlas of Quartz Sand Surface Textures*. Cambridge University Press, Cambridge.

Krinsley, D.H. and McCoy, F.W., 1978. Aeolian quartz sand and silt. In *Scanning Electron Microscopy in the Study of Sediments*, W.B. Whalley (Ed.). GeoAbstracts, Norwich, pp. 249–260.

Krinsley, D.H. and Trusty, P., 1985. Environmental interpretation of quartz grain surface textures. In *Provenance of Arenites*, G.G. Zuffa (Ed.). D. Reidel Publishing, Rotterdam, pp. 213–229.

Laird, K.R., Fritz, S.C., Maasch, K.A. and Cumming, B.F., 1996. Greater drought intensity and frequency before AD 1200 in the northern Great Plains, USA. *Nature*, **384**, 552–554.

Laity, J., 1992. Ventifact evidence for Holocene wind patterns in the east-central Mojave Desert. *Zeitschrift für Geomorphologie*, Supplementband **84**, 1–16.

Lancaster, N., 1988. Development of linear dunes in the southwestern Kalahari, southern Africa. *Journal of Arid Environments*, **14**, 233–244.

Lancaster, N., 1993. Development of Kelso Dunes, Mojave Desert, California. *National Geographic Research and Exploration*, **9**, 444–459.

Lancaster, N., 1995. *Geomorphology of Desert Dunes*. Routledge, London.

Lancaster, N., 1997. Response of aeolian geomorphic systems to minor climate change: examples from the southern Californian deserts. *Geomorphology*, **19**, 333–347.

Lancaster, N. and Tchakerian, V.P., 1996. Geomorphology and sediments of sand ramps in the Mojave Desert. *Geomorphology*, **17**, 151–165.

Livingstone, I. and Thomas, D.S.G., 1993. Modes of linear dune activity and their palaeo-environmental significance: an evaluation with reference to southern African examples. In *The Dynamics and Environmental Context of Aeolian Sedimentary Systems*, K. Pye (Ed.). Geological Society Special Publication No. 72, pp. 37–46.

Livingstone, I and Warren, A., 1996. *Aeolian Geomorphology*. Longman, London.

Machette, M.N., 1985. Calcic soils of the southwestern United States. In *Soils and Quaternary Geology of the Southwestern United States*, D.W. Weide (Ed.). Geological Society of America Special Paper 203, pp. 1–22.

Madole, R.F., 1994. Stratigraphic evidence of desertification in the west-central Great Plains within the past 1000 yr. *Geology*, **22**, 483–486.

Madole, R.F., 1995. Spatial and temporal patterns of late Quaternary aeolian deposition, eastern Colorado, USA. *Quaternary Science Reviews*, **14**, 155–177.

Magaritz, M. and Enzel, Y., 1990. Standing-water deposits as indicators of late Quaternary dune migration in the northwestern Negev, Israel. *Climatic Change*, **16**, 307–318.

Margolis, S.V. and Krinsley, D.H., 1974. Processes of formation and environmental occurrence of microfeatures on detrital quartz grains. *American Journal of Science*, **274**, 449–464.

McClure, H.A., 1976. Radiocarbon chronology of late Quaternary lakes in the Arabian Desert. *Nature*, **263**, 755–756.

McDonald, E.V., McFadden, L.D. and Wells, S.G., 1995. The relative influences of climate change, desert dust, and lithologic control on soil-geomorphic processes on alluvial fans, Mojave Desert, California. In *Ancient Surfaces of the East-Mojave Desert*, R.E. Reynolds and J. Reynolds (Eds). San Bernardino County Museum Association Quarterly 42, pp. 35–42.

McFadden, L.D., Wells, S.G. and Jercinovich, M.J., 1987. Influence of eolian and pedogenic processes on the origin and evolution of desert pavements. *Geology*, **15**, 504–508.

Meek, N., 1997. Paleoclimatic implications of the Mojave River system. *Association of American Geographers, Abstracts with Programs*, **93**, 175.

Mehringer, P.J., Jr. and Wigand, P.E., 1986. Holocene history of Skull Creek dunes, Catlow Valley, southeastern Oregon, USA. *Journal of Arid Environments*, **11**, 117–138.

Muhs, D.R. and Maat, P.B., 1993. The potential response of aeolian sands to Greenhouse Warming and precipitation reduction on the Great Plains of the United States. *Journal of Arid Environments*, **25**, 351–361.

Muhs, D.R., Stafford, T.W., Jr., Been, J. *et al.*, 1997a. Holocene eolian activity in the Minot dune field, North Dakota. *Canadian Journal of Earth Sciences*, **34**, 1442–1459.

Muhs, D.R., Stafford, T.W., Jr., Swinehart, J.B. *et al.*, 1997b. Late Holocene eolian activity in the mineralogically mature Nebraska Sand Hills. *Quaternary Research*, **48**, 162–176.

Nanson, G.C., Chen, X.Y. and Price, D.M., 1992. Lateral migration, thermoluminescence chronology, and colour variation of longitudinal dunes near Birdsville in the Simpson Desert, Australia. *Earth Surface Processes and Landforms*, **17**, 807–820.

O'Hara, S.L., 1997. Human impacts on dryland geomorphic processes. In *Arid Zone Geomorphology: Process, Form and Change in Drylands*, D.S.G. Thomas (Ed.). John Wiley, London, pp. 639–658.

Olson, C.G., Nettleton, W.D., Porter, D.A. and Brasher, B.R., 1997. Middle Holocene aeolian activity on the High Plains of west-central Kansas. *Holocene*, **7**, 255–261.

Orme, A.R. and Tchakerian, V.P., 1986. Quaternary dunes of the Pacific coast of the

Californias. In *Aeolian Geomorphology*, W.G. Nickling (Ed.). Allen and Unwin, Boston, pp. 149–175.

Pease, P.P., 1998. Aeolian sediment transport, Sultanate of Oman and the Arabian Sea. Unpublished DPhil Thesis, Texas A&M University.

Pease, P.P., Tchakerian, V.P. and Tindale, N.W., 1998. Aerosols over the Arabian Sea: geochemistry and source areas for aeolian desert dust. *Journal of Arid Environments*, **39**, 477–496.

Petit-Maire, N. and Riser, J. (Eds), 1983. *Sahara ou Sahel? Quaternaire Recent du Bassin de Taoudenni, Mali.* Imprimerie Lamy, Marseille.

Pokras, E.M. and Mix, A.C., 1985. Eolian evidence for spatial variability of late Quaternary climates in tropical Africa. *Quaternary Research*, **24**, 137–149.

Prospero, J.M., Charlson, R.J., Mohnen, V. *et al.*, 1983. The atmospheric aerosol system: an overview. *Review Geophysics and Space Physics*, **21**, 1607–1629.

Pye, K., 1982. Morphological development of coastal dunes in a humid tropical environment, Cape Bedford and Cape Flattery, North Queensland. *Geografiska Annaler*, **A64**, 212–227.

Pye, K., 1987. *Aeolian Dust and Dust Deposits.* Academic Press, London.

Pye, K. and Tsoar, H., 1990. *Aeolian Sand and Sand Dunes.* Unwin Hyman, London.

Quade, J., 1986. Late Quaternary environmental changes in the upper Las Vegas Valley, Nevada. *Quaternary Research*, **26**, 340–357.

Quade, J., Forester, R.M., Pratt, W.L. and Carter, C., 1998. Black mats, spring-fed streams, and late-glacial-age recharge in the southern Great Basin. *Quaternary Research*, **49**, 129–148.

Rendell, H.M. and Sheffer, N.L., 1996. Luminescence dating of sand ramps in the eastern Mojave Desert. *Geomorphology*, **17**, 187–197.

Rendell, H.M., Yair, A. and Tsoar, H., 1993. Thermoluminescence dating of periods of sand movement and linear dune formation in the northern Negev, Israel. In *The Dynamics and Environmental Context of Aeolian Sedimentary Systems*, K. Pye (Ed.). Geological Society Special Publication No. 72, pp. 69–74.

Rendell, H.M., Lancaster, N. and Tchakerian, V.P., 1994. Luminescence dating of late Quaternary aeolian deposits at Dale Lake and Cronese Mountains, Mojave Desert, California. *Quaternary Science Reviews*, **13**, 417–422.

Rognon, P., 1987. Late Quaternary climatic reconstruction for the Maghreb (north Africa). *Palaeogeography, Palaeoclimatology, Palaeoecology*, **58**, 11–34.

Rognon, P. and Williams, M.A.J., 1977. Late Quaternary climatic change in Australia and North Africa: A preliminary interpretation. *Palaeogeography, Palaeoclimatology, Palaeoecology*, **21**, 285–327.

Sarnthein, M., 1978. Sand deserts during glacial maximum and climatic optimum. *Nature*, **272**, 43–46.

Sarnthein, M. and Koopman, B., 1980. Late Quaternary deep-sea record of northwest African dust supply and wind circulation. *Palaeoecology of Africa*, **12**, 239–253.

Sarnthein, M., Tetzlaff, G., Koopman, B. *et al.*, 1981. Glacial and interglacial wind regimes over the eastern sub-tropical Atlantic and northwest Africa. *Nature*, **293**, 193–196.

Sirocko, F., Lange, H. and Erlenkeuser, H., 1991. Atmospheric summer circulation and coastal upwelling in the Arabian Sea during the Holocene and the last glaciation. *Quaternary Research*, **36**, 72–93.

Smart, P. and Tovey, N.K., 1981. *Electron Microscopy of Soils and Sediments: Examples.* Oxford University Press, Oxford.

Smith R.S.U., 1982. Sand dunes in North American deserts. In *Reference Handbook on the Deserts of North America*, G.L. Bender (Ed.). Greenwood Press, Westport, CT, pp. 481–526.

Spaulding, W.G., 1991. A middle Holocene vegetation record from the Mojave Desert of North America and its paleoclimatic significance. *Quaternary Research*, **35**, 427–437.

Stokes, S. and Breed, C.S., 1993. A chronostratigraphic re-evaluation of the Tusayan Dunes, Moenkopi Plateau and southern Ward Terrace, northeastern Arizona. In *The Dynamics* and *Environmental Context of Aeolian Sedimentary Systems*, K. Pye (Ed.). Geological Society Special Publication No. 72, pp. 75–90.

Stokes, S. and Swinehart, J.B., 1997. Middle-and late-Holocene dune reactivation in the Nebraska Sand Hills, USA. *Holocene*, **7**, 263–272.

Stokes, S., Thomas, D.S.G. and Washington, R., 1997. Multiple episodes of aridity in southern Africa since the last interglacial period. *Nature*, **388**, 154–158.

Stokes, S., Haynes, G., Thomas, D.S.G. *et al.*, 1998. Punctuated aridity in southern Africa during the last glacial cycle: the chronology of linear dune construction in the northeastern Kalahari. *Palaeogeography, Palaeoclimatology, Palaeoecology*, **137**, 305–322.

Tchakerian, V.P., 1989. Late Quaternary aeolian geomorphology, east-central Mojave Desert, California. Unpublished DPhil Thesis, University of California, Los Angeles.

Tchakerian, V.P., 1991. Late Quaternary aeolian geomorphology of the Dale Lake sand sheet, southern Mojave Desert, California. *Physical Geography*, **12**, 347–369.

Tchakerian, V.P., 1992. Aeolian geomorphology of the Dale Lake sand ramp. In *Old Routes to the Colorado*, R.E. Reynolds (Ed.). San Bernardino County Museum Association Special Publication 92-2, San Bernardino, California, pp. 46–60.

Tchakerian, V.P., 1994. Paleoclimatic interpretations from desert dunes and sediments. In *Geomorphology of Desert Environments*, A.D. Abrahams and A.J. Parsons (Eds). Chapman and Hall, London, pp. 631–643.

Tchakerian, V.P., 1997. North America. In *Arid Zone Geomorphology: Process, Form and Change in Drylands*, D.S.G. Thomas (Ed.). John Wiley, London, pp. 523–541.

Thomas, D.S.G., 1987. Discrimination of depositional environments using sedimentary characteristics in the Mega Kalahari, central southern Africa. In *Desert Sediments: Ancient and Modern*, L. Frostick and I. Reid (Eds). Geological Society of London Special Publication, 35. Blackwell Science, Oxford, pp. 293–306.

Thomas, D.S.G., 1997. Reconstructing ancient arid environments. In *Arid Zone Geomorphology: Process, Form and Change in Drylands*, D.S.G. Thomas (Ed.). John Wiley, London, pp. 577–605.

Thomas, D.S.G. and Shaw, P.A., 1991. 'Relict' desert dune systems: interpretations and problems. *Journal of Arid Environments*, **20**, 1–14.

Thomas, D.S.G., Stokes, S. and Shaw, P.A., 1997. Holocene aeolian activity in the southwestern Kalahari Desert, southern Africa: significance and relationships to late-Pleistocene dune-building events. *Holocene*, **7**, 273–281.

Thomas, J.V., Kar, A., Kalath, A.J. *et al.*, 1998. Late Pleistocene and Holocene history of aeolian accumulation in the Thar Desert, India. *1998 International Conference on Aeolian Research (ICAR-4) Abstracts*, **81**.

Tsoar, H. and Møller, J.T., 1986. The role of vegetation in the formation of linear sand dunes. In *Aeolian Geomorphology*, W.G. Nickling (Ed.). Allen & Unwin, Boston, pp. 75–95.

Tsoar, H. and Illenberger, W., 1998. Reevaluation of sand dunes' mobility indices. *Journal of Arid Land Studies*, **7S**, 265–268.

Valero-Garcés, B.L., Laird, K.R., Fritz, S.C. *et al.*, 1997. Holocene climate in the northern Great Plains inferred from sediment stratigraphy, stable isotopes, carbonate geochemistry, diatoms, and pollen at Moon Lake, North Dakota. *Quaternary Research*, **48**, 359–369.

Walker, T.R., 1979. Red color in dune sands. *US Geological Survey, Professional Paper*, **1052**, 61–81.

Warren, A., 1976. Morphology and sediments of the Nebraska Sand Hills in relation to Pleistocene winds and the development of aeolian bedforms. *Journal of Geology*, **84**, 685–700.

Warren, A., 1988. The dunes of the Wahiba Sands. *Journal of Oman Studies, Special Report*, **3**, 131–160.

Warren, A. and Allison, D., 1998. The palaeoenvironmental significance of dune size hierarchies. *Palaeogeography, Palaeoclimatology, Palaeoecology*, **137**, 289–303.

Wasson, R.J., 1984. Late Quaternary paleoenvironments in the desert dunefields of Australia.

In *Late Cainozoic Paleoclimates of the Southern Hemisphere*, J.C. Vogel (Ed.). Balkema, Rotterdam, pp. 419–432.

Wasson, R.J., Rajagaru, S.N., Misra, V.N. *et al.*, 1983. Geomorphology, late Quaternary stratigraphy and paleoclimatology of the Thar Desert. *Zeitschrift für Geomorphologie* Supplementband **45**, 117–152.

Wells, G.L., 1983. Late-glacial circulation over central North America revealed by aeolian features. In *Variations in the Global Water Budget*, F.A. Street-Perrott, M. Beran and R. Ratcliffe (Eds). Reidel, Dordrecht, pp. 317–330.

Wells, S.G., McFadden, L.D. and Dohrenwend, J.C., 1987. Influence of late Quaternary climatic changes on geomorphic and pedogenic processes on a desert piedmont, eastern Mojave Desert, California. *Quaternary Research*, **27**, 130–146.

Wells, S.G., Anderson, R.Y., McFadden, L.D. *et al.*, 1989. *Late Quaternary paleohydrology of the eastern Mojave River drainage, southern California: quantitative assessment of the late Quaternary hydrologic cycle in large arid watersheds*. New Mexico Water Resources Institute Report No. 242.

Wells, S.G., McFadden, L.D. and Shultz, J.D., 1990. Eolian landscape evolution and soil formation in the Chaco dune field, southern Colorado Plateau. *Geomorphology*, **3**, 517–546.

Wells, S.G., Brown, W.J., Enzel, Y. *et al.*, 1994. A brief summary of the late Quaternary history of pluvial Lake Mojave, eastern California. In *Geological Investigations of an Active Margin*, S.F. McGill and T.M. Ross (Eds). Geological Society of America Cordilleran Section Guidebook 94, pp. 182–188.

Wiggs, G.F.S, Thomas, D.S.G., Bullard, J.E. and Livingstone, I., 1995. Dune mobility and vegetation cover in the southwest Kalahari Desert. *Earth Surface Processes and Landforms*, **20**, 515–529.

Williams, M.A.J., 1975. Late Quaternary tropical aridity synchronous in both hemispheres? *Nature*, **253**, 617–618.

Williams, M.A.J., 1982. Quaternary environments in northern Africa. In *A Land Between Two Niles*, M.A.J. Williams and D.A. Adamson (Eds). Balkema, Rotterdam, pp. 43–63.

Wilson, I.G., 1972. Universal discontinuities in bedforms produced by the wind. *Journal of Sedimentary Petrology*, **42**, 667–669.

Wintle, A.G., Lancaster, N. and Edwards, S.R., 1994. Infrared stimulated luminescence (IRSL) dating of late-Holocene aeolian sands in the Mojave Desert, California, USA. *Holocene*, **4**, 74–78.

Zimbelman, J.R., Williams, S.H. and Tchakerian, V.P., 1995. Sand transport paths in the Mojave Desert, southwestern United States. In *Desert Aeolian Processes*, V.P. Tchakerian (Ed.). Chapman and Hall, London, pp. 101–129.

13 Luminescence Dating of Aeolian and Coastal Sand and Silt Deposits: Applications and Implications

ASHOK K. SINGHVI[1] and ANN G. WINTLE[2]
[1] *Earth Science and Solar System Division, Physical Research Laboratory, Ahmedabad, India*
[2] *Institute of Geography and Earth Sciences, University of Wales, Aberystwyth, UK*

This review aims to provide a flavour of the new results that luminescence dating is providing in respect of palaeoclimatic reconstruction. Current luminescence dating procedures are reviewed with emphasis on the recently developed optical dating methods. New ideas concerning the thermal stability of various luminescence signals and the effects of using laboratory preheat procedures to isolate geologically stable signals are discussed. A new approach for reworked (incompletely zeroed) loess is mentioned. The advantages of single aliquot measurements are presented and an example is given of its recent extension to the measurement of single grains of quartz.

Almost 500 dates (both TL and IRSL) have been selected from dated loess deposits around the world and used to construct a probability density plot. Such a plot shows the episodic nature of the loess record for the last 100 ka. A similar analysis is presented for OSL dates on quartz sands from the Kalahari Desert. Dates for other desert areas of Africa, and also India, are reviewed.

This paper also presents a comprehensive listing of dating studies on coastal sands from Australia, mainly dated using the TL signal from quartz. The fossil dune ridges in SE Australia have provided a testing ground for several luminescence signals. These dunes, associated with high sea-level stands during the past interglacials, have been dated using both the 375°C TL peak of quartz and the IRSL signal from the feldspar microliths within quartz. Some dates for isolated marine sand outcrops from the New South Wales coast have been shown to give ages that are unlikely given their elevation relative to present sea-level. The possibility that they are the result of a tsunami is discussed.

At the other end of the time-scale, IRSL dates for sand spit development in Canada and Ireland are discussed. The dates show that sand movement and subsequent stabilisation occurred within the last 1000 years.

INTRODUCTION

In order to compare the deep-sea oxygen isotope record of climate change with that found in the palaeoclimatic records on land, precise and accurate dating methods are required for terrestrial sediments. These would permit the testing of possible synchroneity of response in different parts of the world to a particular climatic

Aeolian Environments, Sediments and Landforms. Edited by A.S. Goudie, I. Livingstone and S. Stokes.
© 1999 John Wiley & Sons, Ltd.

perturbation. There is no universally applicable method for the dating of terrestrial sediments. Radiocarbon dating is limited in its time range (40–50 ka), by the need for calibration and by the necessity for preservation of appropriate organic materials; in addition, radiocarbon dating is often carried out on material from a different sedimentary unit. On a longer time-scale, and where a quasi-continuous sequence of aeolian sediment is available, palaeomagnetic measurements can be applied, as for the loess deposits in China. More recently cosmogenic isotopes, such as ^{26}Al and ^{10}Be, have been measured for quartz grains from desert sands (Nishizumi et al., 1993); the in situ production of these isotopes has the potential to provide information on sand movement.

In the context of these other methods, reviewed by Stokes (1997) for use in aeolian environments, luminescence dating offers several advantages, such as:

1. dating of the most recent depositional event;
2. use of constituent minerals for the dating analysis (a fact that eliminates any ambiguity in respect of sample-stratum correlation);
3. a continuous dating range of a few decades to a few hundred ka; and
4. the prospect of developing a self-consistent chronology by dating various stratigraphic units in a profile; these units can even represent diverse depositional environments.

The first systematic study used the thermoluminescence signal from deep-sea sediments (Wintle and Huntley, 1979). The development of optical dating during the last decade has enlarged the range of applicability and led to improvements in both precision and accuracy (Huntley et al., 1985; Aitken, 1998). Reports of luminescence dating applications can be found in Prescott and Robertson (1997), and Singhvi and Krbetschek (1996). The methodological details can be found in Aitken (1985; 1998) and Wintle (1997).

THE PHENOMENA

Thermoluminescence (TL) refers to the thermally stimulated emission of light from minerals. This light is emitted subsequent to exposure of the mineral to ionising radiation (α, β and γ). Irradiation causes ionisation in the material resulting in a production of free charges (electrons) in the crystal lattice. Most charges recombine instantaneously but a few charges are trapped at lattice defects. The binding energy of a charge at a defect site is dependent on the charge environment surrounding the defect and can result in mean residence times for the trapped charge (for storage at ambient temperature) which range from a few seconds to a few million years. The application of heat evicts the trapped charges, some of which radiatively recombine to produce thermoluminescence. A simple proportional relationship exists between the luminescence intensity and the amount of radiation exposure. This proportionality depends on the number of available trapping sites and the storage temperature; at high doses the traps become filled and saturation in the growth curve of luminescence vs dose is observed. The colour of the emitted light depends exclusively on the type of luminescence centres and in some cases it is possible to identify the lattice defect using the spectrum of the emitted light. For natural minerals such as quartz and feldspars, luminescence emission ranges from near visible ultraviolet (~300 nm) to red (~700 nm) (Krbetschek et al., 1997). The emission of luminescence occurs only once on heating. Multiple heating of a sample without irradiation does not yield luminescence, since the first heating evicts all the trapped charges.

It is also possible to evict the trapped charges optically, e.g. by illuminating the sample with daylight or selected-wavelength light. This results in Optically Stimulated Luminescence (OSL); more specific terms, such as Green Light Stimulated Luminescence (GLSL) and Infrared Stimulated Luminescence (IRSL), are generally used to indicate the colour of the excitation light (see Duller, 1996a) and are often grouped as optical dating techniques. TL and GLSL or IRSL may or may not originate from the same defects and thus may have different behaviour, such as sensitivity to daylight. A typical TL record is a plot of the intensity of luminescence vs the temperature when the material is heated at a known (generally linear) heating rate. This is called a glow curve and in general, the higher the temperature, the higher is the stimulation energy and this implies a higher stability at ambient temperature. In optical dating, the luminescence intensity is plotted as a function of the illumination time to give an OSL decay curve. Unlike the TL glow curve, where the temperature of the light emission denotes its stability, the OSL decay curve does not provide any information on the thermal stability of the trapped charge that gives rise to the OSL signal. In general, additional experiments are made to establish the stability of the OSL signal. On the other hand, lack of heating during the OSL measurement offers several advantages, such as:

(i)   possible applicability to materials that undergo a phase change on heating;
(ii) possible probing of deep traps that cannot be readily accessed due to inter-ference from incandescence, which is unavoidable at higher temperatures.

However, the luminescence detection range needs to be narrower in OSL measurements, so as to ensure that the stimulation light is not observed along with the OSL signal. A further advantage is that, unlike thermal stimulation, optical stimulation can be made using a sequence of short light exposures, a property that allows for multiple read-out from the same sample aliquot. As discussed below, this advantage has been exploited in the development of single aliquot methods (Duller, 1994), which have even been applied to single grains (Murray and Roberts, 1997).

THE DATING METHOD

Luminescence dating requires: (i) natural minerals that are capable of registering radiation doses; (ii) a weak but constant environmental radiation flux; (iii) signals that have high thermal stability; and (iv) a zeroing mechanism, e.g. heating to 400°C, or sufficient daylight exposure to evict trapped charge, and thus reduce the luminescence signal to zero or a near-zero residual value. Luminescence dating is applicable to: (i) heated objects ranging from burnt pottery to sediments contact baked from lava flows; and (ii) to daylight bleached sediments where the daylight exposure results in optical bleaching of the trapped charges. Both these types of event ensure that at the time of burial the samples have zero or close to zero trapped charged population. Dating can also be used for samples that were formed *in situ*, such as calcite, which would have zero trapped charge at the time of crystallisation. Once buried, the minerals are exposed to radiation from the decay of naturally-occurring radioactive elements, *viz.* ^{238}U, ^{232}Th, ^{40}K, and also cosmic rays. Consequently, reacquisition of trapped charge is initiated and this continues unabated until excavation. The long mean lives of these radionuclides imply that the radiation flux is almost constant over time-scales of millions of years, provided that no disequilibrium is present in the ^{238}U decay chain. The age of burial can be estimated using a simple age equation

age = luminescence acquired since burial / annual rate of luminescence acquisition

In practice the equation is more complex, since the radiation flux comprises alpha particles, beta particles, gamma rays and cosmic rays. Each of these have a different path length that varies from a few μm to a few tens of cm, and induce luminescence with different efficiency. A more general version of the age equation is

$$age = P/(aD_\alpha + D_\beta + D_\gamma + D_c)$$

where *a* is an alpha efficiency factor which depends on the sample and defect centre population. Values of *D* are component dose rates arising from each of the radiation types. *P*, the palaeodose, is the equivalent laboratory beta dose that can induce an intensity of luminescence identical to that in the natural sample (also called the equivalent dose, ED or D_e). Thus for a date to be obtained, two types of measurement are used, *viz.* a luminescence measurement to estimate *P* and a radioactivity assay to obtain the annual dose appropriate for the grain size used for the estimation of *P*.

Sample homogeneity, or knowledge of the lack of it, is crucial for the estimation of the environmental dose rate. The gamma ray range of 30 cm implies that homogeneity within 30 cm of the samples is desirable. In extremely inhomogeneous strata with different facies, field gamma measurement is recommended, using either a portable gamma ray spectrometer or using a TL dosimeter such as Al_2O_3:C or CaF_2:nat or $CaSO_4$:Dy that can be buried on site for a known period. The dose acquired can provide an estimate of the annual environmental gamma dose averaged over the spatial inhomogeneity in the strata. Al_2O_3 can provide a reasonable dose estimate for a typical burial time of few days (Bøtter-Jensen, 1999a).

Factors in Luminescence Dating

Grain Size

Since different types of radiation have different ranges, it is important to recognise the contribution of each radiation type. The range of alpha particles is a few tens of μm; this implies that, while fine silt particles receive full exposure to alpha irradiation, only the outer layer of a sand-sized grain is irradiated by alpha particles. This has led to dating being applied to two different grain sizes, *viz.* the fine grain technique using the 4–11 μm size fraction and the coarse grain technique employing mineral grains of a few hundred μm. In the latter the outer layer is etched away; thus it is reasonable to assume that the remainder of the grain has received only the dose coming from the beta, gamma and cosmic radiation. The choice of technique is dictated by the mean grain size of the sediment.

Palaeodose Estimation

In general the estimation of the palaeodose is non-trivial on account of several complicating factors, such as non-linearity of the luminescence *vs* dose growth curve both at low and at high doses (supralinearity and saturation), estimation of alpha efficiency in the case of fine grain samples, and changes in sensitivity of the sample during laboratory measurement. In the *additive dose method* various identical portions of the sample are used and given varying amounts of laboratory doses. A plot of luminescence intensity *vs* laboratory dose gives the growth curve which is extrapolated to zero, or the initial residual luminescence intensity, to obtain *P*. Such a procedure implicitly assumes that the nature of the luminescence *vs* dose growth curve is known. This works well for samples below the saturation region, but considerable error can occur for samples in or near to the saturation region. For such samples an interpolation approach, *the regeneration method* is often used. In this case the luminescence is removed appropriately by either thermal or optical stimulation and a luminescence *vs* dose growth curve is constructed by irradiation of identical portions. The natural intensity of luminescence is matched to this curve and the palaeodose is obtained, with the proviso that the optical bleaching/ stimulation/thermal annealing did not alter the luminescence sensitivity. This needs additional experiments on the pre- and post-annealing/bleaching treatments. Occasionally a combined approach is employed; termed the *Australian slide technique*, this approach uses a combination of the additive and the regenerated growth curves in which the curves are made to slide over each other along the dose axis until an optimum overlap is reached. At each slide interval the sensitivity can be varied to optimise the overlap and thus allow for sensitivity changes (Prescott *et al.*, 1993). The amount of horizontal movement to provide the best match gives the palaeodose *P*. For reasonable precision, it may be often necessary to measure up to 50 aliquots of the sample.

Annual Dose Estimation

Estimation of annual dose implies measurement of the individual concentrations of radioelements, ascertaining the status of their radioactive equilibrium and their alpha efficiency. Techniques such as ZnS(Ag) alpha counting, alpha spectrometry, gamma spectrometry, beta counting etc. are used depending on the material to be dated. An important aspect is the average moisture content of the sample. Water is devoid of radioactivity but attenuates the radiation flux effectively. Thus for given radioelement concentrations, the net radiation flux that is delivered to the sample is less for a wet sample as compared with a dry sample. The fractional change in dose rate (and hence the age) nearly equals the fractional change in water content of the sediment. Another factor that merits attention is the internal beta dose for mineral grains which have internal radioactivity, e.g. K-feldspar which can have up to 14 per cent potassium in the crystal lattice. The cosmic ray flux will change with latitude, altitude and depth. While the latitude and altitude variations can be accounted for, the depth dependence can not be calculated in the case of an accreting system with periods of sedimentation and quiescence. Since the depth itself is a time-varying function and time is the quantity being measured, an average depth is used for the computation.

Initial Luminescence Level: The Zeroing

In any dating application it is crucial to establish either that the pre-existing signal was zero or that it was reset to zero at the time of deposition. For samples heated to up to 400°C or more, the thermal zeroing is complete and the initial luminescence level (TL or OSL) is zero. Laboratory experiments have indicated that the optical bleaching of mineral luminescence occurs rapidly to start with and progressively slows down as the bleaching progresses such that the signal asymptotically approaches a low but finite residual value. Bleaching of TL occurs at a significantly slower rate when compared with OSL. There is also a dependence of the bleaching rate on the exact spectrum of the light and this also controls the level of the residual. In the case of water-laid sediments, the incoming daylight signal is modified by the water column and the sediment load; turbulence affects the net illumination flux reaching the mineral grains. For these materials procedures such as the 'spectrum-restrained partial bleach' methods (Berger et al., 1990) and more recently the differential partial bleach methods (Singhvi and Lang, 1998) have been used. These methods use an estimation of palaeodoses under different illumination conditions with the premise that, if the laboratory illumination is tuned such that if only those traps that were emptied during the depositional event are sampled, a realistic estimate of palaeodose can be expected. The differential partial bleach adopts the approach of isolating the most readily bleachable signal by analysing signal intensity from adjacent intervals in an OSL decay curve and analysing them as a partial bleach data set.

Stability of Signal

It is important to establish whether the luminescence signal used for the dating analysis has the necessary thermal stability over the dating range. For TL, the thermal stability is indicated by the temperature of the TL emission. However, when observing ultraviolet emission from fine grains from loess deposits, an under-estimation of ages occurred, such that an upper limit of 100–150 ka was placed on TL ages obtainable from such samples using the regeneration method (Debenham, 1985). In IRSL dating of loess, a similar upper limit was seen (Musson and Wintle, 1994). Other studies have indicated that ultraviolet emission in K-feldspars have a lower stability compared with the blue emission. Recent studies suggest that the underestimation arises due to a luminescence centre being produced as a consequence of daylight exposure and laboratory irradiation but with a medium term mean life (30–40 ka). This implies that standard trap kinetics analysis would still yield an activation energy indicating high stability, but the loss of medium term luminescence centres shows up in the regeneration analysis as an increased sensitivity for samples irradiated in the laboratory and recorded immediately (Rao, 1997). Another cause of age underestimation is athermal, or anomalous, fading; this was found for the TL of volcanic feldspar, even for the high temperature signal for which sufficient stability was expected. Preheating procedures have been suggested in order to reduce the effect of anomalous fading, to isolate traps with adequate thermal stability, to redistribute charge and to equalise sensitivity change. Wintle (1997) has discussed these procedures for TL and OSL of both quartz and feldspar. A recent study by Wintle and Murray (1999) has suggested that for the OSL of quartz, the largest effect of preheating is to alter the luminescence efficiency.

Methodological Aspects

Luminescence measurement systems are based on a chamber that houses the sample, the optical excitation source/heater plate and a luminescence detection unit. A survey of the existing instrumentation and technology is provided by Bøtter-Jensen (1997). Optical detection is usually made with a bi-alkali photo-multiplier tube (EMI 9635 QA) with a quartz window (to extend the optical response in the ultraviolet) linked to a standard photon counting system coupled to a multi-channel counting system and a PC. A sequential record of the photon count with elapsed time is then available. The time can be related to the decay time for OSL or to temperature in TL measurement (via the heating rate controller). The excitation source for IRSL is usually a set of diodes emitting at ∼880 nm where the stimulation of IRSL emission is most efficient, although some studies at 940 nm have also been made. GLSL excitation may be performed with monochromatic light at 514 nm from an Ar–ion laser. Other alternatives are a Kr–ion laser, which additionally provides the possibility of using other monochromatic wavelengths up to the red end of the spectrum. More often a halogen lamp, or Xenon lamp, assembly is used with a series of optical filters to constrain the illumination spectrum. Care is needed to ensure that the illumination is uniform over the entire sample, and the illumination flux remains constant through a measurement cycle. More recently bright blue light emitting diodes have been used for stimulation of quartz OSL (Bøtter-Jensen et al., 1999b).

Spectral information for the TL/OSL emission bands and stability of the respective signals is important for optimising the dating analysis, though it must be borne in mind that considerable spectral variations exist in spectra of same mineral but from different provenance (Krbetschek et al., 1997). Thus a universally applicable detection system cannot be constructed. Irradiation is usually carried out using a $^{90}Sr/^{90}Y$ beta source, and ^{241}Am alpha sources in a vacuum irradiator assembly.

Preparation of the samples includes a cleaning of the sample with HCl and H_2O_2 in subdued light (the colour being dependent on the material being processed). The fine grains are subsequently dispersed using a deflocculating agent, such as sodium oxalate or Calgon, followed by a separation of fine silt sized grains using Stokes'

Law. Coarse grains are sieved and density-separated using heavy liquids, such as sodium polytungstate.

Two types of analysis protocols *viz. multiple-aliquot* and *single-aliquot* analysis are used. In the multiple aliquot analysis, several identical portions of the sample are used; their luminescence output is normalised using either the sample weight as the normalising factor or the OSL from a short illumination (the *natural normalisation* or the *'short shine' normalisation*, respectively). The palaeodose estimate can be obtained from additive dose growth curves, constructing growth curves by recording the luminescence output of samples as received (N) and samples additionally irradiated in the laboratory (N+betas). A correct extrapolation of the growth curves provides the palaeodose. For well-bleached sediments, additional experiments are carried out to provide the initial residual luminescence that is subtracted; for OSL and IRSL, this is close to zero. Extrapolation for many samples is non-trivial and for these a regeneration growth curve is made by optically bleaching the samples and then irradiating them in the laboratory to construct a growth curve on which the natural TL (or OSL) intensity is superimposed to provide the palaeodose. This dose is acceptable provided that additional measurements are made to establish that the laboratory bleaching did not alter the luminescence sensitivity. For multiple aliquot procedures, appropriate curve fitting is vital and several good software packages are available commercially. The nature of the growth curve and the fitting procedure used must be examined to permit an estimation of the reliability of the palaeodose (Grün and Brumby, 1994; Felix and Singhvi, 1997).

As mentioned earlier, it is also possible to stimulate optically for a very short time, e.g. 0.1 s, and obtain an OSL (or IRSL) signal that is representative of the trapped charge population which could be emptied by stimulation over a longer period of time, e.g. 100 s. It is thus possible to construct an OSL growth curve for a *single aliquot* of sample containing around 1000 grains. This can be achieved either by adding dose, or by bleaching the natural sample and regenerating the growth curve, procedures known as the *single-aliquot additive-dose* and *single-aliquot regeneration* methods, respectively (Duller, 1994). Such procedures require heating after the laboratory irradiations, as well as before measurement of the natural OSL. A preheat is needed to ensure that, after laboratory irradiation, charge is redistributed as it would have been in nature when irradiation occurred much more slowly and at ambient temperature. An even more important requirement is that either the preheat results in the same luminescence efficiency for both natural and laboratory irradiated samples (Murray et al., 1997) or any luminescence efficiency changes can be monitored (Wintle and Murray, 1999). For feldspar IRSL age determination, the single-aliquot additive-dose procedure has been widely adopted; it usually involves 24 discs, with six being used to quantify loss of stable charge on each preheat cycle and eighteen discs being used to give 18 independent values of the palaeodose (Duller, 1994; Wintle et al., 1998).

Following the initial development of a single-aliquot additive-dose procedure for quartz (Murray et al., 1997), a single-aliquot regeneration procedure has been developed (Murray and Roberts, 1998) and, with a few modifications, this procedure is likely to be widely adopted. The same single-aliquot procedures can be applied to single grains (Murray and Roberts, 1997). This new approach has helped to solve a problem of age overestimation found when TL was used to date quartz grains from an excavated rock shelter at Jinmium in northern Australia (Fullagar *et al.*, 1996). These sediments form part of a sand apron that has been formed by raindrop-impacted sheet flow; however, close to the rock face the sand contained crumbly fragments of bedrock. When single grain OSL measurements were made on one of the sand units, a considerable scatter in equivalent dose was obtained, indicating inadequate exposure to light before burial, an interpretation supported by a re-analysis of the TL data (Spooner, 1998). Taking those grains with lower equivalent doses, a younger age was obtained (Roberts *et al.*, 1998a). A spread in values of equivalent dose has also been obtained for aeolian deposits from Allen's Cave in southern Australia (Murray and Roberts, 1998); this was attributed to inhomogeneity of the beta dose rate, as the sediment was cemented and contained clay (Olley *et al.*, 1997). At two other archaeological sites in northern Australia, Roberts *et al.* (1998b) also found scatter in the single grain analyses for un-cemented and well-bleached aeolian sands. The precise cause of this scatter is unknown, but the ages obtained using the mean of the palaeodoses gave ages in agreement with those from TL and multiple grain, single aliquot OSL measurements. Clearly we can gain much information from such measurements, and new equipment has been developed that permits the OSL signal to be obtained for a number single grains mounted on a standard sized disc (Duller *et al.*, 1999).

LOESS

A considerable number of luminescence dating studies have been carried out on the loess/palaeosol sequences which are considered to be the longest continental archives of climatic change (see also Chapter 10) and more recent studies have indicated their potential in reconstructing past atmospheric circulation patterns (Biscaye *et al.*, 1997). The majority of the effort has been expended on the dating of loess/palaeosol sequences in China and Europe. More recently there have been studies on loess from central Asia (Zhou *et al.*, 1995) and the USA (Berger and Busacca, 1995, Richardson *et al.*, 1997). A review of loess dating prior to 1990 was given by Wintle (1990).

Initial optimism of the applicability of luminescence dating to loess was tempered by the observation of an underestimation of ages obtained using regeneration of the luminescence signals. A mean life of about 150 ka was calculated for TL and IRSL ages (Debenham, 1985; Musson and Wintle, 1994). Although efforts were made to overcome the age underestimation using a 'strong thermal wash' (Zöller and Wagner, 1989), and different detection windows (blue instead of UV) have been tried, no satisfactory physical explanation has been put forward. Hence, no universal methodology has been developed to obtain reliable ages beyond these limits, although in some circumstances older ages in agreement with the proposed stratigraphy have been obtained when additive dose procedures are used (Berger, 1994). It is now clear that the regeneration ages are underestimated on account of a sensitivity change. In a recent study (Rao, 1997; Singhvi and Rao, in preparation), the age underestimation of the luminescence signal is suggested to be due to the decay of sensitivity with time. Rao (1997) proposed a model that enables the deduction of a sample-dependent correction for the decay. Though the method has yielded plausible ages for samples from China and Tadjikistan, more data are needed before a general recommendation for the correction of the ages is made.

An interesting case study relates to a loess/palaeosol profile at Jiuzhoutai in the central loess plateau in China (Fang et al., 1997). This is perhaps one of the most detailed sequences with $\sim$40 m of loess between palaeosols S_0 and S_1. Within the loess profile, evidence of the Blake event is reported and the TL dates cited by Fang et al. (1997) are in agreement with this age assignment (111–118 ka), in contrast to the age underestimation observed in most other profiles. Samples from Jiuzhoutai were recently examined by Singhvi et al., (pers. comm.) and an age underestimation effect was found for both IRSL and TL dates. It is difficult to find the cause of this discrepancy since basic data on the TL dates were not provided in the original publication. Singhvi et al. (pers. comm.) found the growth curve of the loess became non-linear at absorbed doses in excess of 100 Gy and systematic overestimation of the palaeodose can occur if sufficient rigour in the curve fitting procedures is not adopted. Singhvi et al. (pers. comm.) used the evidence of the Blake event to calculate the mean life of the TL and IRSL signals which agreed with previously suggested values of 150 and 70 ka, respectively (Xie and Aitken, 1992).

It is essential that a TL or OSL date is provided with maximum possible details on the methodology so that the pedigree and the applicability of the date is ascertained. Equally important is the location of the sample in the loess/palaeosol sequence. A-horizons experience bioturbation and consequently in situ solar resetting of the luminescence clock occurs. Solar exposure ceases with the removal of the sediment from the sediment–atmosphere interface and its subsequent burial by fresh sediment. The luminescence age on a sample from the A-horizon thus reflects the age of soil formation. On the other hand, the B- and C-horizons do not experience any pedoturbation and thus samples from these horizons provide the age of deposition of the parent loess. Thus luminescence ages on palaeosol horizons without a micromorphological identification of the soil horizons are likely to be meaningless.

In spite of these limitations, a good example of the use of luminescence dating is provided in the Carpathian basin. The loess/palaeosol sequences along the River Danube in Central Europe are considered to be the longest continental archives of climate change, with the exception of those in China. Based on radiocarbon dates on material from palaeosols and the soil micromorphology, it was suggested by Pécsi (1985) that the Mende Base (MB) soil was formed during the last interglacial. On the other hand, using a comparison with deep-sea oxygen isotope data and the presence of the Brunhes–Matuyama boundary, a substantially older age of $\sim$220 ka was assigned to the overlying F_2 soil (Bronger's notation, Bronger, 1976) and 450 ka to the MB soil (Kukla, 1977). Luminescence dating of these sequences by different laboratories provided consistent results (Figure 13.1) and demonstrated that the Mende Base soil does not represent the last interglacial soil. Instead it can be concluded that the soil MF_2 (Pécsi's notation), the upper part of Bronger's F_2 soil, represents the last interglacial soil. The earlier chronology based on radiocarbon dating provided ages that were far too young.

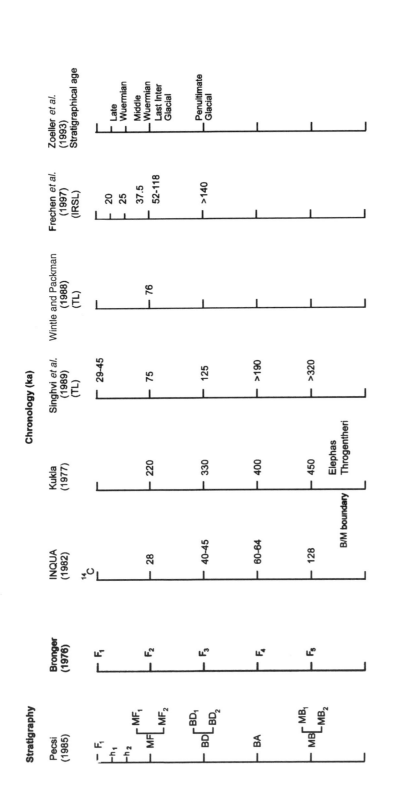

Figure 13.1 A synopsis of the stratigraphy and chronology of loess-palaeosol sequences along the River Danube in Central Europe. The earliest interpretation (Pécsi) was that the Mende Base (MB) soil was formed during the last interglacial. The results of the luminescence studies of Wintle and Packman (1988), Singhvi et al. (1989), Zöller et al. (1994) Frechen et al. (1997) all suggest that MF_2 represents the last interglacial, and that the MB soil is from a much earlier interglacial

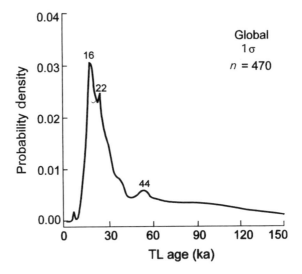

Figure 13.2 A global synthesis of ~470 luminescence ages from loess/palaeosol sequences using the procedure developed by Venkatesan and Ramesh (1993). In this analysis a Gaussian curve for each age was constructed such that the age itself defined the peak and the full width at half maximum was given by the experimental error (1σ). Each of the Gaussian curves was then plotted on a time axis to generate the probability curve. The TL dates were from published literature and are tabulated in Rao (1996). These include about 370 dates from Europe, 40 from China, 35 from the USA, 20 from New Zealand, and a few others. Only methodologically sound dates were used and, in the case of multiply dated horizons, a single mean age was used. Also, only ages on loess and soil B and C-horizons were used. More details on this analysis procedure are being published elsewhere. It should be noted that, whilst the clustering of the ages at particular times is significant, the height of a peak reflects the number of analyses and has no climatic connotation

More recent dating work on closely spaced samples at the Paks and Mende profiles also indicates that, notwithstanding age underestimation, the loess record has considerable time breaks, indicating long periods with no loess accumulation (Frechen *et al.*, 1997). Recent work at Jiuzhoutai also shows loess accumulation as episodic with long periods of hiatus in between (Singhvi *et al.*, pers. comm.). Both sets of data are in agreement with other suggestions for the episodic nature of loess accumulation (Rao, 1997). In a global synthesis of luminescence ages using a probability density approach, Singhvi *et al.* (1998) demonstrated that the age distribution had a peak at about 17 ka with a width of about 5 ka (Figure 13.2). Analysis of this curve and of regional sub-sets suggests that the most recent phase of loess accumulation peaked after the glacial maximum and was of limited duration (Rao, 1997; Singhui *et al.*, 1998).

This observation from global synthesis, and determinations from closely-sampled and dated individual sites in Europe and China, coupled with the recognition of large hiatuses in loess records, calls into question the use of other methods based on the constancy of dust particle and magnetic particle fluxes, and their tuning with the oxygen isotope records.

DESERT SANDS

Deserts comprise about 30 per cent of the land area between the horse latitudes and provide unique records of the interplay between geology and climate change with minimal biospheric interference. Expansion and contraction of deserts, in response to climatic change, control the surface albedo changes on land; dating of these changes are important in climate reconstruction and help provide realistic boundary conditions for global circulation models. However until the demonstration of the absolute dating of deserts sands using luminescence (Singhvi *et al.*, 1982) no method was available to date aeolian deposits. In view of the long distance transport of sand grains by saltation, it is reasonable to assume that desert sand grains are totally bleached and consequently they provide ideal material for dating. The past few years have seen a considerable amount of effort expended on the dating of desert sands; recent reviews of the literature were published by Wintle (1993) and Stokes (1997).

The climatic issues that have been addressed include:

(i)   The antiquity of aeolian dynamism in the deserts (onset of aridity).
(ii)   The climate favouring the sand accumulation and its phase relationship with the global climatic cycles (especially since a direct correlation of glacial aridity and desert expansion has been suggested by Sarnthein (1978)).
(iii) The timing of aeolian accumulation phases, the dune mobility rates and factors controlling them, and past periods of dune stability.

Stokes *et al.* (1997) obtained OSL dates on quartz grains from a large number of cores across the Kalahari, giving a regional perspective. They concluded that dune accumulation has been episodic, with the peak occurring after the period of maximum glacial aridity. Figure 13.3 summarises the dates in a histogram constructed by generating a Gaussian curve for each independent age, such that the peak in the maximum of the Gaussian occurs at the age itself and the experimental error on the age is the full width at half maximum. All Gaussians, each representing a single OSL date, were then superimposed on a time axis to obtain a probability curve (Figure 13.3). The data indicate several peaks in the sand accumulation record, occurring at 115–95 ka, 46–41 ka, 26–20 ka and 16–9 ka. A comparison of the data from the north-eastern Kalahari and the total data set for the Kalahari as a whole, is also presented (Figure 13.3(b)); this indicates increased sand accumulation during the late Holocene in the south-western Kalahari. A mention must be made of the radioactivity assay for these sands from the Kalahari. They are quartz rich and, as such, have extremely low levels of radioactivity; up to 30 per cent of the total dose is provided by cosmic rays. The values of U, Th and per cent K_2O are near the detection limits of the measurement method (neutron activation analysis) and this increases the error term for the annual dose rate. The question surrounding the distribution of radioelements within the sands is an important one. If the radioelements were concentrated in only a few grains, then, clearly, some quartz

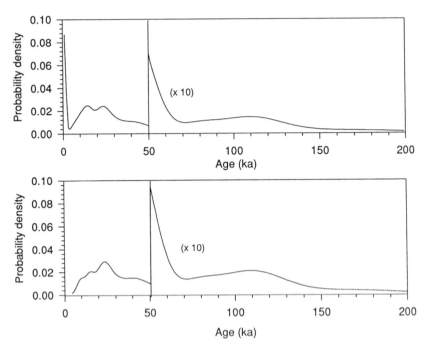

Figure 13.3 Distribution of luminescence ages for aeolian sands in the Kalahari (from Stokes *et al.*, 1997): (a) total data set; (b) data excluding SW Kalahari

grains would have received more dose during burial than others. This situation would be revealed by making single grain palaeodose determinations.

Other studies in Africa include the dating of aeolian sediments from linear dunes in the Lantewa dune field in Nigeria; from their data Stokes and Horrocks (1998) infer that dune-building peaked during the period 18–13 ka. This post-dates the period of maximum aridity, but lends support to other records of aridity in the region. A few studies in the Sahara (e.g. Stokes *et al.*, 1998) are also available, but are insufficient to attempt climatic reconstruction.

Luminescence dating of aeolian sands in the Thar Desert has highlighted several important points. Firstly, the TL ages of > 150 ka at different sites in the Thar settled a long-standing controversy as to whether the onset of desertification was related to human activities. Secondly, the Holocene record suggests multiple episodes of dune accumulation that correlate well with the records of changes in the hydrology of the lakes in the region, preceding them by a few centuries. Studies also indicate that a peak in aeolian accumulation occurred significantly after the glacial epoch and was of a limited duration, coinciding with the timing of the re-establishment of the south-west monsoon (Thomas *et al.*, 1999). In addition, the rate of dune migration was obtained using a section through a 7 m high transverse dune (Kar *et al.*, 1998). Ages were obtained from the top and bottom of each of the sand units identified in the field, using the IRSL signal from sand-sized potassium-rich feldspars and a multiple aliquot approach. The radioactivity content of these dunes is approximately a factor of 10 higher than those reported for the Kalahari (Stokes *et al.*, 1997). Sand deposition occurred around 6000 years ago, 2000 years ago, between 2000 and 600 years ago and, most recently, from 600 years ago. Kar *et al.* (1998) used the dates, and the visible dune crest positions, to calculate dune mobility. They concluded that during the recent phase of sand movement, the dune crest had been moving at about 3 cm per annum, at least 10 times as fast as during the earlier periods.

The luminescence dating of desert sands has provided a robust application of the method, since the basic premise of complete bleaching is satisfied. Comparisons with other quartz sands suggest that samples with a multiple history of bleaching prior to deposition are better suited to OSL dating. A related application has been to pedogenic carbonates found within a dune matrix (Singhvi *et al.*, 1996). In this case, the dose–rate experienced by the quartz grains changes as a result of the deposition of the calcium carbonate. The difference in the palaeodose of quartz or feldspar taken from the part of the dune that has experienced precipitation of carbonate and that which has not, can be used as a measure of the time that has passed since the carbonate was deposited

COASTAL SANDS

The first TL dates for coastal dunes were obtained by Singhvi *et al.* (1986) for red sands found on the coast of Sri Lanka. Since then coastal dunes have provided excellent sites for exploring both the upper and lower limits of luminescence dating techniques. The best testing ground for the upper dating limit has been the series of fossil dune ridges that run parallel to the coast in the south-east of South Australia, south of Adelaide. The ridges represent times of high sea level from previous interglacial periods, and slight tectonic uplift has resulted in them being left in the landscape with a spacing of about 10 km. The Brunhes–Matuyama transition has been located between two ridges near Naracoorte (Figure 13.4), about 70 km inland (Idnurm and Cook, 1980). These dunes are constructed of carbonate-cemented quartz grains and it is the 375°C TL peak in the 90–125 μm quartz grains that is used as the chronological tool. Initial TL studies by Huntley *et al.* (1985) showed that the absolute signal intensities increased with the inferred age of the dune ridge. Although the 375°C TL peak is not bleached as fast as the 325°C TL peak, it shows reproducible growth characteristics and is amenable to the 'Australian Slide' method of analysis. Figures 13.5(a) and (c) show the growth response of the TL at 360°C for two dune samples (SESA-61 and SESA-72) giving dates of 118 ± 4 ka and 196 ± 12 ka (Huntley *et al.*, 1994). In each case, the additive dose data have been shifted along the dose axis until they coincide with the regenerated data. For these samples no sensitivity changes are seen between the additive dose and regenerated data sets, and the fitted curve is taken as the sum of an exponential and a linear term. Using this approach, good agreement was obtained with ages based on assignation of the ridges to particular oxygen isotope stages back to about 500 ka (Huntley *et al.*, 1993a, 1994); the full comparison data set was presented by Prescott and Robertson (1997). The confidence given by these studies enabled Murray-

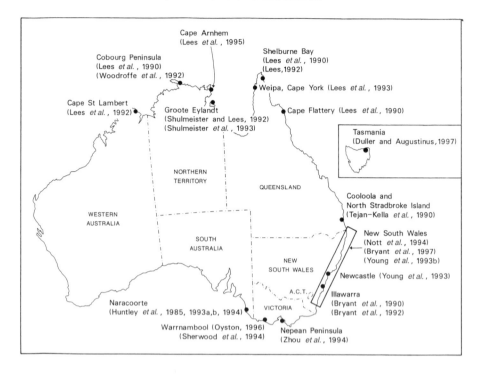

Figure 13.4   Sites of coastal dunes in Australia (including Tasmania inset) for which luminescence ages have been obtained

Wallace *et al.* (1997) to obtain ages for three dune ridges further south and helped them to constrain the extent of Quaternary uplift in the area.

These quartz-rich carbonate-cemented dunes are also likely to provide the best environment in which to test the upper limit, since the dose rate is remarkably low and thus saturation is approached at a slower rate. Values of U and Th are both about 1 ppm on the basis of in situ gamma spectrometry, neutron activation and thick-source alpha counting (Huntley *et al.*, 1994). Potassium is also low (<0.15% K on the basis of *in situ* gamma spectrometry, atomic absorption analyses and X-ray fluorescence spectrometry), since there are no feldspars and little clay. This leads to time-averaged dose rates of about 0.5 Gy ka^{-1} or less; in this situation cosmic rays will contribute significantly ($\sim$0.14 Gy ka^{-1}) to the environmental dose rate. However, it should be noted that incorrect ages will be calculated if the dose rate has changed through time, as the result of carbonate dissolution and reprecipitation. In this case, the present day measurement of the dose rate will be too low compared with the average dose rate in the past, depending on when the carbonate was precipitated.

Two other luminescence measurements have been made on samples from the dune ridges. OSL measurements reported by Huntley *et al.* (1996) showed that the natural signal for quartz from the Woakwine ridge, linked to the last interglacial (oxygen isotope stage 5e), was within 50 per cent of the saturation level as deter-

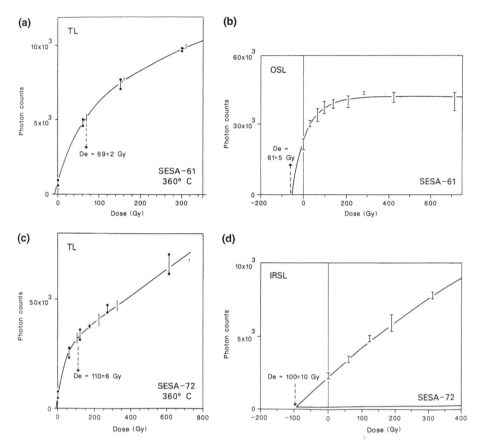

Figure 13.5 Growth curves for luminescence signals for quartz from two beach ridges in SE South Australia: (a) and (c) 360°C TL growth curves for added (x) and regenerated (•) doses, shifted using the 'Australian slide' approach for determination of the equivalent dose D_e, for two samples (SESA-61 and SESA-72) of different age (redrawn from Huntley *et al.*, 1994); (b) OSL growth curve for SESA-61 (118±4 ka) showing saturation (redrawn from Huntley *et al.*, 1996); (d) IRSL growth curve for SESA-72 (196112 ka) (redrawn from Huntley *et al.*, 1993b)

mined by the response to laboratory irradiation (Figure 13.5(b)). This apparently early saturation of the initial part of the OSL signal is likely to prevent its application to dune ridge samples significantly older than 120 ka. However, the data suggested that thermal stability was not a problem. The data in Figure 13.5(b) can be compared with the 360°C TL data for the same sample (Figure 13.5(a)).

The other signal was that obtained on infrared stimulation of the quartz grains. This IRSL signal was thought to be derived from small (~10 μm) feldspar inclusions within the quartz grains (Huntley *et al.*, 1993b). The growth curve for the additive dose data for sample SESA-72 is shown in Figure 13.5(d). The IRSL response to added dose (up to 300 Gy) can be compared with the 360°C response from the same quartz grains (Figure 13.5(c)). IRSL measurements on younger sands did not show anything other than a linear response for doses from 0 to 100 Gy, thus permitting the extrapolation used to obtain the equivalent dose. The IRSL growth can also be compared with the quartz OSL growth (Figure 13.5(b)) for the younger sample; it appears that the IRSL signal would provide an optical dating signal which could be used to extend the age range to beyond 100 ka. Indeed, the IRSL ages agree with the TL ages back to 200 ka, but for older dunes they show underestimation compared with the quartz TL ages (Huntley et al., 1993b; Prescott and Robertson, 1997). The agreement with the TL results below 200 ka suggests that anomalous fading is not likely to be the cause of this underestimation. How-ever, thermal fading, with a lifetime of about 1 million years for the temperature range experienced at that location, would explain the observed discrepancy. The results of these studies should be borne in mind when dating beyond 120 ka is contemplated.

TL dating using the 375°C TL peak of quartz has been used in almost all luminescence dating studies on dunes from the Australian coast (Figure 13.4). Besides the studies mentioned above, the Adelaide laboratory has obtained four TL dates for coastal dune sands at Cooloola and North Stradbroke Island, just north of the New South Wales border with Queensland (Tejan-Kella et al., 1990). Two similar studies by the La Trobe University laboratory were carried out near Warrnambool in Victoria; two calcarenites were dated at Point Ritchie, as part of a multiple dating method study (Sherwood et al., 1994) and Oyston (1996) obtained TL dates which linked the dunes to times of high sea level, e.g. 500 ka, 330 ka, 240 ka, 200 ka, as well as the last interglacial and the Holocene.

The Wollongong University luminescence laboratory has dated sands from Victoria and along the New South Wales coast. Zhou et al. (1994) reported 11 TL dates for coastal aeolianites on the Nepean Peninsula in Victoria. For the New South Wales coast, Bryant et al. (1997) reported the dating of isolated remnants of marine sands with about 20 samples from barrier deposits dating to about 125 ka, the last interglacial (5e), or to ~96 ka (5a), and another four marine sands dating to around 240 ka (stage 7). In addition, two more recent phases of barrier accretion from 50 to 80 ka and at about 32 ka were identified, suggesting times when local sea level was close to the present level. These sands, found at about 2–3 m above present sea level, present a conundrum for a region that is considered to have been tectonically stable over the late Pleistocene. Tsunami deposition was identified as the cause of a 25 ka age for sand from four other sites where sand layers were also found at about 2 m above present sea level.

Dating of recent tsunami deposits on the Pacific coast of North America has been reported by Huntley and Clague (1996); comparison with radiocarbon dating suggested that all the grains were well bleached whilst on the inter-tidal flat. However, it is possible that the sands from New South Wales, dated to 55 and 32 ka, represent average TL ages provided by the measurement of a mixture of grains, some of which have been exposed to light on the surface when the sea level was about 100 m lower around 25 ka, and others picked up from older underlying sands. Luminescence studies of mixtures of grains have been reported by Lamothe et al. (1994) who obtained IRSL signals from single feldspar grains derived from glaciomarine deposits and for quartz from archaeological sites in Australia, as discussed earlier.

Fortunately, aeolian dune sands are less likely to present any similar problem of grain admixture. About 50 TL dates have been obtained for aeolian sands along the New South Wales coast (Bryant *et al.*, 1990, 1992, 1994; Nott *et al.*, 1994; Nott and Price, 1991; Young *et al.*, 1993a,b). Extensive TL studies have also been reported for coastal dune fields along the north coast of Australia, from Cape St. Lambert in northern Western Australia (Lees *et al.*, 1992), Cape Arnhem (Lees *et al.*, 1995) and the Cobourg Peninsula in the Northern Territory (Woodroffe *et al.*, 1992) to Cape York in Queensland (Lees *et al.*, 1992, 1993; Lees, 1992). Other dates have reported by Lees *et al.* (1990), Shulmeister and Lees (1992) and Shulmeister *et al.* (1993).

Duller and Augustinus (1997) used four samples from linear dunes in the north-eastern part of Tasmania to compare several different sets of luminescence measurements. This included a modern back beach dune for which an OSL age of 14 ± 6 years was obtained, demonstrating that these quartz grains were well bleached at deposition. The multiple aliquot OSL data sets showed considerable scatter, but agreed with the single aliquot dose determinations, which also showed scatter. The TL measurements are affected by the presence of a large residual signal, as observed after bleaching in the laboratory, and provide a good example of the advantage of using OSL dating methods. One of the linear dunes appears to have been emplaced about 40 thousand years ago, prior to the last glacial maximum. This interpretation contrasts with the conclusions of Thom *et al.* (1994) who, on the basis of a compilation of radiocarbon and TL dates, suggest that dunes were forming during the period from 25 to 15 ka.

In New Zealand, some Holocene ridges near Leithfield, North Canterbury, have been dated using the 375°C TL from quartz (Shulmeister and Kirk, 1996). A coastal dune sand from south-western North Island, where there was good age control from a well-dated ash, had previously been dated at the same laboratory (Shepherd and Price, 1990). Four charcoal dates gave a mean age of $22\,590 \pm 230$ conventional ^{14}C yr for the ash layer, below which was the sand sample with a TL date of 24.2 ± 3.7 ka. In addition, Duller (1992) reported comparisons of TL and IRSL ages for potassium feldspars for a number of dune sands and has more recently obtained ages for the Koputaroa dunes found in south-west North Island (Duller, 1996b). These studies show the advantage of using more than one luminescence signal, as a check on the efficiency of zeroing at deposition.

In the Northern Hemisphere there have been far fewer luminescence studies of coastal dunes. Pye *et al.* (1995) used multiple aliquot OSL of quartz to obtain Holocene ages for coastal sands in Northwest England. However, the OSL sensitivity is far less than that found for quartz from hot, arid environments such as Australia, where the grains have been through many cycles of erosion and deposition over millions of years. This has led to IRSL measurement procedures being developed for potassium feldspar separates. Ollerhead *et al.* (1994) used multiple aliquot IRSL procedures on potassium feldspars to determine the stages of development of a coastal spit in New Brunswick, Canada. Ages as young as 5 ± 30 (or 12 ± 2) years were found for the modern dune once an appropriate measurement and analysis technique was developed to deal with the IRSL signal caused by charge transfer, found after the sample was preheated. The oldest dunes on the spit were found to have ages of no more than 800 years; this was in agreement with the spit accretion rate calculated on the basis of historical evidence.

Remnant basal sands and dunes on a spit across Dingle Bay in Southeast Ireland were dated using a single aliquot procedure for feldspars IRSL (Wintle *et al.*, 1998). The ages indicated that all the sand had been mobile within the past 600 years, and no remnant of sand was found which related to the Holocene high sea level. Similar young ages have been obtained using the same procedure for coastal dunes on the east coast of England (Knight *et al.*, 1998). Single aliquot IRSL measurements were also used to date dunes along the coast of south-western France (Clarke *et al.*, 1999). The dates show a major phase of activity in the late Holocene (3600–3300 years ago) with widespread remobilisation of sand in the seventeenth century and are being used to reinterpret the pre-existing dune chronology which had been based on dune morphology.

CONCLUSIONS

From the applications discussed in the previous sections, it can be seen that luminescence dating can provide reliable ages for aeolian sediments ranging in age from a few hundred to a few hundred thousand years. In arid regions, the family of luminescence techniques is particularly useful, as there is often a lack of material suitable for other dating methods. Although some successful comparisons have been made with other dating methods, radiocarbon ages on material from loess/palaeosol sequences have been shown to give rise to severe underestimation of the periods of deposition. Also, only by obtaining closely spaced luminescence dates is it possible to determine periods when no dust was deposited.

In coastal environments in Australia, dating has been carried out on both aeolian and marine sands. Some dates confirm our knowledge of Holocene and last interglacial sea levels, but others go against our knowledge of sea level lowering during oxygen isotope stage 4 to 2. Whether this is explainable in terms of more recent tsunami deposits, or mixtures of grains from deposits of different ages, or from changes in environmental dose rate through time needs to be investigated.

Fundamental research in the last five years, using natural materials such as the south-eastern South Australian dune ridges, has led to the knowledge that the OSL signal of quartz is limited by saturation, rather than thermal stability. The converse is true for the IRSL from feldspar inclusions in the same quartz grains. Fundamental research has also led to the development of more precise age determinations using single aliquot measurements on multiple grains. It has also led to improved accuracy for all OSL measurements, as we determine the optimum preheat procedure to use, based on our understanding of sensitivity changes (particularly for older sands) and charge transfer (particularly for younger sands). In turn, this understanding has led to the development of single aliquot procedures being adopted for single grain measurements, application of which can solve problems caused by the mixing of old and young grains.

REFERENCES

Aitken, M.J., 1985. *Thermoluminescence Dating.* Academic Press, London.

Aitken, M.J., 1998. *Introduction to Optical Dating.* Oxford University Press, Oxford.

Berger, G.W., 1994. Thermoluminescence dating of sediments older than ~100 ka. *Quaternary Geochronology (Quaternary Science Reviews),* **13**, 445–455.

Berger, G.W. and Busacca, A.J., 1995. Thermoluminescence dating of late Pleistocene loess and tephra from eastern Washington and southern Oregon and implications for the eruptive history of Mount St. Helens. *Journal of Geophysical Research,* **100** (B11), 22 361–22 374.

Berger, G.W., Luternauer, J.L. and Clague, J.J., 1990. Zeroing tests and application of thermoluminescence dating to Fraser River delta sediments. *Canadian Journal of Earth Sciences,* **27**, 1737–1745.

Biscaye, P.E., Grousset, F.E., Revel, M. *et al.*, 1997. Asian provenance of glacial dust (stage 2) in the Greenland Ice Sheet Project 2 ice core, Summit, Greenland. *Journal of Geophysical Research – Oceans,* **102** (C12), 26 765–26 781.

Bøtter-Jensen, L., 1997. Luminescence techniques: instrumentation and methods. *Radiation Measurements,* **27**, 749–768.

Bøtter-Jensen, L., Banerjee, D., Junger, H. and Murray, A.S., 1999a. Retrospective assessment of environmental dose rates using optically stimulated luminescence from Al_2O_3: C and quartz. *Radiation Protection Dosimetry,* **84**, 537–542.

Bøtter-Jensen, L., Mejdahl, V. and Murray, A.S., 1999b. New light on OSL. *Quaternary Geochronology (Quaternary Science Reviews),* **18**, 303–309.

Bronger, A., 1976. Zur Quartären Klima- und Landschaftsentwicklung des Karpatenbeckens auf (paläo-) pedologischer und bodengeographischer Grundlage. *Kieler Geographische Schriften,* **45**, 1–268.

Bryant, E.A., 1992. Last interglacial and Holocene trends in sea-level maxima around Australia; implications for modern rates. *Marine Geology,* **108**, 209–217.

Bryant, E.A., Young, R.W., Price, D.M. and Short, S.A., 1990. Thermoluminescence and uranium-thorium chronologies of Pleistocene coastal landforms of the Illawarra region, New South Wales. *Australian Geographer,* **21**, 101–112.

Bryant, E.A., Young, R.W., Price, D.M. and Short, S.A., 1992. Evidence for Pleistocene and Holocene raised marine deposits, Sandon Point, New South Wales. *Australian Journal of Earth Sciences,* **39**, 481–493.

Bryant, E.A., Young, R.W., Price, D.M. and Short, S.A., 1994. Late Pleistocene dune chronology: near-coastal New South Wales and eastern Australia. *Quaternary Science Reviews,* **13**, 209–223.

Bryant, E.A., Young, R.W. and Price, D.M., 1997. Late Pleistocene marine deposition and TL chronology of the New South Wales, Australian coastline. *Zeitschrift für Geomorphologie,* **41**, 205–227.

Clarke, M.L., Rendell, H.M., Pye, K. *et al.*, 1999. Dune development on the Aquitaine coast, southwestern France. *Zeitschift für Geomorphologie Supplementband,* submitted.

Debenham, N.C., 1985. Use of UV emissions in TL dating of sediments. *Nuclear Tracks and Radiation Measurements,* **10**, 717–724.

Duller, G.A.T., 1992. Comparison of equivalent doses determined by thermoluminescence and infrared stimulated luminescence for dune sands in New Zealand. *Quaternary Science Reviews,* **11**, 39–43.

Duller, G.A.T., 1994. Luminescence dating using single aliquots: new procedures. *Quaternary Geochronology (Quaternary Science Reviews),* **13**, 149-156.

Duller, G.A.T., 1996a. Recent developments in luminescence dating of sediments. *Progress in Physical Geography,* **20**, 127–145.

Duller, G.A.T., 1996b. The age of the Koputaroa dunes, southwest North Island, New Zealand. *Palaeogeography, Palaeoclimatology, Palaeoecology,* **121**, 105–114.

Duller, G.A.T. and Augustinus, P., 1997. Luminescence studies of dunes from north-eastern Tasmania. *Quaternary Geochronology (Quaternary Science Reviews)*, **16**, 357–365.

Duller, G.A.T., Bøtter-Jensen, L., Kohsiek, P. and Murray, A.S., 1999. A high-sensitivity optically stimulated luminescence scanning system for measurement of single sand-sized grains. *Radiation Protection Dosimetry*, **84**, 325–330.

Fang, X-M., Li, J-J., Van der Voo, R. *et al.*, 1997. A record of the Blake Event during the last interglacial paleosol in the western Loess Plateau of China. *Earth and Planetary Science Letters*, **146**, 73–82.

Felix, C. and Singhvi, A.K., 1997. Study of non-linear luminescence-dose growth curves for the estimation of paleodose in luminescence dating: results of Monte Carlo simulations. *Radiation Measurements*, **27**, 599–609.

Frechen, M., Horvath, E. and Gabris, G., 1997. Geochronology of Middle and Upper Pleistocene loess sections in Hungary. *Quaternary Research*, **48**, 291–312.

Fullagar, R.L.K., Price, D.M. and Head, L.M., 1996. Early human occupation of northern Australia: archaeology and thermoluminescence dating of Jinmium rock-shelter, Northern Territory. *Antiquity*, **70**, 751–773.

Grün, R. and Brumby, S., 1994. The assessment of errors in past radiation doses extrapolated from ESR/TL dose-response data. *Radiation Measurements*, **23**, 307–315.

Huntley, D.J. and Clague, J.J., 1996. Optical dating of tsunami-laid sands. *Quaternary Research*, **46**, 127–140.

Huntley, D.J., Hutton, J.T. and Prescott, J.R., 1985. South Australian sand dunes: a TL sediment test sequence, preliminary results. *Nuclear Tracks and Radiation Measurements*, **10**, 757–758.

Huntley, D.J., Hutton, J.T. and Prescott, J.R., 1993a. The stranded beach-dune sequence of south-east South Australia: A test of thermoluminescence dating, 0–800 ka. *Quaternary Science Reviews*, **12**, 1–20.

Huntley, D.J., Hutton, J.T. and Prescott, J.R., 1993b. Optical dating using inclusions within quartz grains. *Geology*, **21**, 1087–1090.

Huntley, D.J., Hutton, J.T. and Prescott, J.R., 1994. Further thermoluminescence dates from the dune sequence in the southeast of South Australia. *Quaternary Science Reviews*, **13**, 201–207.

Huntley, D.J., Short, M.A. and Dunphy, K., 1996. Deep traps in quartz and their use for optical dating. *Canadian Journal of Physics*, **74**, 81–91.

Idnurm, M. and Cook, P.J., 1980. Palaeomagnetism of beach ridges in South Australia and the Milankovitch theory of ice ages. *Nature*, **286**, 699–702.

Kar, A., Felix, C., Rajaguru, S.N. and Singhvi, A.K., 1998. Late Holocene growth and mobility of a transverse dune in the Thar Desert. *Journal of Arid Environments*, **38**, 175–185.

Knight, J., Orford, J.D., Wilson, P. *et al.*, 1998. Facies, age and controls on recent coastal sand dune evolution in North Norfolk, Eastern England. *Journal of Coastal Research*, **26**, 154–161.

Krbetschek, M.R., Götze, J., Dietrich, A. and Trautmann, T., 1997. Spectral information from minerals relevant for luminescence dating. *Radiation Measurements*, **27**, 695–748.

Kukla, G.J., 1977. Pleistocene land-sea correlations I. Europe. *Earth Science Reviews*, **13**, 307–374.

Lamothe, M., Balescu, S. and Auclair, M., 1994. Natural IRSL intensities and apparent luminescence ages of single feldspar grains extracted from partially bleached sediments. *Radiation Measurements*, **23**, 555–561.

Lees, B.G., 1992. Geomorphological evidence for Late Holocene climatic change in northern Australia. *Australian Geographer*, **23**, 1–10.

Lees, B.G., Lu, Y.C. and Head, J., 1990. Reconnaissance thermoluminescence dating of northern Australian coastal dune systems. *Quaternary Research*, **34**, 169–185.

Lees, B.G., Lu, Y.C. and Price, D.M., 1992. Thermoluminescence dating of dunes at Cape St. Lambert, East Kimberleys, northwestern Australia. *Marine Geology*, **106**, 131–139.

Lees, B.G., Hayne, M. and Price, D. 1993. Marine transgression and dune initiation on western Cape York, northern Australia. *Marine Geology*, **114**, 81–89.

Lees, B.G., Stanner, J., Price, D.M. and Lu, Y., 1995. Thermoluminescence dating of dune podzols at Cape Arnhem, northern Australia. *Marine Geology*, **129**, 63–75.

Murray, A.S. and Roberts, R.G., 1997. Determining the burial time of single grains of quartz using optically stimulated luminescence. *Earth and Planetary Science Letters*, **152**, 163–180.

Murray, A.S. and Roberts, R.G., 1998. Measurement of the equivalent dose in quartz using a regenerative-dose single-aliquot protocol. *Radiation Measurements*, **29**, 503–515.

Murray, A.S., Roberts, R.G. and Wintle, A.G., 1997. Equivalence dose measurement using a single aliquot of quartz. *Radiation Measurements*, **27**, 171–184.

Murray-Wallace, C.V., Belperio, A.P., Cann, J.H. *et al.*, 1997. Late Quaternary uplift history, Mount Gambier region, South Australia. *Zeitschrift für Geomorphologie*, **106**, 41–56.

Musson, F.M. and Wintle, A.G., 1994. Luminescence dating of the loess profile at Dolni Vestonice, Czech Republic. *Quaternary Geochronology (Quaternary Science Reviews)*, **13**, 411–416.

Nishizumi, K., Kohl, C.P., Arnold, J.R. *et al.*, 1993. Role of in-situ cosmogenic radionuclides [10]Be and [26]Al in the study of diverse geomorphic processes. *Earth Surface Processes and Landforms*, **18**, 407–425.

Nott, J.F. and Price, D.M., 1991. Late Pleistocene to early Holocene aeolian activity in the upper and middle Shoalhaven catchment, New South Wales. *Australian Geographer*, **22**, 168–177.

Nott, J., Young, R., Bryant, E. and Price, D., 1994. Stratigraphy vs. pedogenesis: problems of their correlation within coastal sedimentary facies. *Catena*, **23**, 199–212.

Ollerhead, J., Huntley, D.J. and Berger, G.W., 1994. Luminescence dating of sediments from Buctouche Spit, New Brunswick. *Canadian Journal of Earth Sciences*, **31**, 523–531.

Olley, J.M., Roberts, R.G. and Murray, A.S., 1997. Disequilibria in the uranium decay series of the sedimentary deposits at Allen's Cave, Nullarbor Plain, Australia: Implications for dose-rate determination. *Radiation Measurements*, **27**, 433–443.

Oyston, B., 1996. Thermoluminescence dating of quartz from Quaternary aeolian sediments in southeastern Australia. Unpublished PhD Thesis. La Trobe University, Bundoora, Australia.

Pésci, M., 1985. Chronostratigraphy of Hungarian loesses and the underlying subaerial formation. In *Loess and the Quaternary: Chinese and Hungarian Case Studies*, M. Pésci (Ed.). Akadémiai Kiadó, Budapest, pp. 33–49.

Prescott, J.R. and Robertson, G.B., 1997. Sediment dating by luminescence: A review. *Radiation Measurements*, **27**, 893–922.

Prescott, J.R., Huntley, D.J. and Hutton, J.T., 1993. Estimation of equivalent dose in thermoluminescence dating – the Australian slide method. *Ancient TL*, **11**, 1–5.

Pye, K., Stokes, S. and Neal, A., 1995. Optical dating of aeolian sediments from the Sefton coast, northwest England. *Proceedings of the Geologists' Association*, **106**, 281–292.

Rao, M.S., 1997. Studies on the physical basis of luminescence chronology and its applications. Unpublished PhD Thesis. Nagpur University, Nagpur, India.

Richardson, C.A., McDonald, E.V. and Busacca, A.J., 1997. Luminescence dating of loess from Northwest United States. *Quaternary Geochronology (Quaternary Science Reviews)*, **16**, 403–415.

Roberts, R.G., Bird, M., Olley, J. *et al.*, 1998a. Optical and radiocarbon dating at Jinmium rock shelter in northern Australia. *Nature*, **393**, 358–362.

Roberts, R., Yoshida, H., Galbraith, G. *et al.*, 1998b. Single-aliquot and single-grain optical dating confirm thermoluminescence age estimates at Malakanunja II rock shelter in northern Australia. *Ancient TL*, **16**, 19–24.

Sarnthein, M., 1978. Sand deserts during glacial maximum and climatic optimum. *Nature*, **272**, 396–398.

Shepherd, M.J. and Price, D.M., 1990. Thermoluminescence dating of late Quaternary dune sand, Manawatu/Horowhenua area, New Zealand: a comparison with ^{14}C age determinations. *New Zealand Journal of Geology and Geophysics*, **33**, 535–539.

Sherwood, J., Barbetti, M., Ditchburn, R. *et al.*, 1994. A comparative study of Quaternary dating techniques applied to sedimentary deposits in southwest Victoria, Australia. *Quaternary Geochronology (Quaternary Science Reviews)*, **13**, 95–110.

Shulmeister, J. and Lees, B.G., 1992. Morphology and chronostratigraphy of a coastal dunefield; Groote Eylandt, northern Australia. *Geomorphology*, **5**, 521–534.

Shulmeister, J., Short, S.A., Price, D.M. and Murray, A.S., 1993. Pedogenic uranium/thorium and thermoluminescence chronologies and evolutionary history of a coastal dunefield, Groote Eylandt, northern Australia. *Geomorphology*, **8**, 47–64.

Shulmeister, J. and Kirk, R.M., 1996. Holocene history and a thermoluminescence based chronology of coastal dune ridges near Leithfield, North Canterbury, New Zealand. *New Zealand Journal of Geology and Geophysics*, **39**, 25–32.

Singhvi, A.K., Banerjee, D., Ramesh, R. *et al.*, 1996. A luminescence method for dating 'dirty' pedogenic carbonates for paleoenvironmental reconstruction. *Earth and Planetary Science Letters*, **139**, 321–332.

Singhvi, A.K., Bluszcz, A., Murthy, C.S.R. and Rao, M.S., 1998. Synthesis of luminescence chronology of loess accumulation episodes and implications for global land-sea correlation. First IGBP PAGES Open Science Meeting, London, Abstracts, p. 118.

Singhvi, A.K., Bronger, A., Sauer, W. and Pant, R.K., 1989. Thermoluminescence dating of loess-palaeosol sequences in the Carpathian Basin (east Central Europe); a suggestion for a revised chronology. *Chemical Geology (Isotope Geosciences Section)*, **73**, 307–317.

Singhvi, A.K., Deraniyagala, S.U. and Sengupta, D., 1986. Thermoluminescence dating of Quaternary red-sand beds; a case study of coastal dunes in Sri Lanka. *Earth and Planetary Science Letters*, **80**, 139–144.

Singhvi, A.K. and Krbetschek, M.R., 1996. Luminescence dating: a review and a perspective for arid zone sediments. *Annals of Arid Zone*, **35**, 249–279.

Singhvi, A.K. and Lang, A., 1998. Improvements in infra-red dating of partially bleached sediments – the 'differential' partial bleach technique. *Ancient TL*, **16**, 63–71.

Singhvi, A.K., Sharma, Y.P. and Agrawal, D.P., 1982. Thermoluminescence dating of sand dunes in Rajasthan, India. *Nature*, **295**, 313–315.

Spooner, N.A., 1998. Human occupation at Jinmium, northern Australia: 116 000 years ago or much less? *Antiquity*, **72**, 173–178.

Stokes, S., 1997. Dating of desert sequences. In *Arid Zone Geomorphology*, D.S.G. Thomas (Ed.). John Wiley, Chichester, pp. 607–637.

Stokes, S., Thomas, D.S.G. and Washington, R., 1997. Multiple episodes of aridity in southern Africa since the last interglacial period. *Nature*, **388**, 154–158.

Stokes, S., Haynes, G., Thomas, D.S.G. *et al.*, 1998. Punctuated aridity in southern Africa during the last glacial cycle: the chronology of linear dune construction in the northeastern Kalahari. *Palaeogeography, Palaeoclimatology, Palaeoecology*, **137**, 305–322.

Stokes, S. and Horrocks, J., 1998. A reconnaissance survey of the linear dunes and loess plains of northwestern Nigeria: granulometry and geochronology. In *Quaternary Deserts and Climatic Change*, A.S. Alsharhan, K.W. Glennie, E. Whittle and R. Kendall (Eds). Balkema, Rotterdam, pp. 165–174.

Tejan-Kella, M.S., Chittleborough, D.J., Fitzpatrick, R.W. *et al.*, 1990. Thermoluminescence dating of coastal sand dunes at Cooloola and North Stradbroke Island, Australia. *Australian Journal of Soil Research*, **28**, 465–481.

Thom, B., Hesp, P. and Bryant, E., 1994. Last glacial 'coastal' dunes in Eastern Australia and implications for landscape stability during the Last Glacial Maximum. *Palaeogeography, Palaeoclimatology, Palaeoecology*, **111**, 229–248.

Thomas, J.V., Var, A., Kalath, A.J. *et al.*, 1999. Late Pleistocene and Holocene history of aeolian accumulations in the Thar Desert, India. *Zeitschrift für Geomorphologie*, **116**, 1–14.

Venkatesan, T.R. and Ramesh, R., 1993. Consideration of analytical uncertainties while plotting histograms. *Journal of the Geological Society of India*, **41**, 313–317.

Wintle, A.G., 1990. A review of current research on TL dating of loess. *Quaternary Science Reviews*, **9**, 385–397.

Wintle, A.G., 1993. Luminescence dating of aeolian sands – an overview. In *The Dynamics and Environmental Context of Aeolian Sedimentary Systems*, K. Pye (Ed.). Geological Society (London) Special Publication, 72, 49–58.

Wintle, A.G., 1997. Luminescence dating: laboratory procedures and protocols. *Radiation Measurements*, **27**, 769–817.

Wintle, A.G. and Huntley, D.J., 1979. Thermoluminescence dating of a deep-sea sediment core. *Nature*, **279**, 710–712.

Wintle, A.G. and Murray, A.S., 1999. Luminescence sensitivity changes in quartz. *Radiation Measurements*, **30**, 107–118.

Wintle, A.G. and Packman, S.C., 1988. Thermoluminescence ages for three sections in Hungary. *Quaternary Science Reviews*, **7**, 315–320.

Wintle, A.G., Clarke, M.L., Musson, F.M. *et al.*, 1998. Luminescence dating of recent dunes on Inch Spit, Dingle Bay, southwest Ireland. *The Holocene*, **8**, 331–339.

Woodroffe, C.D., Bryant, E.A., Price, D.M. and Short, S.A., 1992. Quaternary inheritance of coastal landforms, Cobourg Peninsula, Northern Territory. *Australian Geographer*, **23**, 101–15.

Xie, J. and Aitken, M.J., 1982. The hypothesis of mid-term fading and its trial on Chinese Loess. *Ancient TL*, **9**, 21–25.

Young, R.W., Bryant, A.E., Price, D.M. *et al.*, 1993a. Theoretical constraints and chronological evidence of Holocene coastal development in central and southern New South Wales, Australia. *Geomorphology*, **7**, 317–329.

Young, R.W., Bryant, E.A. and Price, D.M., 1993b. Last interglacial sea levels on the south coast of New South Wales. *Australian Geographer*, **24**, 72–75.

Zhou, L.P., Dodonov, A.E. and Shackleton, N.J., 1995. Thermoluminescence dating of the Orkutsay loess section in Tashkent region, Uzbekistan, Central Asia. *Quaternary Science Reviews*, **14**, 721–730.

Zhou, L.P., Williams, M.A.J. and Peterson, J.A., 1994. Late Quaternary aeolianites, palaeosols and depositional environments on the Nepean Peninsula, Victoria, Australia. *Quaternary Science Reviews*, **13**, 225–239.

Zöller, L. and Wagner, G.A., 1989. Strong or partial thermal washing in TL-dating of sediments? In Long and Short Range Limits in Luminescence Dating. *Research Laboratory for Archaeology and the History of Art. Occasional Publication* No. 9.

Zöller, L., Oches, E.A. and McCoy, W.D., 1994. Towards a revised chronostratigraphy of loess in Austria with respect to key sections in the Czech Republic and in Hungary. *Quaternary Geochronology (Quaternary Science Reviews)*, **13**, 465–472.

14 Conclusions

STEPHEN STOKES
School of Geography, University of Oxford, UK

In compiling this edited volume, a key intention was to provide a forum for active researchers in the range of sub-disciplines which contribute to knowledge of the nature and utilisation of aeolian sediment systems to present up to date assessments of the state of each field. In scanning the preceding chapters it is clear that we have very much attained this goal. Each chapter has systematically visited both historical beginnings and the most contemporary investigations and ideas in their fields. They collectively chronicle aeolian research as an area which is undergoing rapid change and development and one that has expanded greatly over the past two decades. Additionally, many chapters have served to emphasise the considerable applied as well as academic significance of such research. This applicability is apparent not only in the discussions of on- and off-site impacts of dust erosion and deposition, but in the consideration of the nature and evolution of coastal and inland aeolian systems and how this impacts on their management. Interestingly, in Chapter 5, Thomas further points out that in addition to the clearly observable anthropogenic impacts of housing and recreational development, and industrial impacts on coastal dune systems, that it is increasingly necessary to contemplate longer time-scales, and wider issues of management of many inland semi-arid and sub-humid areas. Such management will be forced under even greater scrutiny in many regions where anthropogenically-induced climate changes are anticipated to result in reduced precipitation and soil moisture, and enhanced temperature seasonally or annually (e.g. Muhs and Maat, 1993).

As stated, a strong theme through most papers is the dramatic developments in research activity and its diversity over the past one or two decades. Whether relating to enhancements in instrumentation for measuring wind field parameters or sediment transportation or in the application of increasingly refined strategies of luminescence dating, aeolian research has come into a new age of investigation incorporating an ever broader cross-section of engineering and applied physical and geophysical techniques and equipment. Indeed the future may identify this final few decades of the century as the period in which aeolian-related research underwent a further metamorphosis from outlined in the introductory chapter. This new period is increasingly seeing integrated research being undertaken without singularly stressing the previously predominantly reductionist approaches of past research. Such an approach is typified by the conceptual models described by Kocurek (Chapter 11) in which an understanding of the long-term formation of aeolian sand bodies within the rock record is interpreted from a combination of logical geological inferences, empirically-based models of sediment transport and availability and well-dated later Quaternary aeolian sequences which are employed both to understand the styles of construction and modification of large aeolian sand bodies.

Over the last two decades climatic studies and palaeoenvironmental research in other depositional settings have increasingly identified greater levels of variability with high degrees of precision. To evaluate the responsiveness of aeolian systems to such changes and to incorporate palaeoenvironmental data from aeolian systems into future GCM (General Circulation Model) runs there will clearly be a need for greater understanding of aeolian system reactions to abrupt climatic events and climatic fluctuations. Again a combined approach would appear to be the way forward; one which integrates an appreciation of landscape sensitivity and dune reactivation over annual and decadal time-scales, and links this variability to longer term patterns of activity on broad spatial scales. In this respect it can be anticipated that use of the mobility index developed by Lancaster (1988) in tandem with instrumental and historical records of climate changes will provide more quantified indexes of causation for past periods of enhanced dune activity. Indeed, the end-point of such approaches, which link contemporary climatic and dune variability with mobility of well dated Quaternary sequences and provide deterministic explanations for such patterns via GCM experimental data, is already being realised in ongoing research in southern Africa (e.g. Stokes *et al.*, 1999, in press).

CHALLENGES FOR THE COMING DECADES

While the chapters presented within this volume demonstrate the great strides which have been made in understanding aeolian systems, a number of important outstanding questions and challenges remain.

- There is a need to grasp the combined power of mathematical complexity theory and the paradigm of geomorphic self organisation on a range of scales. Initial work by Werner (1990, 1995) has demonstrated the great potential of predicting gross dune form using such methods. Further application should fill the gaps which have persisted between studies at the grain by grain and larger scales.
- While a number of aspects of sediment transportation have been elucidated there is still little understanding of the initiation and development of aeolian dunes.
- Sediment transport research still requires techniques which allow precise and direct measurement of shear stress and other parameters.
- There remains a need to refine the Mobility Index concept further, possibly linking it with the Drift Direction and Drift Potential approaches developed in the late 1970s (Fryberger, 1979) and develop truly predictive models of future dune activity and dust mobility.
- There is also a need to incorporate more widely newly developed approaches and techniques. In this regard while developments in luminescence dating have been mentioned in a number of places, the full utilisation of the method to

construct objectivity sampled chronologies of dune systems is still a long way off. Likewise, application of new Global Positioning System (GPS) surveying techniques (Stokes, 1999b) will provide an added dimension to sediment transportation studies at the bedform scale.

With the increasing incorporation of new technologies and paradigms, past uncertainties are developing into tractable problems which should be solvable within the next one or two decades. Past and present researchers have provided a strong basis from which these future developments should flourish. Let us hope that future studies and ideas pertaining to aeolian sediments and palaeoenvironments collect less dust than some of our sediment traps!

REFERENCES

Fryberger, S.G., 1979. Dune form and wind regime. *US Geological Survey, Professional Paper*, **1052**, 137–169.

Lancaster, N., 1988. Development of linear dunes in the southwestern Kalahari, southern Africa. *Journal of Arid Environments*, **14**, 233–244.

Muhs, D. and Maat, P.B., 1993. The potential response of aeolian sands to Greenhouse Warming and precipitation reduction on the Great Plains of the United States. *Journal of Arid Environments*, **25**, 351–361.

Stokes, S., Washington, R. and Preston, A., 1999. Late Quaternary Evolution of the Central and Southern Kalahari: Environmental Responses to Changing Climatic Conditions. In *Late Cenozoic Environments and Hominid Evolution: a tribute to Bill Bishop*, P. Andrews and P. Banham (Eds). London, Geological Society, pp. 247–268.

Stokes, S., Goudie, A.S., Ballard, J., Gifford, C., Samieh, S., Embabi, N. and El-Rashidi, O.A., 1999. Accurate dune displacement and morphometric data using Kinematic GPS. *Zeitschrift für Geomorphologie*, Supplemental **116**.

Werner, B.T., 1990. A steady-state model of wind-blown sand transport. *Journal of Geology*, **98**, 1–17.

Werner, B.T., 1995. Eolian dunes: Computer simulations and attractor interpretations. *Geology*, **32**, 1107–1110.

Index